T0332623

SUSTAINABLE COMMUNITIES ON A SUSTAINABLE PLANET: THE HUMAN–ENVIRONMENT REGIONAL OBSERVATORY PROJECT

Scientists and policymakers have come to realize that localities are central to addressing the causes and consequences of global environmental change. Despite this realization, there has been no systematic effort to monitor global change in local places. The goal of the Human–Environment Regional Observatory project (HERO) was to develop the infrastructure necessary to scrutinize and understand the local dimensions of global change, emphasizing the interactions between people and their environment that make them vulnerable to global change.

This book presents the philosophy behind HERO, the methods used to put that philosophy into action, the results of those actions, and the lessons learned from the project. HERO used three strategies: it developed research protocols and data standards for collecting data; it built a web-based networking environment to help investigators share data, analyses, and ideas from remote locations; and investigators field-tested these concepts by applying them in diverse biophysical and socioeconomic settings – central Massachusetts, central Pennsylvania, south-western Kansas, and the US–Mexico border region of Arizona.

The book highlights the unique focus of HERO on how to think about and act on complex, integrative, and interdisciplinary global change science at local scales. It is a valuable resource for global change scientists concerned with collaborating and comparing case studies across time and place.

BRENT YARNAL is Professor and Associate Head of Geography at The Pennsylvania State University. His research and teaching interests bridge the physical and social sciences, and integrate climate change, natural hazards, land-use change, water resources, and the use of environmental information in decision-making. His research focuses on vulnerability to and adaptation planning for present and future climate change, local and regional greenhouse gas emissions inventories and mitigation planning, and the role of climate information in water resource decision-making. He has authored and contributed to numerous books and journal articles, and was an editor of *Climate Research – Interactions of Climate with Organisms, Ecosystems and Human Societies* from 1996 to 2001.

COLIN POLSKY is Associate Professor in the Graduate School of Geography at Clark University, Worcester, MA. He is a geographer specializing in the human dimensions of global environmental change. He has explored ways to blend quantitative and qualitative methods for the study of social and ecological vulnerability to environmental changes in the Arctic, the US Great Plains, and central and eastern Massachusetts. This research requires the blending of statistical techniques (such as empirical downscaling and spatial econometrics) with insight gained from qualitative methods (such as interviews and participant observation). Professor Polsky has an extensive range of publications on the subject.

JAMES O'BRIEN is a Principal Lecturer in the School of Geography, Geology and the Environment at Kingston University, London. His research and teaching interests include geographic information systems (GIS) enterprise and research, GIS software development, internet GIS, GIS and natural hazards, spatial databases, and geographic semantics. He has co-presented papers at international conferences on geographic semantics and the role of GIS in education.

SUSTAINABLE COMMUNITIES ON A SUSTAINABLE PLANET: THE HUMAN–ENVIRONMENT REGIONAL OBSERVATORY PROJECT

Edited by

BRENT YARNAL
The Pennsylvania State University

COLIN POLSKY
Clark University

JAMES O'BRIEN
Kingston University

CAMBRIDGE
UNIVERSITY PRESS

University Printing House, Cambridge CB2 8BS, United Kingdom

One Liberty Plaza, 20th Floor, New York, NY 10006, USA

477 Williamstown Road, Port Melbourne, VIC 3207, Australia

4843/24, 2nd Floor, Ansari Road, Daryaganj, Delhi - 110002, India

79 Anson Road, #06-04/06, Singapore 079906

Cambridge University Press is part of the University of Cambridge.

It furthers the University's mission by disseminating knowledge in the pursuit of education, learning and research at the highest international levels of excellence.

www.cambridge.org
Information on this title: www.cambridge.org/9781108445740

First published 2009
First paperback edition 2017

A catalogue record for this publication is available from the British Library

ISBN 978-0-521-89569-9 Hardback
ISBN 978-1-108-44574-0 Paperback

Contents

Contributors

Ola Ahlqvist, Assistant Professor of Geography, Ohio State University

Sarah Assefa, undergraduate student, Clark University

Andrew Comrie, Professor of Geography, Dean of the Graduate School and Associate Vice President for Research, University of Arizona

Kate Del Vecchio, undergraduate student, Clark University

Mark Gahegan, Professor of Geography and Associate Director of the GeoVISTA Center, The Pennsylvania State University

John Harrington, Jr., Professor of Geography, Kansas State University

Lisa M. Butler Harrington, Professor of Geography, Kansas State University

Rachel M. K. Headley, Research Scientist, US Geological Survey EROS Data Center

Troy Hill, Masters student, Yale University

Diana Liverman, Professor of Geography and Director of the Environmental Change Institute, University of Oxford

Max Lu, Associate Professor of Geography, Kansas State University

Junyan Luo, Ph.D. student, The Pennsylvania State University

Alan MacEachren, Professor of Geography and Director of the GeoVISTA Center, The Pennsylvania State University

Laura Merner, undergraduate student, Clark University

Rob Neff, Assistant Professor of Geography, University of Maryland–Baltimore County

James O'Brien, Principal Lecturer, Kingston University

William Pike, Research Scientist, Pacific Northwest National Laboratory

Colin Polsky, Associate Professor of Geography, Clark University

Robert Gilmore Pontius, Jr., Associate Professor of Geography, Clark University

Cynthia Sorrensen, Assistant Professor of Geography, Texas Tech University

Isaac Tercero, undergraduate student, Clark University

B. L. Turner II, Gilbert F. White Professor of Environment and Society, Arizona State University

Jessica Whitehead, South Carolina Sea Grant Consortium

Brent Yarnal, Professor and Associate Head of Geography, The Pennsylvania State University

Chaoqing Yu, Chinese Institute of Water Resources and Hydropower Research, Beijing

Part I

1

Infrastructure for observing local human–environment interactions

BRENT YARNAL, JOHN HARRINGTON, JR., ANDREW COMRIE, COLIN POLSKY, AND OLA AHLQVIST

The vision: sustainable communities on a sustainable planet

Imagine a world where nature and society coexist in a healthy symbiosis, where human impacts on the environment are minimal, and where communities are safe from natural and technological hazards. Imagine a time when scientists can monitor such sustainable human–environment interactions, when they can interactively share and compare data, analyses, and ideas about those interactions from their homes and offices, and when they can collaborate with local, regional, and international colleagues and stakeholders in a global network devoted to the environmental sustainability of their communities and of the planet.

We contend that to build the sustainable world portrayed above, it is necessary to develop an infrastructure[1] that will support such an edifice. Consequently, this chapter introduces our ideas about the infrastructure needed to realize this vision and how the Human–Environment Regional Observatory project (HERO) attempted to take the initial steps to develop that infrastructure. The chapter also demonstrates that HERO addressed several major growth areas of twenty-first-century science – complex systems, interdisciplinary research, usable knowledge/usable science, and transdisciplinarity – as integral parts of its infrastructure development. The chapter ends by laying out the rationale behind and structure of this book.

Achieving the vision: infrastructure development and HERO

Infrastructure for monitoring global change in local places

To paraphrase the American politician Tip O'Neill,[2] "all global change is local." On the one hand, anthropogenic global environmental change is the accumulated result of billions of individual actions occurring at billions of specific locations. On the other hand, people experience the biophysical and socioeconomic impacts of global

Sustainable Communities on a Sustainable Planet: The Human–Environment Regional Observatory Project, eds. Brent Yarnal, Colin Polsky, and James O'Brien. Published by Cambridge University Press.

environmental change in identifiable places. Efforts to implement adaptations to those impacts, as well as to implement actions to mitigate the human causes of global environmental change, take place locally. Thus, a critical – but until recently, missing – element of the global change research agenda is the integrated[3] study of global change in local places (Kates and Torrie 1998; Wilbanks and Kates 1999).

This book asserts that to develop sustainable communities and a sustainable Earth, it is essential to monitor global change in local places. Why is it important to conduct such monitoring?[4] The many sustainability indicator projects under way demonstrate that monitoring helps communities gauge their progress toward (or regression from) sustainability (e.g., Farrell and Hart 1998). Monitoring shows which actions are improving the local human–environment dynamic; it points to areas of strength and weakness in local human–environment relations. It identifies emerging vulnerabilities to nature or abuses of nature and how fast they are developing. Moreover, if monitoring detects the source of the vulnerability or abuse, it may suggest ways to diminish or eliminate the problem. Importantly, monitoring enables a community to set goals and to determine how far it is from reaching those goals.

Scientists should monitor global change for similar reasons. At all human levels – international, national, and local – monitoring would help gauge progress on adapting to and mitigating global change. Monitoring would tell scientists what adaptation and mitigation strategies are working and which ones are not working. It would identify when and where global change problems are developing and would suggest how urgently society should address the problems. It would enable international bodies, nations, and communities to set goals and measure advancement towards those goals.

Alas, today's global change monitoring efforts emphasize the global scale. They tend to be disjointed and piecemeal, especially at local scales. We believe that the opportunity exists to promote a coordinated monitoring effort that focuses on global change in local places. It affords the scientific community the chance to implement the infrastructure needed to support effective global change monitoring. What should that infrastructure look like?

The Intergovernmental Panel on Climate Change (IPCC) provides one model of the infrastructure needed to monitor global change – in this case, climate change (e.g., IPCC 2007a, b, c). There are problems with this model, however. First, the IPCC provides five-year snapshots of the state of the climate and related environment, but tracks few clearly defined indicators of climate change and even fewer indicators of climate change impacts. Instead, the process relies on a larger suite of unique case studies; in fact, the main job of IPCC scientists is to synthesize diverse case studies and to judge subjectively what they mean *in toto*. The IPCC process does a better job of tracking the socioeconomic activities that cause climate change because national databases of socioeconomic activity tend to be superior to natural science and human–environment databases. Second, the IPCC focuses on global and continental scales – not on the community scale where people ultimately cause, experience, and respond to climate change. Third, IPCC scientists communicate while compiling the five-year reports, but most agree that communications are cumbersome and influenced by international politics and would benefit from continuous discourse. Finally, although scientists form formal networks to conduct an IPCC assessment, these networks disperse after each assessment, with scientists going back to their organizations to re-engage in research and with governments reforming the networks with new people and ideas for the next assessment. There is no mechanism to maintain an ongoing, worldwide network of researchers joined by common interests, collaborating in real time, and free from political constraints. In order to construct a rich picture of climate change, it is necessary to embrace and accumulate the results and perspectives captured by many local, regional, and global studies. Moreover, to integrate such studies, there needs to be support for formality and logical inference, for diverse opinions and approaches, and for the use of imprecise and contested knowledge.

Thus, science needs a new, alternative model for an infrastructure to monitor global change. That model must enable scientists to monitor the ongoing causes and consequences of environmental change across a continuum of scales, including – and with special emphasis on – the local scale. The model must facilitate convenient real-time sharing of data, analyses, and ideas among scientists working in regions and locales around the planet. It must foster a sense of community, purpose, and intellectual freedom among scientists who study global change and who share the goal of sustainable communities on a sustainable planet.

One important aspect of this envisioned infrastructure is the development of research protocols[5] and data standards for scientists working on global change in local places. Such protocols should be flexible, accommodating a broad spectrum of potential users from diverse geographic areas and with varying resources and training (Tran and Wu 2001). The protocols should be dynamic, incorporating new technologies, methodologies, models, data, and intellectual paradigms over time. They should be standardized so that comparisons are possible, showing how processes influencing global change vary over space and time. In addition, global change research protocols should balance data (e.g., quantitative versus qualitative), models (e.g., deterministic versus stochastic), and scope (e.g., multiple spatial and temporal scales). Clearly, there is a tension among the competing concepts of flexibility, dynamism, standardization, and balance, making the development of research protocols for monitoring global change in local places a non-trivial task.

While protocols have been slow to develop, there has been significant progress on international and national data standards, especially in the realm of geospatial data.

Most of these efforts involve governments at international or national levels, such as the European Committee for Standardization (CEN 2007) or the United States Federal Geographic Data Committee (2007). Taking the lead on geospatial standards for industry, the Open Geospatial Consortium (OGC) aims at increasing the interoperability[6] of hardware and software involving spatial information and location – that is, OGC facilitates communication among geographic information systems, vendor brands, data sources, and computing platforms (OGC 2007). Its members include public and private companies, universities, government agencies, and other organizations interested in building geospatial interoperability. Notably, OGC sponsored the Geospatial Information for Sustainable Development Initial Capability Pilot (GISD-ICP) in summer and fall 2002. This pilot project demonstrated how geospatial information standards can enhance sustainable development efforts and showed why such standards are critical at the local level. This pilot was just the beginning – science must go much farther to scale data standards from the global and national levels to the community level. In sum, the efforts of governments and industry reduce the need to develop data standards for global change monitoring by providing clear guidelines for data storage.

Understanding global change in local places cannot happen in isolation. Scientists who monitor this problem must share their data, methods, and ideas so that they can build a picture that helps them know which characteristics are local, which are regional, and which are truly global. The World Wide Web has made it increasingly possible for scientists around the world to know what other scientists are doing. Most Websites, however, do not promote dynamic intellectual interchange or capture the excitement of dynamic communication. In contrast, a collaboratory[7] uses the interconnectivity of the Web to link scientists in near real time, if not real time (MacEachren 2000, 2001). The concept of the collaboratory goes beyond email and instant messaging to include such dimensions as Web-based video conferencing, electronic Delphi tools, and portals that allow scientists and others to share databases, maps, graphs, notebooks, and workspaces interactively. Pilot collaboratories are being developed around the world, but none have realized their potential because of technical, security, and other issues. Only one, introduced in the next section, has focused on global change in local places.

Finally, an essential part of infrastructure aimed at studying and monitoring global change in local places is a network of scientists who will adopt research protocols and data standards and who will engage each other in a collaboratory. There are already hundreds of local-area research and monitoring sites around the world that focus on environmental change and issues of sustainability; only a few sites concentrate on global change in local places. In all cases, however, these sites function largely independently, collecting unique data in unique ways, thus making cross-site comparison more or less impossible now and in the future. Scientists working at these sites sometimes are aware of the work of colleagues at other research and monitoring sites through Web searches, published papers, and conferences, but more often they are unaware of parallel efforts; rarely, if ever, do they coordinate their efforts to leverage natural, but untapped, synergies with their colleagues. It is crucial, therefore, that an international network of researchers develops so that science has an ongoing dialogue about consistent, verifiable, and comparable records of global change in local places over time and over space.

The HERO project

The goal of the Human–Environment Regional Observatory project was to develop a prototype of the infrastructure and concepts needed to understand and monitor global change in local places and to prove that the infrastructure and concepts worked. HERO did not seek to create networks of researchers, but to enable the development of such networks. To reach its goal, the project had three strategies. First, HERO developed research protocols and data standards for collecting human–environment data to facilitate the studying and monitoring of global change at individual sites and to enable cross-site comparisons and generalizations. Second, HERO built a collaboratory to help investigators share data, analyses, and thoughts from remote locations and to collaborate in answering common research questions. Third, HERO tested these ideas by applying the protocols, standards, and networking environment at four proof-of-concept research sites to investigate land-use-induced vulnerability to hydroclimatic variation and change.

The research design came directly from the strategies outlined above and had two essential components. The first component was the Web-based HERO Intelligent Networking Environment (HEROINE, located at The Pennsylvania State University), which had two tasks. One was to develop ways to handle the heterogeneous quantitative and qualitative, biophysical and socioeconomic data generated in local human–environment research. Current approaches to data interoperability, mentioned briefly above, are largely about data communication between different computers in different places. HERO needed to go beyond those approaches to address the issues involved in linking human understanding with formal systems.[8] Consequently, we developed computational methods for modeling knowledge about the conceptual understanding of human–environment interactions and the process of decision-making.

These methods provided a foundation for HEROINE's second task: to build a collaboratory where researchers from around the world could share data; analyze, visualize, and compare those data; and interact with one another while working at their local sites. Approaches developed and tested in the HERO collaboratory included an electronic notebook for posting data of any type (e.g., numbers, words, audio, and video) and of any format at a central repository for instant access by all researchers in the network, no matter where they were located. Codex, a searchable, sharable, dynamic Web portal later replaced the original electronic notebook and allowed researchers to represent different perspectives about information resources. Codex aimed at enabling a global network of scientists to capture, compare, and compute their different approaches to science; it was not limited to data, but included methods, work practices, and the situated nature of any human–environment enquiry (Gahegan and Pike 2006). Another approach was an electronic Delphi tool designed to support remote group decision-making and consensus building through an anonymous, iterative process. A third approach was Web-based video conferencing, which allowed collaborators to interact through their computers regardless of their physical location by seeing and hearing each other, as well as by sharing presentations, documents, and real-time work.

The second component of the HERO research design consisted of proof-of-concept testing. To provide a real-world context for developing the infrastructure, the project focused on the question, "How does changing land use affect the vulnerability of people and places to hydroclimatic variation and change?" Less formally, the question is, "How does land-use change influence vulnerability to droughts and floods?" HERO addressed this question at four HERO proof-of-concept testing sites (HEROs) in diverse biophysical and socioeconomic settings. The four HEROs were located along a decreasing east–west precipitation gradient starting in central and eastern Massachusetts, carrying through central Pennsylvania to southwestern Kansas, and ending in the Arizona–Sonora border region (Figure 1.1). Researchers from these HEROs came from the geography departments at Clark University, The Pennsylvania State University, Kansas State University, and The University of Arizona, respectively. The researchers collected data using the same protocols, stored and shared their data using the same data standards, and interacted through the HERO collaboratory. A vital part of this interaction involved implementing the collaboratory to develop the protocols and standards and to improve tools through group interaction.

To answer the research question posed above, the four HEROs focused on assessing vulnerability and land-use/land-cover change. The vulnerability assessments started by developing a framework for defining vulnerability. Many frameworks exist and some are complementary, but others are not. As a result, HERO team members adopted an iterative approach to frame, operationalize, and report the results of the vulnerability research. The iterations allowed the conceptual framework to adapt to the varying biophysical and socioeconomic contexts of each area as identified by the initial scoping exercise and by later vulnerability analyses.

Vulnerability studies focus on a particular place, at a specific time through its three dimensions – exposure, sensitivity, and adaptive capacity. An understanding

Figure 1.1. Location of HERO study sites.

of place, including the physical characteristics of the landscape and the political and social milieu of the population (Jianchu *et al.* 2005), is essential to analyzing vulnerability. Understanding local vulnerability required examination of local context and knowledge. To develop this understanding, the HEROs engaged in historical research to determine local human–environment interactions and associated land-use/cover changes for each site. After determining vulnerability at the individual sites, cross-site comparison searched for commonalities and differences across the regions.

HERO contended that if society is to study and monitor the local dimensions of global environmental change in the future, then it is essential to develop a cadre of young scientists trained in this research area today. Consequently, an important element of the HERO infrastructure was the training of young scientists. In addition to the many postdoctoral and graduate students who participated in the HERO project, the HERO Research Experiences for Undergraduates (REU) program engaged advanced undergraduate students in field, laboratory, and archival research on human–environment interactions and, especially, global change in local places (Yarnal and Neff 2007). The program followed a cooperative learning model to foster an integrated approach to geographic research and to build collaborative research skills. The program hosted 12–16 students annually, who first engaged in an intensive two-week short course and then formed three- or four-person teams to conduct six weeks of research at the four HEROs. The student teams used the HERO collaboratory to work together across sites and to integrate their research and findings.

Although HERO personnel continually reached out to other national and international networks to share their vision of infrastructure and the collaboratory, the four sites and their faculty, staff, postdoctoral researchers, graduate students, and undergraduate students formed the small network that produced the research covered in this book. Throughout the five years of the project, these researchers turned to a single, overarching question for guidance on infrastructure development:

- How do we understand and monitor local human–environment interactions across space and time?

To answer that all-encompassing question, they addressed three more-focused guiding questions:

- How do we collaborate across space and time?
- How do we build networks of human–environment collaborators?
- How do we benefit from collaboration across academic generations?

The book will return to these questions in Chapter 15.

Addressing complex environmental systems via transdisciplinarity

The vision expressed in the opening paragraph of this chapter asked the reader to imagine a time when scientists could help contribute to community and planetary sustainability by collaborating amongst themselves and with various stakeholders. This vision intersects major growth areas in science over the last several years: complex systems, interdisciplinary research, usable knowledge/usable science, and transdisciplinarity. All of these areas contribute to what the United States National Science Foundation (NSF) calls complex environmental systems (NSF 2003). HERO embraced this approach to science and the concepts behind it.

Complex systems may or may not be complicated, but they are certainly interdependent and integrated. Human–environment systems, which integrate interdependent social and biophysical systems, are therefore complex. HERO, which by its title focused on developing systems to assess, monitor, and study local human–environment interactions, therefore tackled a complex topic.

Interdisciplinary research involves "research by teams or individuals that integrates information, data, techniques, tools, perspectives, concepts, and theories from two or more disciplines or bodies of specialized knowledge to advance fundamental understanding or to solve problems whose solutions are beyond the scope of a single discipline or area of research practice" (National Academy of Sciences 2004). HERO researchers were all from the discipline of geography, but they represented the specialized knowledge of the three of the four research areas of that discipline: human geography (social science), physical geography (physical science), and geographic information science (information science). Reaching the goal of HERO – to explore the possibility of developing the infrastructure needed to build a network of sites devoted to understanding and monitoring human–environment interactions across space and time – was well beyond the scope of any one of these areas of research practice. Instead, investigators worked together in geography's fourth research area – human–environment geography – to build and test the prototype HERO network.

Complex systems, such as integrated human–environment systems, and interdisciplinary research, which integrates the content of multiple bodies of specialized knowledge, work together synergistically. Klein (2004) thinks that the convergence of interdisciplinarity and complexity is part of the larger cultural process of postmodernism in which domains of expertise have become more permeable. She finds that a central feature of postmodernism is the reversal of reductionist tendencies and increasing hybridization, interdependence, and cooperation of science and scientists.

This breakdown of barriers goes beyond the separation among academic disciplines to include greater permeability in the barrier between the academy and larger society, with the related ideas of *usable knowledge* and *usable science* symbolizing the transformation. Usable knowledge is knowledge that generates tools, materials, and ideas that people apply to problem solving and decision making. In human–environment interactions, usable knowledge is "incorporated into the decision making processes of all stakeholders [to enhance] their ability to avoid, mitigate, or adapt to stressors in their environment" (Lemos and Morehouse 2005). Similarly, usable science provides information that specifically addresses societal needs. It is socially distributed, application oriented, and subject to multiple accountabilities (Nowotny *et al.* 2003); rather than being subject only to peer review, science produced for use by society is accountable to the users it aims to serve (Dilling 2007). HERO actively sought not only to build the prototype of an infrastructure to help human–environment scientists break down barriers between areas of scientific specialization, but between those scientists and stakeholders, working together to produce knowledge they could use to develop more sustainable communities.

Transdisciplinarity fuses the concepts of complexity, interdisciplinarity, and usable knowledge/usable science. Klein (2004) cites five keywords that capture the essence of transdisciplinary science: beyond disciplinarity, problem-oriented, practice-oriented, process-oriented, and participatory. She finds that transdisciplinary problem identification does not originate with scientists, but instead with stakeholders, and these real-world problems are not neatly structured, but complex and messy. Moreover, she stresses that transdisciplinarity does not subscribe to reductionist assumptions about how systems work and their components relate, does not operate in the absence of stakeholder and community inputs, and does not suppose that science delivers final, precise estimates with certainty.

The project had the financial and moral support of NSF and the National Oceanic and Atmospheric Administration (NOAA) to take tentative steps into transdisciplinary science. At the time the project received funding in 2000, both agencies were seeking to move their human–environment science into the transdisciplinary realm (see NSF 2003). HERO contributed to NSF initiatives in coupled natural–human systems, coupled biological–physical systems, and people and technology; it also promised to build capacity to address complex environmental challenges through HERO educational programs, scientific outreach, infrastructure development, and technical advances. HERO had the opportunity to move human–environment science into uncharted territory and thereby advance the study of complex environmental systems.

Structure of this book

This book presents an in-depth view of HERO. It is divided into six parts that follow a logical progression and are meant for sequential reading. After this introductory chapter (Part I), Part II discusses the HERO Intelligent Networking Environment and related geospatial technologies and applications. Its three chapters include a look at the theory behind collaboration (Chapter 2), the HERO collaboratory (Chapter 3), and two applications that use local knowledge for representing and reasoning (Chapter 4). Part III sets the scholarly and field contexts for later analyses in its three chapters. The first of these, Chapter 5, offers an overview of the methods HERO used to assess local context and vulnerability, whereas Chapter 6 introduces the historical–environmental context of the four HEROs and Chapter 7 gives a satellite-based overview of the land use and land cover analyses used at these sites. Part IV consists of three chapters on vulnerability. The first of these chapters (Chapter 8) presents the methodological approaches used to assess vulnerability and the second and third chapters (Chapters 9 and 10) apply those methodologies in the four HERO regions. Readers who want to know more about the regions instead of following the intended chapter sequence can jump to Part V, which features four chapters that describe the local human–environment interactions in the HERO regions – central Massachusetts, central Pennsylvania,[9] southwestern Kansas, and the southern Arizona–northern Sonora border lands. The book concludes with Part VI, which reviews the HERO vision; answers the guiding questions and discusses sometimes hard, but necessary lessons learned; and expresses the need for HEROs.

Notes

1. Infrastructure can be defined as the foundation or basic framework of a system or organization. An alternative definition of infrastructure is the resources – such as the rules, software, or personnel – required for an activity.
2. Tip O'Neill was Speaker of the House of the United States Congress, 1977–1987. His most famous quote was, "all politics is local."
3. By integrated, we mean studies that include the dynamic interaction of biophysical and socioeconomic processes contributing to global change. Necessarily, such studies require both biophysical and social scientists.
4. In this context, monitoring is the act of observing, recording, or keeping track of something in order to set a baseline and to establish variations around that baseline over time.
5. Research protocols are guidelines that specify how a research process should work or how scientists should apply a methodology or suite of methodologies to a particular problem.
6. Interoperability refers to the connecting of data, people, and diverse systems. In its technical sense, interoperability describes the standardization of computer systems or software programs to enable the automatic and accurate exchange of data by using common file formats and protocols. In its broader sense, interoperability refers to the management of organizational procedures and cultures for maximizing opportunities for exchange and reusing information. There are several types of interoperability spanning a broad continuum of meaning, including technical, semantic, political/human, intercommunity, legal, and international. See Miller (2000) for details.
7. A collaboratory is a Web-based environment aimed at fostering remote collaboration among scientists.
8. Formal systems are systematic rules that define how to manipulate some defined set of symbols. Formal systems consist of the set of symbols for constructing formulae (an "alphabet"), a set of rules to guide the construction of the formulae (a "grammar"), acknowledged goals of the formulae or principles on which the rules are based ("axioms"), and reliable methods of evaluating the validity of arguments in the formulae ("rules of inference"). A common example of a formal system would be the rules for a card game.
9. The names Susquehanna River Basin (SRB-) HERO and Central Pennsylvania HERO are used interchangably throughout this book, depending on context.

References

Dilling, L., 2007. Towards science in support of decision making: characterizing the supply of carbon cycle science. *Environmental Science and Policy* **10**(1): 48–61.

European Committee for Standardization (CEN), 2007. Accessed at www.cenorm.be.

Farrell, A., and M. Hart, 1998. What does sustainability really mean? The search for useful indicators. *Environment* **40**(9): 4–9, 26–31.

Gahegan, M., and W. Pike, 2006. A situated knowledge representation of geographical information. *Transactions in GIS* **10**(5): 727–749.

IPCC, 2007a. Global climate projections. In Climate Change 2007: *The Physical Science Basis*, eds. S. Solomon, D. Qin, M. Manning, Z. Chen, M. Marquis, K. B. Averyt, M. Tignor, and H. L. Miller. Cambridge: Cambridge University Press.

IPCC, 2007b. *Climate Change 2007: Impacts, Adaptation and Vulnerability*, eds. M. L. Parry, O. F. Canziani, J. P. Palutikof, P. J. van der Linden, and C. E. Hanson. Cambridge: Cambridge University Press.

IPCC, 2007c. *Climate Change 2007: Mitigation*, eds. B. Metz, O. R. Davidson, P. R. Bosch, R. Dave, and L. A. Meyer. Cambridge: Cambridge University Press.

Jianchu, X., J. Fox, J. B. Volger, Z. Peifang, F. Yongshou, Y. Lixin, Q. Jie, and S. Leisz, 2005. Land-use and land-cover change and farmer vulnerability in Xishuangbanna Prefecture in southwestern China. *Environmental Management* **36**(3): 404–413.

Kates, R. W., and R. Torrie, 1998. Global change in local places. *Environment* **40**(2): 5, 39–41.

Klein, J. T., 2004. Prospects for transdisciplinarity. *Futures* **36**(4): 515–526.
Lemos, M. C., and B. J. Morehouse, 2005. The co-production of science and policy in integrated climate assessments. *Global Environmental Change Part A* **15**(1): 57–68.
MacEachren, A. M., 2000. Cartography and GIS: facilitating collaboration. *Progress in Human Geography* **24**: 445–456.
MacEachren, A. M., 2001. Cartography and GIS: extending collaborative tools to support virtual teams. *Progress in Human Geography* **25**: 431–444.
Metz, B., O. Davidson, R. Swart, and J. Pan (eds.), 2001. *Climate Change 2001: Mitigation*. Cambridge: Cambridge University Press.
Miller, P., 2000. Interoperability: what is it and why should I want it? *Ariadne* **24**. Accessed at www.ariadne.ac.uk/issue24/interoperability.
National Academy of Sciences, 2004. *Facilitating Interdisciplinary Research*. Washington, D. C.: National Academies Press.
National Science Foundation (NSF), 2003. *Revolutionizing Science and Engineering through Cyberinfrastructure*, Report of the NSF Blue-Ribbon Advisory Panel on Cyberinfrastructure. Accessed at www.communitytechnology.org/nsf
Nowotny, H., P. Scott, and M. Gibbons, 2003. Mode 2′ revisited: the new production of knowledge. *Minerva* **41**(3): 179–194.
Open Geospatial Consortium (OGC), 2007. Open GIS Consortium. Accessed at http://www.opengis.org.
Tran, L. and S.-Y. Wu, 2001. HERO project's methodology for protocol development: HERO–MEPRODE. Draft report available at http://hero.geog.psu.edu/products/metainfopaperrevised2.pdf.
United States Federal Geographic Data Committee, 2007. Standards. Accessed at www.fgdc.gov/standards/st.andards.html.
Wilbanks, T. J., and R. W. Kates, 1999. Global change in local places: how scale matters. *Climatic Change* **43**: 601–628.
Yarnal, B., and R. Neff, 2007. Teaching global change in local places: the HERO Research Experiences for Undergraduates program. *Journal of Geography in Higher Education* **31**(3): 413–426.

Part II

2

Theory: computing with knowledge to represent and share understanding

MARK GAHEGAN, WILLIAM PIKE, AND JUNYAN LUO

Introduction

The cornerstone of HERO's technological research was our effort to build a HERO collaboratory, which Pike *et al.* describe in Chapter 3. Another area of HERO technological research attempted to link human understanding and formal systems, such as databases, analyses, and models. Ahlqvist and Yu demonstrate two ways that HERO explored this linkage in Chapter 4.

This chapter lays the conceptual foundations for the technologically focused work of Chapters 3 and 4. It concentrates on computing with knowledge structures and on knowledge sharing between participants who may not be co-located. The chapter is organized around the following five questions:

- Why is a conceptual understanding of collaborative work in general, and HERO work in particular, important, and what advantages does it offer?
- What is the nature of concepts that human–environment scientists create and use in their attempts to understand and model Earth's complex environmental systems?
- How can computational systems represent concepts? What languages and reasoning systems can facilitate concept representation and exploit its structure?
- How can a community of collaborators share conceptual understanding?
- What roles might conceptual tools play in an evolving national cyberinfrastructure for human–environment sciences?

In the end, the chapter shows that before we can begin to collaborate we must be able to answer each of the questions above. The answers to these questions enable us to develop a collaboratory infrastructure for the sharing of meaning, concepts, information, and ultimately knowledge. The theoretical elements that enable us to represent concepts, convey meaning and therefore establish a common framework and a means of exchanging knowledge and information are introduced in this chapter. These theoretical elements are then applied with the infrastructure presented in

Sustainable Communities on a Sustainable Planet: The Human–Environment Regional Observatory Project, eds. Brent Yarnal, Colin Polsky, and James O'Brien. Published by Cambridge University Press.

Chapter 3 and the system of representing concepts introduced in Chapter 4 to build the HERO collaboratory.

Mediating exchange with computational systems

Since their advent, computers have been used by scientists to help distribute information. This sharing was initially by punched cards, magnetic tapes, or disks, and more recently by databases and files connected across the Internet. Had the HERO project been conducted 20 years ago, the focus would undoubtedly have been on the sharing of data, since at that time data sharing was a huge problem pervading the Earth sciences. Now this problem is largely solved, and reliable standards for data exchange are offered by the Open Geospatial Consortium (OGC: www.opengeospatial.org/) and the Open-source Project for a Network Data Access Protocol (OPeNDAP: www.opengeospatial.org/), among others.

Figure 2.1 shows that the problems of exchange of data, information, and knowledge are in fact part of a continuum of abstraction. Importantly, the various levels of this continuum represent significant steps in the progress of computer and information science. Solutions for most of the levels now form the basis of widely used protocols and standards. Yet, the top two levels – those of semantics[3] and pragmatics[4] – remain unsolved, and solutions are elusive because these two levels strive

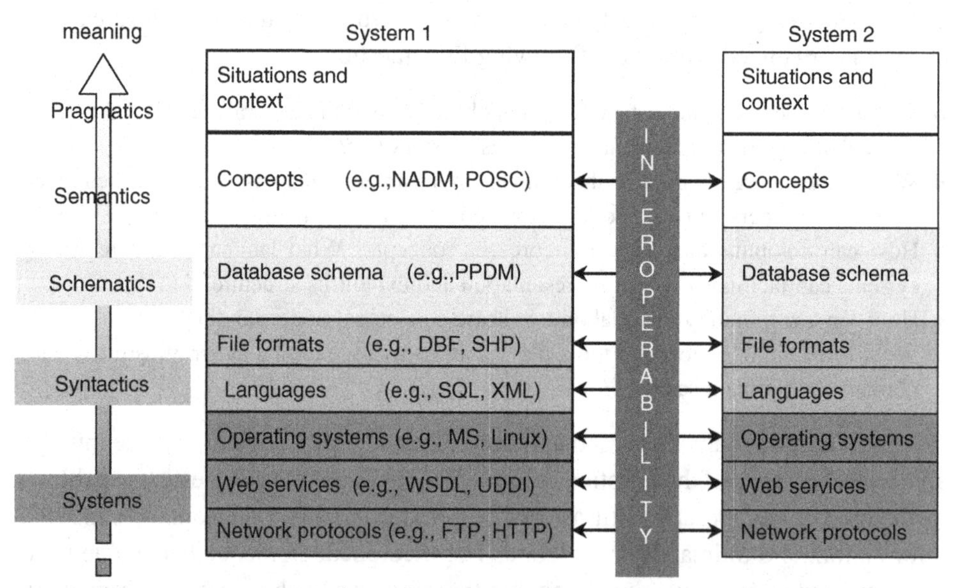

Figure 2.1. Interoperability[1] between two systems expressed as layers. Shading represents the degree of abstraction, with white being the most abstract. Maturity of standards is greatest for the lower-level descriptions of data in terms of systems, schematics, and syntactics.[2] Figure based on Brodaric and Gahegan (2006).

to represent different aspects of meaning or knowledge that are by nature difficult to describe, subjective, and contextual.

Why concentrate on knowledge?

Why is a conceptual understanding of collaborative work in general, and HERO work in particular, important, and what advantages does it offer?

In science (and life), we often go to great lengths to get our ideas understood. In a research setting, such efforts typically take the form of written articles, presentations, and informal conversations. But in collaborative science, and particularly as science communities begin to coalesce around cyberinfrastructures and digital libraries, the results of research usually become divorced from their originators. Sharing our science outcomes effectively involves more than simply making them available – we must also help others to understand them and to use them correctly in their own work. In the context of HERO, each of the four investigative teams tried to understand human–environment interactions within a particular place. These teams wanted ultimately to share their results and their emerging knowledge with each other so that a national, and perhaps international, picture of important aspects of human–environment interaction, such as vulnerability to natural hazards, could be assembled.

Our knowledge is often hard-won. It pervades and influences both how we see the world and how we act in it (including the science we do), and yet we take little care of that knowledge. We are finicky about how we treat numbers, and have strong, universally understood methodologies for the computation and reporting of statistics. In contrast, we let knowledge languish, leaving it partially represented or implied in a dense written record where it can be difficult to find or disentangle. In practice, its fingerprints are all over our work; hypotheses, methods, data, results, conclusions all bear the subtle marks of the understanding we bring with us.

Ultimately, the usefulness of our data to future generations of scientists will depend not only on their ability to decode the syntax and schemas[5] used to represent the data and to comprehend our semantics described in ontologies,[6] but also on how well they can understand the context and situations that shaped the science that produced the data. Such factors as context and situation therefore need to be made as explicit as possible and folded into the computational schemas that represent the data (Magnani *et al.* 1999; Langley 2000). Clearly, we cannot currently hope to capture all the conceptual understanding of experts (Penrose 1989; Dreyfus 1972), but we can at least make a start.

Although computational systems have made great strides in recent times, Vannevar Bush's[7] vision of intimate support for the research process (Bush

1945) remains elusive. It is not that contemporary computational systems are not up to this task, but that software applications have, by and large, developed as independent, monolithic systems that simply do not support the ideas of capturing and working with knowledge. The emphasis instead is placed on working with data.

Our overall aim here is not to supplant the human–environment scientist by a computer with a perfect representation of human–environment knowledge, but to create representations that enable the scientist to grasp important concepts that underlie datasets, methods, articles, and other resources more effectively and more efficiently. In pursuing this goal in the chapters that follow, we make incremental progress towards this aim. At the time of this writing, however, there is a renewed interest in computational semantics,[8] fueled in part by recent progress in description languages[9] and description logics[10] for computing with knowledge structures. This interest has, in turn, led to the production of several useful semantic descriptions of Earth and environmental science, including classifications of places and descriptive characteristics and processes – good examples of which are available from the NASA Ontology Website (http://sweet.jpl.nasa.gov/ontology/) under a program that aims to describe data products in terms of a Semantic Web of Earth and Environment Terminology (SWEET). Equipped with these new description logics, and semantic domain descriptions,[11] we are able to make progress towards

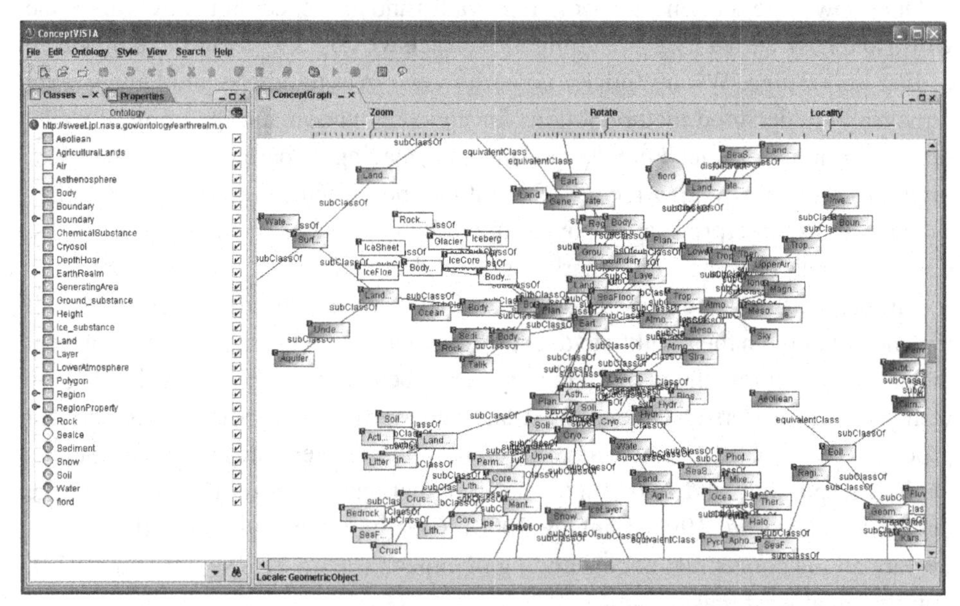

Figure 2.2. Part of NASA's EarthRealm ontology, which forms a descriptive, standardized language for describing Earth's systems and for semantically tagging related resources such as images and maps.

better representation and communication of human–environment meaning between individuals and teams of collaborators.

Figure 2.2 shows an example of part of NASA's EarthRealm ontology, displayed using a concept mapping application called ConceptVista.[12] The left panel in the figure shows a list of concepts (classes) that represent a standardized list of agreed-upon terms for describing aspects of Earth's physical environment. The right panel depicts relationships between these terms in an interactive browser.

The nature of human–environment concepts

What is the nature of concepts that human–environment scientists create and use in their attempts to understand and model Earth's complex environmental systems?

Representing an evolving world

One could assume that better education and improved documentation of the conceptual resources that researchers use, such as classifications and models of processes, could solve the problem of sharing knowledge. Such advances would help, yet ultimately we face the problem that the world around us is continually changing and so is our understanding of the world. This problem of capturing the evolving understanding of an evolving world is certainly not new – Heraclitus, writing around AD 400, recognized the problem thus (quoted from Sowa 2002): "Everything is in flux. But what gives that flux its form is the logos – the words or signs that enable us to perceive patterns in the flux, remember them, talk about them, and take action upon them even while we ourselves are part of the flux we are acting in and on."

In the context of this insight, our task here is to design the words or signs that will enable us to make sense of the flux and to communicate our understanding. Such words or signs connect mentally to the things they stand for (that we express via concepts and their interactions). But, we also need to position these words or signs within the flux – that is, place them within an evolving context.

Whenever we create maps, models, articles, protocols, datasets, or other outcomes, we do so based on our current conceptual understanding. Such products implicitly or explicitly contain aspects of our conceptual understanding, ranging from the categories we construct to describe climate change vulnerability to the theories we create to describe likely future scenarios of climate and society. If we think of ourselves as contributors to the ongoing collections of useful maps, models, articles, protocols, and datasets describing the state of our world, then a pertinent question we could ask ourselves is, "How useful a record will we leave behind for the next generation of scientists?"

If the concepts we employ are well understood by others, and if these concepts are not evolving, then our records may prove to be adequate. However, when we consider that many of the concepts we use to understand the world are evolving, just as the systems they describe also evolve, then we have recognized a significant problem.

During our careers, we have witnessed conceptual changes in such human–environment notions as vulnerability and the various classifications used in land-use mapping, and importantly also in our understanding of the roles these concepts play in broader Earth science. Nevertheless, our datasets and tools do not capture that understanding, even though they are highly dependent upon it. How can future scientists understand our work without a description of this unrepresented and evolving knowledge? Perhaps it is possible for them to construct an understanding from articles we wrote and presentations we made. The stakes are high, however, given the importance of understanding our changing world; do we want to hang our work on such a flimsy structure? We are in a situation where the human–environment dynamic is changing, perhaps faster than ever before (e.g., National Research Council 2001). Leaving a more complete account of our efforts to record the state of the Earth's systems and human–environment interactions will allow future generations of scientists to understand these changes better and perhaps be able to respond to them more wisely.

The image in Figure 2.3 shows a concept map of the emerging conceptual understanding of HERO researchers that was captured during discussions of protocols for describing climate change vulnerability during an all-hands project meeting in Arizona, 2004. Note that vulnerability (the central node) is composed of the three major dimensions: sensitivity, exposure, and adaptive capacity (dark gray oval nodes). These dimensions can be further decomposed into additional components (square dark gray, light gray and white nodes). During the life of the project, ideas about vulnerability changed considerably as new perspectives were taken into account and as new situations were encountered in the field. Concept maps such as Figure 2.3, for example, could help other researchers understand how HERO team members conceived of vulnerability at the time a particular vulnerability map was created.

What aspects of meaning should we represent?

Representing our mental concepts, and the relationships between them, poses many challenges. Perhaps first among these challenges is the problem of which aspects of a concept are useful to represent. There are many different perspectives that might be taken to answer the question, "What does this concept mean?" or , "How should this concept be represented?" (Brodaric *et al.* 2000).

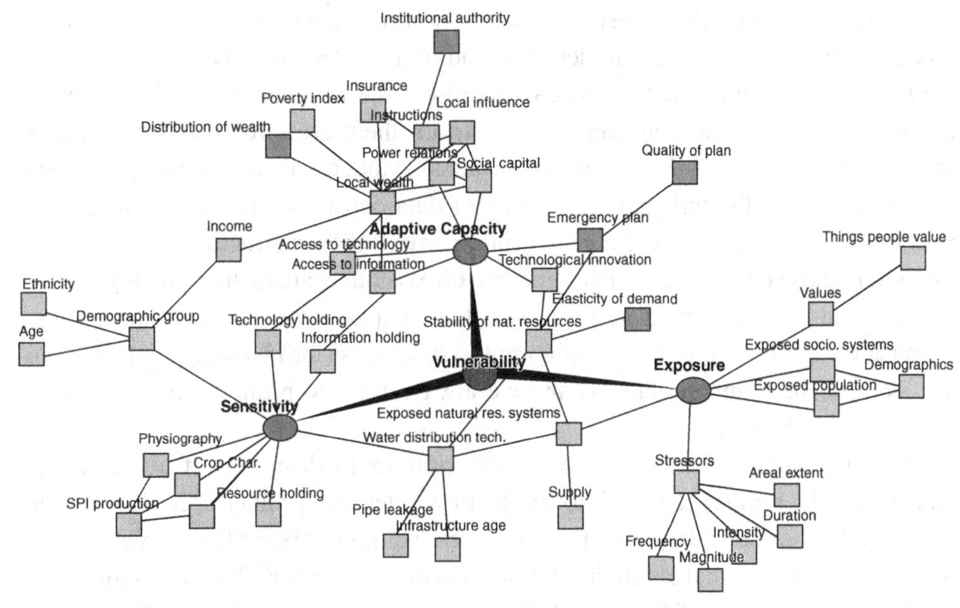

Figure 2.3. A concept map showing the notion of vulnerability to climate change, captured during discussions of protocols for vulnerability during a HERO all-hands meeting in Arizona, 2004. See text for further details.

Some scientists tend to concentrate on questions relating to "what," "where," and "when," but other scientists focus on questions of "who," "how," and "why." Accordingly, we could base our conceptual descriptions on:

(1) Data used to synthesize a concept
(2) Methods used to form a concept
(3) Tasks that use a concept
(4) Relationships that a concept has to other concepts
(5) Situations surrounding the creation and use of a concept
(6) Researchers who use a concept.

The first three conceptual descriptions would require careful representation of the scientific process and, particularly, of the workflows used to create and deploy concepts (e.g. Gahegan *et al.* 2003). The fourth requires a more semantically oriented set of tools, such as computational ontologies (Denny 2004). The fifth requires us to represent some of the pragmatic aspects that surround concepts, such as when and where they were used (Pike and Gahegan 2007). The sixth involves the representation of social networks that underlie a research community (Berman and Brady 2005). At the one extreme, any one of these perspectives, by itself, provides only a partial account of underlying meaning. At the other extreme, there are some scientific endeavors that require only one or two of these perspectives to be complete. In most cases, however, a healthy mix of these perspectives reduces the risk that concepts may be misunderstood and used inappropriately.

One could assume that the place to represent the conceptual details of, say, a dataset should be in a metadata document, but there are three reasons why this assumption is incorrect. First, current metadata standards focus on data producers and do not typically anticipate the needs of data consumers (Fisher *et al.* 2004). Second, few standards exist for representing any of the six structures in the above list, and in those cases where such standards do exist, they are in their infancy. Third, a traditional text document might not be the right structure for such knowledge because we often need to represent complex structures (for instance, items 4 and 6 are likely to be graph structures) and we always want to promote more effective searching and browsing.

We could apply the six conceptual descriptions to determine if a land-cover dataset would be useful for us to apply to a scientific problem. To answer the question, "What does the dataset mean?" (for example, "What land-cover classes does it contain and what are their relationships one to another?") would suggest the need for descriptions of the underlying classifications or ontologies. To answer, "When and where did the dataset originate?" would require the relevant spatial scale and temporal interval and would demonstrate that geographical metadata still have an important role to play. To answer, "How was the dataset created?" would call for a representation of the various steps taken (such as gathering training data) and methods used in its creation (such as classifiers and accuracy-reporting tools). Note here, however, that deeper questions could reveal insight into methodologies, hypotheses, theories, and metaphysics that are currently not represented. To answer, "How should the dataset be used?" could require good documentation and examples. To answer, "Who created the dataset?" would raise issues of trust and intention, although the information itself could be easily included in metadata descriptions. To answer, "Who has used the dataset?" would connect with the idea of social networks of users and would require documenting each time the resource was used.

The final question, "Why was the dataset made?" is far more open-ended, and consequently far more problematic to represent. It evokes agendas and motivations, which can have personal, social, and political facets, as well as scientific ones. It is not clear how to represent such facets and how much detail would be useful to represent.

Thus, answering some of these questions poses a severe, perhaps impossible challenge to our current representational tools. Nevertheless, answers to many of the above notions can indeed be captured, represented, and communicated – to some extent – via existing or emerging information technologies such as the Codex system, created to help manage the knowledge resources of the HERO team. We briefly describe Codex here; a detailed description follows in Chapter 3.

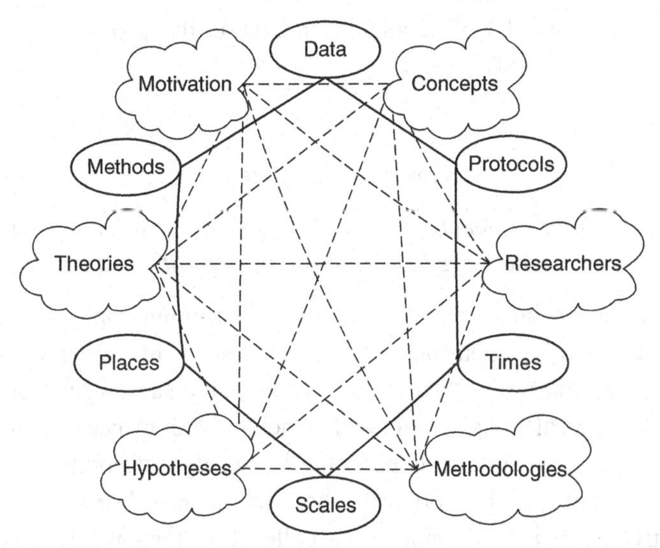

Figure 2.4. A nexus of the various knowledge facets described in the text that together form a conceptualization of the roots of meaning in human–environment research.

In the Codex project, we based our conceptual framework for representing multifaceted knowledge on the nexus structures described in the writings of the philosopher Alfred North Whitehead (1938), as shown in Figure 2.4. We treated each facet of meaning as a concept or as a set of related concepts. Notice in the figure that the concepts representing the more straightforward aspects of "what," "where," "when," and "how" are depicted as ellipses, whereas the harder-to-represent notions relating to "methodology," "motivation," and "actors" are shown as more ephemeral clouds. Although not all possible connections are shown in the figure, the implication is that all conceptual aspects are interwoven into a single web of meaning. To take the analogy further, a subset of these concepts sometimes can carry meaning, but removing too many of them (and the links among them) can cause the structure to become weak or unstable to the point of failure.

By providing a conceptual framework whereby the user could represent (or the system could capture) aspects of meaning, Codex supplied a meta-model from which models of knowledge could be constructed by adopting (and perhaps by further specializing) some pieces and by ignoring others. For example, during the later phases of the HERO project, the Codex knowledge portal provided support for concepts, files, tools, groups of people, places, and tasks. The remaining facets were not included at that time, but could be included in future work. With HERO, we have just begun to study the nature of the connections among the concepts shown in the nexus and to evaluate the utility that each concept adds, so while the nexus is for us a meta-model of understanding in human–environment science, it is still only a straw model. We do not intend Codex and the nexus as the last word in knowledge modeling, but rather one of the first.

Representing concepts

How can computational system represent concepts? What languages and reasoning systems can facilitate concept representation and exploit its structure?

How computers understand a representation is an intriguing topic. There have been many long debates in philosophy and artificial intelligence regarding the possibility of intelligent machines that truly "know" the world in the same way that humans do. Although the theoretical existence of "real" machine intelligence is highly questionable, in practice it is possible to explain the knowledge of computational systems by using the ideas of the philosophical school of functionalism, which regards the intelligent activities of human minds as a collection of mental functions (Putnam 1988). From the functionalist perspective, it is the interaction between mental functions and the environment that directly affects human life. Those functions are unobservable and only knowable through observed behaviors. Therefore, it is possible to understand mental functions by studying the outcomes they demonstrate. While the functionalist view itself is highly debatable in philosophy (Ravenscroft 2005), it provides a convenient theoretical ground for the development of useful computational systems. For example, establishing whether a system "knows" the meaning of the term *climate change* could be simplified to determining whether it responds to this term by highlighting related knowledge structures and performing desired computations with methods and data. If it responds in this fashion, then the system is interpreting the term in a semantically sound way. Thus, we can regard the development of knowledge representation as a process that aims to translate informal knowledge into structures that can be mapped into a functionalist system, in which meaning and reasoning are defined by the actions of the system (Brachman and Levesque 1985).

Formal systems

A formal system can be defined by the following five elements:

A finite set of symbols;

A grammar that strictly defines how to construct well-formed formulae (WFF) using the symbols;

A set of axioms, which are predefined and pre-assumed WFFs;

A set of inference rules, which defines the relations between WFFs, especially how to derive new WFFs from existing ones;

A set of theorems, which include all axioms and all WFFs derived from them.

The term *formal* comes from the philosophical concept *form* (which is in contrast to *content*); here it indicates that the symbols of the system have no meaning beyond their structural semantics, and the reasoning of a formal system is entirely subject to the structure of the system as defined by the above five elements. In other words, formal systems are based on the assumption that the structure of a representation can be separated from its meaning, and that the system provides ways to define the structure. Humans may be able to assign and interpret meaning from this kind of structure, but by itself the structure is simply an empty container (albeit a very carefully defined one).

Formal models

A formal model is a model of some domain (such as human–environment interaction) in which all the basic constructs of the domain are defined as formal symbols and their relationships defined as formal rules. A formal model is usually built on top of a more general formal language – such as the Web Ontology Language (OWL) or Resource Description Framework (RDF) – defined below. A formal model also includes symbols and relational rules dedicated to the targeted domain, so the model can be regarded as the formal language specifically tailored to a targeted domain, and formalization can be regarded as the act of building such a language. The relationships among formal languages, formal models, and knowledge representations can be summarized by the layered architecture depicted in Figure 2.5. Note that Level 1 corresponds to the nexus of meaning shown in Figure 2.4, and Level 2 to the various HERO research resources.

One can think of the above relationships as similar to the case of natural language representation. English, for example, served as the ground language for human communication within HERO (Level 0). The terminology of the human–environment domain(s) then defined the domain-dependent vocabulary that was used to

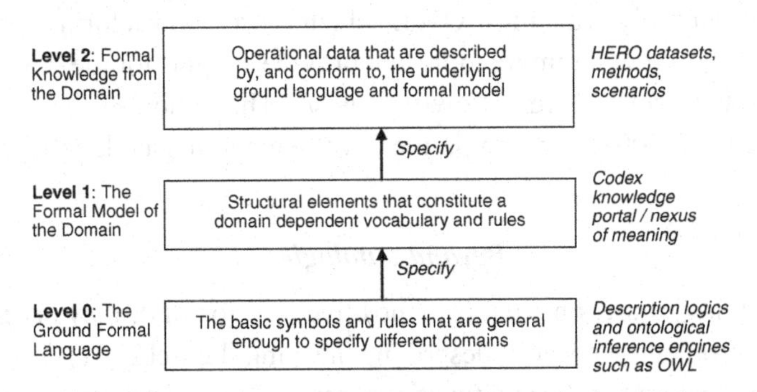

Figure 2.5. A layered view of formal languages, formal models, and formal knowledge; inspired by Uschold and Jasper (1999).

describe human–environment knowledge (Level 1). Finally, textual descriptions about particular human–environment problems, regions, or applications, comprised Level 2, drawing on the vocabulary defined in Level 1.

Ontologies

New research in information science in the area of ontologies is leading towards richer representations of semantics, as they relate to concepts and relationships that researchers use. These semantics can include useful relationships that can be formally defined, such as *part-of*, *instance-of*, and *type-of*. So, for example, we can state formally that a particular geographical place (such as southwestern Kansas) is part of a larger system (such as the industrialized wheat farming), is also an instance of some process occurring (such as aquifer depletion), and that aquifer depletion is a type of human–environment interaction that can leave the resident population at long-term risk of losing industrialized wheat farming.

A computational ontology is a model that is rich in semantic constructs, describing concepts and relationships of significance to a knowledge domain or community. Note that this definition is different from the philosophical meaning of ontology, which is concerned with describing what is true, or what can be, from a metaphysical standpoint. Computational ontologies can represent (and enforce) certain kinds of semantic structures, such as generalization hierarchies and part-of relationships (Guarino 1998; Holsapple 2003). Under various rules and constraints, ontological systems are typically implemented via description logics (Sowa 1999), applied to an underlying graph structures, and supported by computable inferences.

The current interest in ontologies within many research communities is a direct result of their ability, first, to provide a ground formal language with which meaning can be constructed (Level 0) and, second, to define the various concepts, relationships, rules, and constraints by which a particular formal domain model can be synthesized (Level 1). One of the most popular ontological languages and modeling environments currently available is OWL, which is itself both a formal language and a set of application programming tools that provide support for the development of domain models (www.w3.org/TR/owl-features/). The notion of the semantic web (Berners-Lee *et al.* 2001) is based on the development of such languages.

Beyond ontologies

Although ontologies are, on balance, a positive step forward, they are not a silver bullet for the problems we face in describing meaning (Fensel 2001). They approach meaning in a purely symbolic and functional sense. There is debate among ontology researchers as to whether there really are unifying ontologies, which can form a conceptual basis for understanding domain knowledge (such as human–environment interaction) that practitioners can agree on, or whether there are as many ontologies as there are practitioners – or more if you count the fact that understanding changes through time. Ontologies fail in that they do not address how researchers working separately or together apply knowledge (i.e., epistemology), and they do not capture a practical sense of what data, tools, and methods experts use, why they make the choices they do, or what deep conceptual understanding they make their choices from.

In our efforts to support the needs of human–environment researchers – to whom epistemologies are equally as important as ontologies – we have extended the common notion of computational ontologies as a means to represent ontological knowledge. Ironically, we have also used ontologies to describe aspects of epistemology – specifically aspects of pragmatics, workflows, and social networks. Although not designed for this purpose, the abilities of ontological tools to reason with complex relationships and abstract concepts makes them an ideal basis on which to construct systems that implement the nexus shown in Figure 2.4. Consequently, we can define a common semantic language for human–environment research communities, such as HERO, that can be used to:

- Construct concept maps to represent the conceptual understanding underlying resources (e.g., data, tools, methods, or ideas) that the community could share
- Describe (tag) resources with aspects of their meaning
- Provide audit trails that show how resources have been used, and by whom
- Model workflows to show how resources were created.

Chapter 3 provides further details of the Codex knowledge portal that provides this functionality to collaborating researchers. Chapter 4 gives additional insights into the internal structure of concepts and the possible construction of similarity measures between concepts.

Recording and sharing conceptual understanding

How can a community of collaborators share conceptual understanding?

The metaphor of the scientific notebook

A useful metaphor for our purposes is the scientific notebook that provides a full description of experiments, augmenting the results with hypotheses, aims, methods, and conclusions, thus providing all that is needed for a third party to repeat the experiments. Taking a more long-term perspective, such notebooks also serve as narratives that recount the course of a project or scientific career, making them valuable aids to tracing the development of ideas. The keeping of notebooks is still a requirement in many of the laboratory-based sciences, but is not an established practice in the human–environment sciences or geographic information sciences. This lack of a formal record makes it difficult for other researchers to gain insights into our work. That is not to say such aspects of our work are not reported; pieces may appear in project reports to sponsors, published articles, oral presentations, and even scripts and log files created by commercial software. However, in these forms the details are difficult to elucidate and their connection with the results (data) is tenuous or non-existent. In short, they are not accessible enough. Even in the form of the traditional notebook, these details would not be readily searchable without considerable investment in text pre-processing and mark-up.

From this perspective, our research can be seen as developing a modern, computer-based scientific notebook – one that captures and represents a rich characterization of the scientific process by keeping an account of tasks performed by individuals and, thus, a reproducible record of activity. As such it extends earlier efforts that address only the sharing of data and methods (Myers *et al.* 2001).

Imagine that aspects of this computer-based notebook could be shared with other researchers. Not all contents would be shared, but individuals and groups could choose to what degree they grant other researchers access. Such a notebook – if sharable by many scientists – could help provide answers to questions such as:

- Who first introduced this concept?
- What methods and datasets have been used to synthesize or signify this concept?
- Which individuals and groups have applied this concept, and to what problems?
- Do the reported aims of two individuals using the same concept agree?
- Do they further agree with the aims of the originator of the concept?

This kind of sharing could lead to significant productivity gains, since researchers could learn not only from the articles of colleagues, but also from work practices, experiments, conceptual understandings, and other previously hidden details that could save time and effort.

Despite increasing reliance on digital datasets, computational analyses, and networked data libraries, many of the advantages of electronic science, such as automated capture of audit trails that could reveal much about the course of exploration, go unexploited. The "undo" command, coupled with the speed with which analyses can be repeated, has seemingly obviated the need for meticulous records of one's investigation. Electronic science too easily emphasizes products over process, illustrated by the profusion of Web-based data archives. While accessible databases are a commendable first step, we suggest that the contemporary state of computing offers possibilities for much richer information sharing. We argue for a philosophy of e-science that marries the interconnectedness of digital research tools with the introspection enabled by traditional record-keeping. In systems informed by this philosophy, interoperability is not simply a technical characteristic, but a design strategy that promotes effective cooperation between both human and electronic components of the research process. Moreover, by leveraging advanced computational and visualization methods, collections of shared electronic notebooks might provide insights into the process of knowledge construction – how concepts are proposed, debated, adopted, and refined through time by collaborating researchers.

To support the idea of sharing only some aspects of work, and keeping others hidden, there is a need to define levels of sharing via a hierarchy of private, group, and community interfaces to notebooks. At the individual level, a notebook holds personal concepts that others cannot access; at the group level, the notebook represents points of agreement (or disagreement) among collaborating scientists (such as the HERO conceptual model of vulnerability shown in Figure 2.3); and at the community level, the notebook holds broadly shared, discipline-wide concepts (such as the NASA EarthRealm ontology shown in Figure 2.2). In the context of human–environment research, a community notebook could define nationally or internationally agreed concepts leading to shared protocols for land-cover assessment, land-use change, or natural hazard vulnerability. By contrast, a group notebook could represent the shared knowledge of researchers working on a specific problem or location, such as a local watershed.

The Codex knowledge portal represents our attempt to construct such a notebook. In addition to supporting different perspectives on underlying knowledge, it supports shared, but private workspaces where group resources, and their understanding of those resources, can reside. The portal is accessible to all members, wherever they may be, via a Web browser. Figure 2.6 shows a user's personal workspace after login, with lists of his conceptual and data resources – that is, items held personally and items held by groups of which he is a member. See Chapter 3 for more on Codex.

The future

What roles might conceptual tools play in an evolving national cyberinfrastructure for human–environment sciences?

At the time of writing, much money and intellectual effort is being invested in the construction of national and international cyberinfrastructures to support science and social science activities. A sample of these programs includes Network for Earthquake Engineering Simulations (NEES), the Space Physics and Aeronomy Research Collaboratory (SPARC), the National Ecological Observatory Network (NEON), the Geosciences Network (GEON), Chronos, EarthScope, the Science

Figure 2.6. A Codex personal workspace showing concepts and files – two of the access points into the nexus of Figure 2.4.

Environment for Ecological Knowledge (SEEK), the Grid Physics Network (GriPhyN), the International Virtual Data Grid Laboratory (iVDGL), and the High Energy Physics Collaboratory for the ATLAS project (NSF 2003). These efforts represent the beginnings of programs to improve the process of science, and much of the work to date has concentrated on implementation questions and is targeted at the lower levels of cyberinfrastructure organization (Figure 2.7). As more practical problems are being resolved, attention is shifting from the low-level syntax and schema to the upper-level semantics and pragmatics.

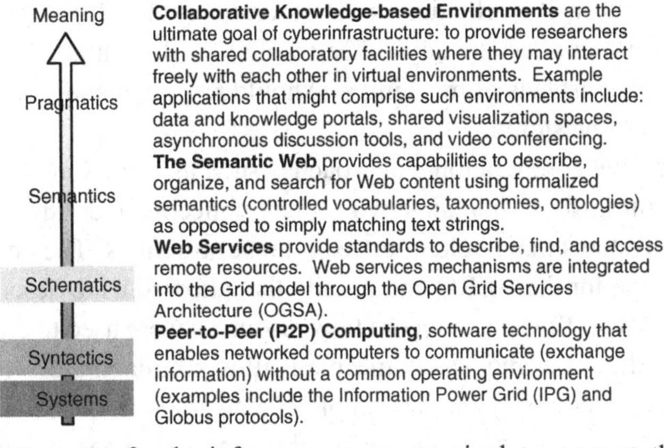

Figure 2.7. Layers of cyberinfrastructure are required to support the idea of collaborative knowledge-based environments.

The vision of what cyberinfrastructure could become, and the roles it could play in science, is much broader than providing a shared supercomputer and massive data store for a community of researchers. There is a recognized need for sharing methods and workflows, for representing cybercommunities, and for fostering knowledge-level understanding (NSF 2003). As pointed out in the introduction, this vision is not new; for instance, it played a central role in some of the writings of Vannevar Bush presented earlier in this chapter. His vision finds new life in these recent efforts to organize and share the fruits of our research labors.

From a historical perspective, the HERO project was among the first of many such projects that aimed to go beyond sharing resources to sharing *understanding* of these resources – otherwise the project's research would have produced nothing more than a digital library. There is still much to do in this regard, but involvement with HERO has led some team members to begin working with other cyberinfrastructure teams, notably the Geosciences Network (GEON: www.geogrid.org) which is experimenting with a Codex-like portal for accessing large collections of geoscientific data.

Conclusions

With each of the questions posed at the start of this chapter answered we have demonstrated the importance of having a conceptual understanding of collaborative work in general, and for the HERO work in particular. We highlighted that by working from a common conceptual understanding we can communicate human–environment meaning between individuals and collaborators more readily. We noted that human–environment concepts have meanings that change and evolve over time and as a result we need to record snapshots of our conceptual understanding as a record for the researchers who will work with our collaboratively developed products (maps, models, diagrams, articles, or any other results) in the future. In doing so we grappled with what aspects of meaning we should represent and how those aspects of meaning should be stored.

We examined the computational approaches such as formal models and ontologies to representing these human–environment concepts and explored languages and reasoning systems that could operate on those structures. The end result was a common language for the HERO community that allowed us to describe and share meaning, model workflows, and record how resources were used to create products. The ability to describe and share meaning allows collaborators to share their conceptual understanding by linking simple objects together to explain more complex concepts (e.g. vulnerability comprises the concepts exposure, sensitivity, and adaptive capacity, which are each made up of simpler concepts).

The theory presented in this chapter forms the basis of building the infrastructure of the collaboratory (Chapter 3). The need for human–environment researchers to be able to share meaning, data, information, and knowledge drove the development of the collaboratory, while the need for conveying meaning and making use of a common framework influenced the tools used within the infrastructure. Chapter 4 presents methods for representing and reasoning with conceptual understanding and how we can achieve a common conceptual understanding. Again the ideas presented here of representing meaning with a view to building an overarching ontology are crucial to enabling this shared understanding. The theory that underpins the infrastructure, tools, and use of the collaboratory permits a consensus building approach responsible for most of the results presented in this book. These results benefited from the collaboratory users being able to negotiate a common understanding, adhering to a shared vocabulary, and using a common set of methods.

Notes

1. Interoperability in a computing context is the concept of seamless data sharing across disparate systems. For data to be interoperable, their meaning must be retained no matter how they are stored. It is possible for the meaning of data to change as a result of the conversion process. More generically when sharing information or knowledge the sharing (or interoperability) can only take place if the two entities exchanging data understand the other's terms, concepts, and language.
2. Syntactics is the formal structure of language, data (e.g. a database structure) or the format of the data such as Arc/INFO, ArcGIS, Microsoft Excel. Syntax specifies the rules of language or how data must be stored (for example the data must only be numeric).
3. Semantics is the relationship between how and what we store (the syntax and schema) of data, and how that data relate to concepts and entities in the real world. For example if we wanted to represent a township's water supply in a database we might store a point with spatial coordinates (e.g. latitude and longitude) and some attribute information (e.g. water quality). That database point might represent a water supply but the concept of what a water supply is will not be recorded in the database.
4. Pragmatics is the study of the aspects of meaning and use of language that are dependent on the speaker, the listener, the context of the discussion, and the goals of the speaker. Pragmatics also considers information that is not supplied to a listener but implied.

5. Schemas are definitions that specify the structure of data. They might be data structures, data models, or the representation of data within a computer.
6. Ontology is defined as "a logical theory that gives an explicit, partial account of a conceptualization" (Guarino 1998). A less formal definition is the study of things that exist. An ontology provides a structured, homogeneous view of the domain and its concepts, rules, and roles (Guarino 1998), allowing a standardized description to be constructed to describe properties of data, methods, process, and concepts.
7. Vannevar Bush is credited with establishing a relationship between the scientific community and government in the USA during World War II, changing the way scientific research was carried out in the USA and leading to the establishment of the National Science Foundation. As Director of the Office of Scientific Research and Development in 1944 Bush had a vision of government-supported scientific research under a unified agency responsible for funding and coordination. While not all of Bush's ideals were realized the NSF was established in the 1950s at his urging.
8. Computational semantics like formal semantics seeks to formalize the meanings of sentences and discourses precisely. Computational semantics differs from philosophical semantics in that instead of aiming to situate meaning within a general understanding of the intentionality of human mental states computational semantics attempts to answer the following questions (from Stone 2000):
   - How can ambiguities and contextual dependencies in the meanings of terms and concepts be represented compactly?
   - How can these meanings be resolved automatically and efficiently?
   - How can semantic representations be related to other computational representations of the world?
   - How can we compute the inferential consequences of semantic representations?
9. Description languages allow for the formal representation of semantics and ontologies. These languages support formal semantics as they allow the relationships that exist between entities in an ontology to be formally defined, contain reasoning operators that allow objects to be compared and methods for exchanging entities between different representations of those entities (between different ontologies).
10. Description logics are a family of knowledge-representation languages upon which description languages are based. The logical operators within description logics define the membership of a concept or term within a group (or class) and allow the comparison of concepts based on their membership and roles within these classes.
11. Semantic tagging aims to join keyword tags to relational structures such as an ontology so that the keyword is placed in the context of a domain. For example vulnerability has many meanings in different contexts but linking the keyword "vulnerability" to an ontology of drought and climate change puts the term into a specific context.
12. ConceptVista was developed at the GeoVISTA Center of The Pennsylvania State University to serve several funded research projects, including HERO. It provides a highly adaptive visualization of conceptual structures and includes many advanced tools for computing with concepts and relationships. Interested readers can go to http://www.geovista.psu.edu/ConceptVISTA, where the application is available for download.

References

Berman, F., and Brady, H., 2005. Final Report: NSF SBE-CISE *Workshop on Cyberinfrastructure and the Social Sciences*. Accessed at www.sdsc.edu/sbe/.

Berners-Lee, T., J. Hendler, and O. Lassila, 2001. The semantic web. *Scientific American* **284**(5): 34–43.

Brachman, R., and H. Levesque (eds.), 1985. *Readings in Knowledge Representation*. Los Altos, CA: Morgan Kaufman.

Brodaric, B., and M. Gahegan, 2006. Representing geoscientific knowledge in cyberinfrastructure: challenges, approaches and implementations. In *GeoInformatics:*

Data to Knowledge, Special Paper No. 397, ed. A. K. Sinha, pp. 1–20. Boulder, CO: Geological Society of America.

Brodaric, B., M. Gahegan, M. Takatsuka, and R. Harrap, 2000. Geocomputing with geological field data: is there a ghost in the machine? In *Proceedings of GeoComputation 2000*, Greenwich, UK, August 23–25, 2000, abstract.

Bush, V. (ed.), 1945. *Science: The Endless Frontier*. Washington, D. C.: United States Government Printing Office.

Denny, M., 2004. Ontology tools survey, revisited: O'Reilly XML.com. Accessed at www.xml.com/pub/a/2004/07/14/onto.html.

Dreyfus, H., 1972. *What Computers Cannot Do: The Limits of Artificial Intelligence*. New York: Harper and Row.

Fensel, D., 2001. *Ontologies: A Silver Bullet for Knowledge Management and Electronic Commerce*. New York: Springer.

Fisher, P., A. Comber, and R. Wadsworth, 2004. Land cover mapping and the need for expanded metadata. *Proceedings of GIScience 2004, 3rd International Conference on Geographic Information Science*, University of Maryland Conference Center, USA, October 20–23, 2004.

Gahegan, M., X. Dai, J. Macgill, and S. Oswal, 2003. From concepts to data and back again: connecting mental spaces with data and analysis methods. *Proceedings of the 7th International Conference on GeoComputation*, Bristol, UK, Sep. 2003. Accessed at www.geocomputation.org/2003/index.html.

Guarino, N. (ed)., 1998. *Formal Ontology in Information Systems*, Vol. 46 of Frontiers in Artificial Intelligence and Applications. Amsterdam: IOS Press.

Holsapple, C. W. (ed.), 2003. *Handbook on Knowledge Management,* Vol. 2, *Knowledge Directions*. Berlin: Springer.

Langley, P., 2000. The computational support of scientific discovery. *International Journal of Human–Computer Studies* **53**: 393–410.

Magnani, L., N. J. Nersessian, and P. Thagard (eds.), 1999. *Model-Based Reasoning in Scientific Discovery*. New York: Kluwer.

Myers, J., E. Mendoza, and B. Hoopes, 2001. A collaborative electronic notebook. *Proceedings of the IASTED International Conference on Internet and Multimedia Systems and Applications*, Honolulu, HI, August 13–16, 2001, pp. 334–338.

National Research Council (NRC), 2001. *Climate Change Science: An Analysis of Some Key Questions*. Washington, D. C.: Committee on the Science of Climate Change, National Research Council.

National Science Foundation (NSF), 2003. *Revolutionizing Science and Engineering through Cyberinfrastructure*, Report of the National Science Foundation Blue Ribbon Advisory Panel on Cyberinfrastructure. Accessed at www.communitytechnology.org/nsf_ci_report/.

Penrose, R., 1989. *The Emperor's New Mind: Concerning Computers, Minds, and the Laws of Physics*. Oxford: Oxford University Press.

Pike, W., and M. Gahegan, 2007. Beyond ontologies: towards situated representations of scientific knowledge. *International Journal of Human–Computer Studies* **65**(7): 674–688.

Putnam, H., 1988. *Representation and Reality*. Cambridge, MA: MIT Press.

Ravenscroft, I., 2005. *Philosophy of Mind: A Beginner's Guide*. Oxford: Oxford University Press.

Sowa, J., 1999. *Knowledge Representation: Logical, Philosophical, and Computational Foundations*. Pacific Grove, CA: Brooks/Cole.

Sowa, J. F., 2002. Signs, processes and language games: foundations for ontology. Accessed at www.jfsowa.com/pubs/signproc.htm.

Stone, M., 2000. Towards a computational account of knowledge, action and inference in instructions. *Journal of Language and Computation* **1**: 231–246.

Uschold, M., and R. Jasper, 1999. A framework for understanding and classifying ontology applications. In *Proceedings of the IJCAI-99 Workshop on Ontologies and Problem-Solving Methods (KRR5)*, Vol. 18, eds. V. and A. Benjamins, B. C. Gomez-Perez, N. Guarino, and M. Uschold, pp. 1–11. Stockholm, Sweden: CEUR-WS.

Whitehead, A. N., 1938. *Modes of Thought.* New York: Macmillan.

3

Infrastructure for collaboration

WILLIAM PIKE, ALAN MACEACHREN, AND BRENT YARNAL

Introduction

In a world connected by networks that enable instant transmission of voices, images, and data, environmental science is changing in ways that bring researchers, students, decision-makers, and citizens closer than ever before. Realizing the potential of this connected world depends on building an infrastructure, both technological and human, that enables effective interaction.

Why is infrastructure necessary? Local actions have global impacts, and global changes have local effects. Understanding the full complexity of environmental problems depends on the ability of researchers, students, decision-makers, and stakeholders to work across the continuum of scales that characterize the causes of and responses to environmental change (Association of American Geographers Global Change in Local Places Research Team 2003; Kates and Wilbanks 2003). A primary goal for a flourishing HERO network would be to build the information resources to support long-term scientific research partnerships needed to understand these changes. For the data collection and analysis efforts of a HERO network to succeed, geospatial technology and methods that are developed and implemented must meet two goals. First, the technology and methods must facilitate context- and task-sensitive encoding of data in, and retrieval of data from, the very large and complex data warehouses that will develop. Second, the technology and methods must support collaboration among scientists at different HEROs as they work together on common problems.

As noted in Chapter 1, monitoring global and local indicators can be an effective way to assess the implications of environmental change and the effectiveness of strategies to manage it (National Research Council 2003). Monitoring identifies emerging vulnerabilities to or abuses of nature and how fast they are developing, and importantly, it enables international bodies, nations, and local communities to set goals and measure progress toward those goals. Despite its promise, monitoring by itself is an incomplete solution: there must be a way to synthesize local

Sustainable Communities on a Sustainable Planet: The Human–Environment Regional Observatory Project, eds. Brent Yarnal, Colin Polsky, and James O'Brien. Published by Cambridge University Press.

environmental change into expressions of global impacts and to translate global processes into locally meaningful terms. Achieving these aims requires a new approach to scientific infrastructure, and specifically to infrastructure that supports scientific collaboration.

Collaboration is not a new idea – scientists, decision-makers, and stakeholders have been working together to solve environmental problems for a long time. Traditional forms of collaboration, however, such as letters, telephone calls, file sharing, email, and journal articles, are not always suited to the needs of present and future environmental monitoring.

To address these needs, one vision for collaborative environmental science uses an infrastructure based on new Internet technologies and the emerging "Semantic Web" that supports interaction among scientists, decision-makers, and stakeholders (Berners-Lee *et al.* 2001). The goal of this infrastructure is not to replace established forms of collaboration, but to augment them with deeper interaction and consensus-building techniques that bring advantages not available with traditional modes of communication. To be successful, however, this infrastructure must be designed around three characteristics of effective scientific collaboration:

- *Continuity: enabling communities to link local studies to larger problems.* Regional and global trends can only be detected if the many groups that study local environmental change have mechanisms for sharing and comparing their research. The communication and synthesis of research results must be a continuous process, as communities update and evaluate the state of their understanding. Environmental research must also keep future generations in mind, giving those who follow records of earlier data and reasoning.
- *Informality: giving diverse groups a role in science.* The research process and decision-making must be open to all who can make important contributions and whose decisions should be informed by the research, including scientists, students, policy-makers, and private citizens. The means of collaboration must allow communication and information-sharing among these parties to occur at many time scales, from real-time chats to long-term data archiving. Overcoming social and political encumbrances can also demand informal ways of interacting.
- *Ubiquity: leveraging Internet connectivity to achieve greater communication.* As Internet connectivity increases, research and decision-making communities no longer need to be isolated geographically. Tools that support continuous, informal communication – from handheld devices used by researchers in the field to Web browsers in remote schools – can be made available worldwide and can be designed to minimize the technical requirements of remote users.

Systems that support this style of interaction are often called "collaboratories," a broad term used to describe efforts that enable geographically dispersed teams of scientists to have access to data repositories, conversation spaces, and even instrumentation (Cerf 1993; Finholt 2002). Collaboratories provide resources that one might find in a physical laboratory, but in a distributed, virtual space. Collaboratories have been built previously to support research in physical sciences (Kouzes *et al.* 1996; Olson *et al.* 1998; Russell *et al.* 2001; Keahey *et al.* 2002; Schissel *et al.* 2002), health sciences (Craver and Gold 2002; Olson *et al.* 2002), and computational science (Kaur *et al.* 2001). A primary focus in initial collaboratory efforts has been on developing and implementing the networking technologies required to make remote connections possible and secure (Kaur *et al.* 2001). A common goal has been to enable remote connections to laboratory instruments (Henline 1998) and to simulation models (Schissel 2002). More generally, emphasis has been on shared technologies to facilitate real-time data collection or control of experiments in the physical or medical sciences.

With some collaboratories moving beyond the experimental stage to everyday use, researchers are asking whether specific collaboratory implementations have been successful, if they have involved a wider range of scientists than traditional collaboration, and how collaboratories influence group work. In a meta-analysis of 10 years of published research, Olson and Olson (2000) identified key aspects of individual and group dynamics that can lead to success or failure of technology-mediated, same- and different-place group work. Specifically, they found that groups starting with much common ground, engaging in loosely coupled work, and having a readiness both for collaboration and for collaborative technology are likely to have the most success in taking advantage of what technology can offer. More recently, Sonnenwald and colleagues (Sonnenwald and Li 2003; Sonnenwald *et al.* 2003) explored aspects of social interaction among science students collaboratively using a remote specialized scientific instrument called the nanoManipulator. They concluded that students with a strong collaborative learning style tended to be more positive about the relative advantage, compatibility, and complexity of the collaboratory system over face-to-face work. Students with a preference for individualistic learning, or students working with another student having that preference, had a somewhat negative reaction to the nanoManipulator (specifically about the degree to which the results of the innovation were easily seen and understood by the group).

In spite of the developments above, only limited progress has been made toward applying or understanding implications of the collaboratory concept to the social sciences, generally, and to the study of human–environment interaction, more specifically (Kuhlman *et al.* 1997). This lack of extension of the collaboratory concept to human–environment science is surprising in relation to developments in US science policy, which has recognized the need for what has been called "mega-collaboration" to address critical global problems (Zare 1997).[1]

A prototype collaboratory

Barriers to interaction across distributed research sites, such as those involved in the HERO project, will slow integration of knowledge required to resolve the challenging research questions that underlie development of a distributed network (Finholt 2002). HERO built a prototype collaboratory to serve as a proof-of-concept for the social and technological infrastructure needed to evaluate environmental change across geographic locations and scales. The four HERO sites represented dramatically different physical and human landscapes and therefore dramatically different types of human–environment relations and environmental change problems. In central Massachusetts, brownfields and urban development pressure the environment. Central Pennsylvania experiences the complex interactions of agrarian, industrial, and post-industrial social groups, whereas southwestern Kansas feels stress from large agricultural operations and associated water demands. The politics of the border region and of water-resource issues are dominant in southern Arizona.

The HERO collaboratory provided a suite of methods that offered targeted support for particular kinds of collaboration. Scientists and others interacted in two ways: same-time (synchronously, or real-time) and different-time (asynchronously). Same-time interaction tends to be of short duration, from a few minutes to a few hours. Different-time interactions cover a wider range of scales, from email messages or threaded discussions that take place on and off for a few days, weeks, or months, to much longer-term storage and reuse of data and methods. The suite of tools included three collaboration techniques that supported work across these scales: real-time conferencing, asynchronous discussions, and data- and knowledge-sharing portals (see Figure 3.1)

Each of these three tools met the requirements of continuity, informality, and ubiquity. For continuity, each was "always on," running continuously on Websites accessible to team members anywhere, allowing users to come and go with ease and creating records for long-term storage. For informality, each was user-driven, so

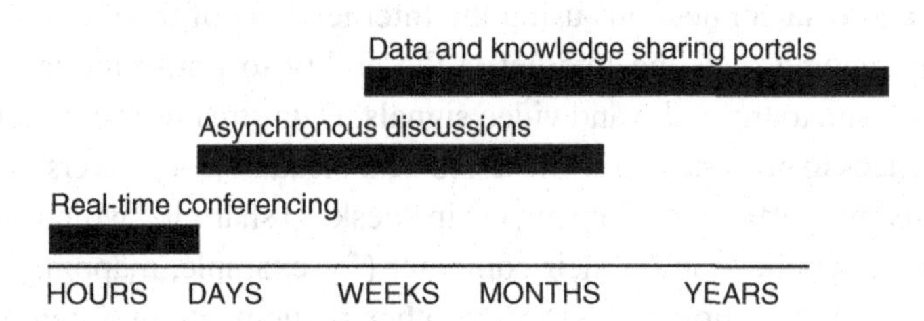

Figure 3.1. Collaboration occurs over a range of timescales. The three components of the HERO human–environment collaboratory are designed to support communities working within and across these scales.

users could initiate the activities that each tool supported without waiting for an invitation from higher up. For ubiquity, each tool required minimal hardware, so that any user with an Internet connection and a Web browser could use it.

The HERO experience yielded important data on the benefits and drawbacks of these three collaboratory tools. It also provided insights into how such tools could become an important part of the future of local environmental change science and decision-making.

Real-time conferencing

One of the keys to creating informal networks of researchers is to help them communicate with each other much like they would in face-to-face meetings. The most obvious way to achieve this communication is through the transmission of voice and image over the Internet. There are many videoconferencing solutions, and some, such as satellite connections between television-equipped studios, do not use the Internet.[2] To make adoption easy, however, the HERO collaboratory emphasized low-cost, existing networks.

Real-time conferencing was the only tool in the HERO collaboratory that used more than a basic personal computer and Internet connection: it required that each participant or group of participants in the same location have access to a Web camera (which is generally inexpensive). In a videoconference, remote collaborators use their Web cameras to connect to a virtual meeting room run on a centralized server. Personal Web cameras lend themselves to spontaneous conversations between two or more colleagues. For scheduled meetings, organizers can invite as many participants as desired. Unlike long-distance telephone conferences, there is no cost beyond that of the existing Internet connection and Web camera.[3]

Collaborating scientists also need to be able to share maps, images, and other graphic depictions of information relevant to discussions and to annotate and manipulate representations presented by others (Brewer *et al.* 2000). To meet these needs, HERO experimented with a range of technologies that support combined video and data connections using the Internet. One of the important features of videoconferencing over the Internet is the ability to transmit supporting data streams in addition to the audio and video signals. Data streams make it possible for videoconferences to become real work sessions, instead of just conversations. Using these data streams, participants can engage in "desktop sharing," which allows them to broadcast any application on their computer (for example, mapping software, a spreadsheet, or presentation slides) to the other participants in a conference (see Figure 3.2). Desktop sharing is not just one-way; it gives participants the ability to control the applications on each other's computer desktops. For example, a participant might explain a statistical model to his or her colleagues, and then allow them

Figure 3.2. In a real-time desktop sharing session, collaborators can manipulate data and applications on each other's computers. Here, nine team members interact with maps of toxic release sites.

to explore the results by changing the values in the model remotely. Case Study 3.1 describes a situation in which videoconferencing and desktop sharing played a vital role in collaborative research.

HERO's decision to manage its own videoconferencing service instead of relying on a commercial service had several benefits to the group and the local environmental change community at large. First, researchers could schedule meetings at will without reliance on an external infrastructure. Second, the service could be made available to HERO researchers around the world without incurring any cost to them or to HERO. Third, HERO could record and archive conference sessions and make them available through its other tools.

Many of the benefits – and many of the problems – of face-to-face communication are replicated in videoconferences. Achieving the fluidity of conversation that comes with real-time conversation without the need for geographic proximity is

Case Study 3.1 Real-time online conferences benefit students

Each summer, 12 to 16 undergraduate students who attended universities in or around the four HERO locations gathered to explore key local environmental change themes across the four study areas. Such themes included creating vulnerability indicators and understanding the sensitivity of drinking-water systems to environmental change. The students spent two weeks together learning theories and methods of local-area global change science and then dispersed to their research sites around the country. Once in the field, they used the HERO collaborative tools to share findings, ask questions, and make cross-site comparisons.

Videoconferences played two roles for the students. First, regularly scheduled meetings helped preserve the social aspects of collaboration. Email is irreplaceable but does not provide the immediate back-and-forth interaction that real-time discussions do. Second, spontaneous videoconferences helped students learn from each other by providing them with a way to talk about (and more important, demonstrate through desktop sharing) problems and insights as they happened. In one such instance, the students were using geographic information systems (GIS) to map census data about their study areas. Rather than create static images of the maps to email to each other, the students gathered in a videoconference to evaluate each others' work. Each team used slightly different methods to create their maps. Through desktop sharing, the teams could manipulate each others' maps "live." That is, students in Massachusetts could remotely manipulate the maps produced by the team in – and running on a computer in – Arizona to make the two sets of maps comparable on the fly, with accompanying audio/video commentary. Instead of trying to create comparable map displays in words alone, the students could communicate by doing.

one of the main advantages of the technique. Desktop sharing may even have benefits over face-to-face meetings because the manipulation of multiple computers by multiple users is difficult to do in a single location (for example, many people standing around one workstation). Still, social behavior rarely changes with the medium; collaborators who monopolize a conversation in real life are just as able to do so in a videoconference. Variability in network quality can also be an issue, with participants on slower networks (as might be the case with researchers scattered around the world) causing degradation in the quality of the communication (such as choppy video and out-of-sync audio) for all participants. Experience with the HERO collaboratory demonstrated that a successful videoconference is one in which the conferencing tool itself recedes from participants' attention, allowing them to focus on their colleagues, not on the communication medium; when poor-quality video or delays in audio transmission interfere with participants' ability to communicate clearly, conferences break down. Current Web-based videoconferencing solutions are not yet able to provide error-free audiovisual communication; poor-quality video plagued all of the products HERO tried, interfering with conferences enough to make collaborators occasionally avoid video altogether. The desktop-sharing tools built into these products proved to be invaluable, however. As a result, HERO researchers generally relied on just the desktop-sharing components of conferencing software. When working on documents collaboratively, for example, they used the traditional telephone conferencing system to narrate. Eliminating the need for cameras and the high-bandwidth connections they require also reduced entry barriers for other researchers and stakeholders.

Collaboratory tools were used to support routine synchronous collaboratory activities among smaller groups of team members (usually four to five people) and full-group meetings among 12–16 HERO Research Experience for Undergraduate (REU) students. For these smaller groups of collaborators, the synchronous collaboratory environment proved to be indispensable. Videoconferences were scheduled frequently for intensive information exchanges and for saving collaborators' time, in comparison to trying to accomplish the same tasks via email or telephone. Activities in these sessions included: exchanging basic understanding of a particular concept (e.g., vulnerability, protocols); discussing the framework for co-authored papers; and engaging in detailed activities such as revising interview scripts prior to field work or figuring out how to compile field data in a GIS after returning from the field.

In addition to supporting HERO's large group scheduled meetings, videoconferencing proved indispensable for small group, synchronous collaboratory activities. These synchronous activities were often most successful if complemented by other communication (e.g., via email) before and after the synchronous session, thus when not completely spontaneous. Before a meeting, email communication was often used to clarify the tasks and to exchange necessary text documents. Ongoing use of collaboratory Codex tools that enable asynchronous collaboration (see below) helped provide context for synchronous exchanges. At the beginning of the meeting, it was useful for an organizer to list a clear agenda to remind participants about the major purpose of the meeting and to suggest a sequence for addressing specific topics. At the end of a session, it was useful to summarize or re-emphasize the major points. After the meeting, participants again used email and collaboratory Codex tools to follow up the meeting topics until goals were reached. For two-person collaboration, the meeting times and agenda could be very flexible. The focus of interaction was usually quite specific and data-sharing functions were often important to problem-solving. Different people could share the same set of data and software to discuss a problem in depth.

Based on feedback over time by our participants, we conclude that the video- and audio-enabled collaboratory environment has many advantages in comparison to text-only chat. Text chats require relatively high typing and reading speeds. Chat can be a tedious form of communication, particularly for those not familiar with the large number of commonly used abbreviations. It typically includes significant time lags between thoughts and expressions. With chat, it is easy to lose key points or misunderstand intentions. In contrast, the video–audio-enabled tools we have used allow participants to clarify ideas and correct misunderstandings in a more natural way. Still, some key differences continue to exist between using the current tools for distributed work and engaging in face-to-face collaboration. In face-to-face situations, body language, mood, and feelings are more easily communicated and impromptu brief interactions and follow up is more practical.

Asynchronous discussions

One of the challenges facing groups of distributed collaborators is the need to make connections across team members' specialized areas of expertise and to create expressions of group opinion. HERO collaborators approached human–environment interactions from a variety of perspectives, among them climate science, land-use analysis, and public policy, so areas of agreement and disagreement about the drivers and effects of environmental change were rarely obvious. As a result, the process of deciding how to monitor environmental change and what parameters were critical in different locations benefited from structured negotiation. Over time, team members argued their own positions, evaluated those of others, and gradually came to improve their understanding of the breadth of a problem and the possibilities for its solution.

Videoconferences and desktop sharing are useful for short bursts of communication and for times when all participants are available. Still, when collaboration needs more time, when participants are not available (such as when they live in different time zones), or when conflicting views may stop the interaction, real-time discussions cannot help. Asynchronous discussions – those in which participants contribute at different times throughout the course of a day, week, or longer – reduce personality conflicts and allow participants time to reflect.

The HERO approach to asynchronous discussion was based on the Delphi method (Turoff and Hiltz 1996). Delphi is a common technique used in decision-making and planning to elicit input from a diverse group of experts, such as those involved in local environmental change problems. Because the method accommodates a wide range of opinions, the goal of Delphi activities is not necessarily to reach consensus, but simply to identify key elements of a problem or points of agreement and disagreement. Delphi activities are anonymous, structured, and iterative, in contrast to the relatively free-flowing discussion that videoconferences support. Developed at the Rand Corporation in the 1940s, the Delphi method was first applied to wartime forecasts by military planners (Dalkey 1969; Linstone and Turoff 1975). In principle, the method works like an asynchronous focus group, run by a moderator who poses questions to a panel, synthesizes feedback, and guides the group toward its goal.

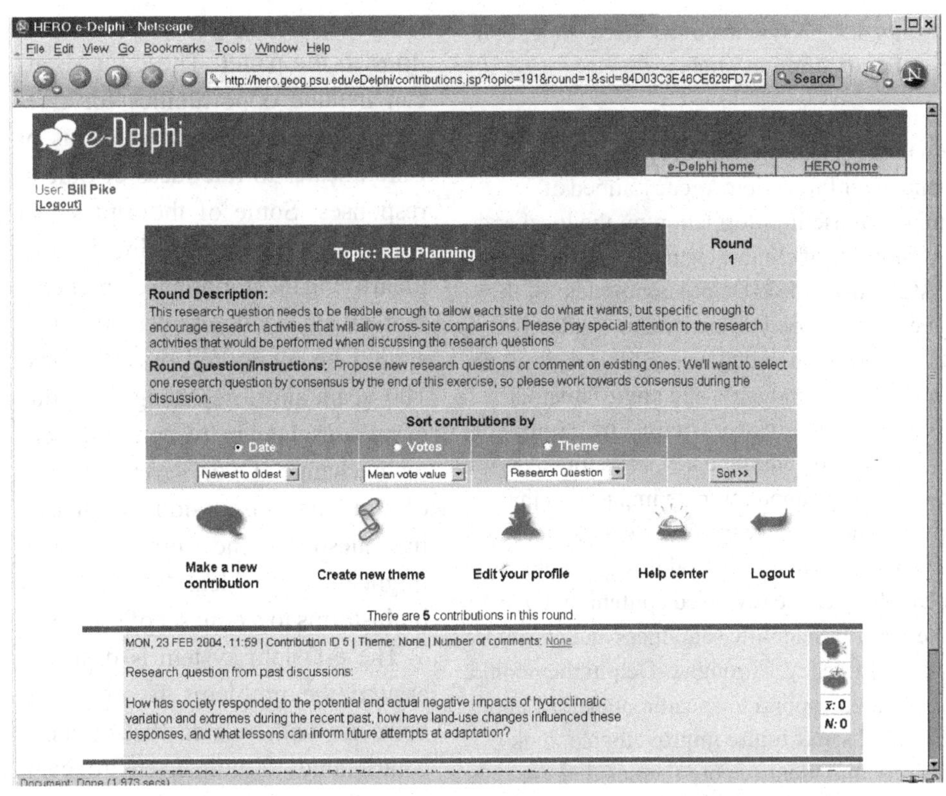

Figure 3.3. The HERO e-Delphi system supports moderated discussion of environmental change involving far-flung participants. Participants can read, comment about, and vote on each others' responses.

Traditionally, Delphi method activities contain several rounds of iteration, each of which might ask the participants to generate new ideas or refine, extend, vote on, or otherwise discuss responses from previous rounds. Recent applications of the method in the environmental sciences include water-resource issues (e.g., Nagels *et al.* 2001; Taylor and Ryder 2003) and climate forecasts (Tapio 2002). Problems in these domains are generally complex and lack an easily defined or readily agreed upon solution. Historically, Delphi activities were conducted through the mail, but HERO implemented a Web-based system called e-Delphi (see Figure 3.3) that made these activities available to anyone via a Web browser.

Over a period of days or weeks, participants in a Web-based Delphi exercise access the system to respond to the moderator's questions, read and discuss how others have responded, and vote on important themes. Users can participate in multiple activities at once, and anyone with Internet access can initiate an e-Delphi activity by inviting a panel of participants; a self-appointed moderator (often the person who initiates the activity) guides the group through its exploration of a

Case Study 3.2 Asynchronous discussions in a diverse climate impacts study

The Consortium of Atlantic Regional Assessments (CARA) – a project sponsored by the US Environmental Protection Agency aimed at disseminating useful information on the local and regional impacts of climate change to stakeholders – successfully used the e-Delphi component of the HERO collaboratory. Participants in CARA's e-Delphi exercise included nearly 90 stakeholders from government, industry, and environmental groups, as well as private citizens. The purpose of the activity was to determine key ways for CARA to refine its message about climate impacts so that it would have meaning for the diverse audiences to consume CARA's research products. Participants in this e-Delphi exercise reviewed content on CARA's Website for both scientific soundness and clarity for a broad constituency. Through e-Delphi they could, at their leisure, respond to specific questions about the content, discuss future improvements, rank suggestions, and identify core themes that the content should address. The primary benefit CARA derived from e-Delphi was the ability to engage many participants from a range of interests, geographic areas, and demographic backgrounds. While in-person meetings, focus groups, or workshops are necessarily limited in who can attend and contribute, the e-Delphi process can be opened to as many people as a moderator wishes. For CARA, it allowed communities often left out of research decisions to participate. The anonymous nature of the activity helped CARA's participants, particularly those who might otherwise have been uncomfortable collaborating with researchers, share their ideas forthrightly. CARA used a voting feature in e-Delphi to help participants select and refine their set of messages on key climate impacts over time. Because they could not be identified, participants were more likely to vote for ideas they felt strongly about rather than those proposed by "important" people.

problem by posing a series of questions to the panel. These questions can include opportunities for free-text responses as well as ballots or Likert scales to rate ideas or others' responses. Some of the topics that were explored using e-Delphi include identifying the important elements of human–environment interaction that should be monitored over the next 100 years and forecasting the value of various human responses to climate change. Over several rounds of structured discussion, summarizing questions, and voting, a panel might refine its ideas from a list of brainstorms to a core set of opinions.

The e-Delphi system is designed around the problems facing collaborators with varied backgrounds contributing to human–environment research (see Case Study 3.2). One of the hallmarks of the Delphi process is anonymity. Each participant posts anonymously, reducing the potential for status conflicts. Anonymity encourages participants to evaluate responses to the moderator's questions on their merits and to engage in commentary on those responses. Since e-Delphi runs over the Web, participants can contribute from their home, office, library, or school. E-Delphi runs from a centralized server that eliminates the need for remote groups to have the technical resources to administer their own versions of the system. The HERO e-Delphi application has been used by research groups around the world that want to leverage the power of the technique, but lack the technical resources to manage their own versions of the tool.

Known "troublemakers" had no more power than other participants did. In face-to-face meetings or videoconferences, particularly those that incorporate a range of perspectives and interests (and the broad nature of environmental change studies makes such a range necessary), status relations and strong opinions can drown out dissent. By contrast, the e-Delphi platform was designed to accommodate this range by instituting certain rules of interaction. These rules – not just anonymity, but also controls on how much, how often, and in what format participants contribute – bring a rare equality among the diverse constituencies interested in environmental change. E-Delphi also preserves a long-term record of discussions that allows future researchers to understand how participants' ideas evolved over time.

Anonymity is the first of four basic tenets of the Delphi method; the others are asynchronicity, controlled feedback, and statistical response (Turoff 1971, 1972). Together, these four characteristics describe an approach to group communication that attempts to overcome some of the shortcomings of face-to-face meetings. Asynchronicity allows Delphi participants to engage in a discussion on their own schedules, where and when they feel it appropriate. A participant might choose to contribute only when the discussion approaches his or her areas of expertise, or after spending more time in reflection than would be possible in real-time meetings. Controlled feedback gives the moderator responsibility to shape the discussion; the moderator may choose how much information to present to the panel, in what form, and how often. Through statistical response, participants' contributions to a Delphi activity may be summarized quantitatively; they may be asked to vote on their support for particular ideas, with the results of this voting used to prioritize discussion in future rounds. In e-Delphi, statistical response is updated continuously in the form of vote results, allowing moderators or participants to keep abreast of the support for various contributions.

An electronic environment for managing Delphi activities also supported HERO's aim to understand the process of geographic knowledge construction and application. Managing all of a research group's Delphi-style interaction through a single environment helps to build a rich knowledge base that reflects the evolution of a community's thinking over time. Every contribution to an e-Delphi exercise is logged, and a body of Delphi activities can be taken together to reveal trends in the concepts that are important to a group of researchers. Elsewhere, Pike and Gahegan (2003) discussed how an evolving text corpus consisting of e-Delphi discussion can be mined for concepts at different levels of abstraction, and how the relationships among these concepts and the participants who discuss them can be visualized. To accomplish this task, the initial HERO e-Delphi tool was extended through addition of several visual and computational methods that allowed the session moderator and the participants to develop a deeper understanding of an evolving moderated

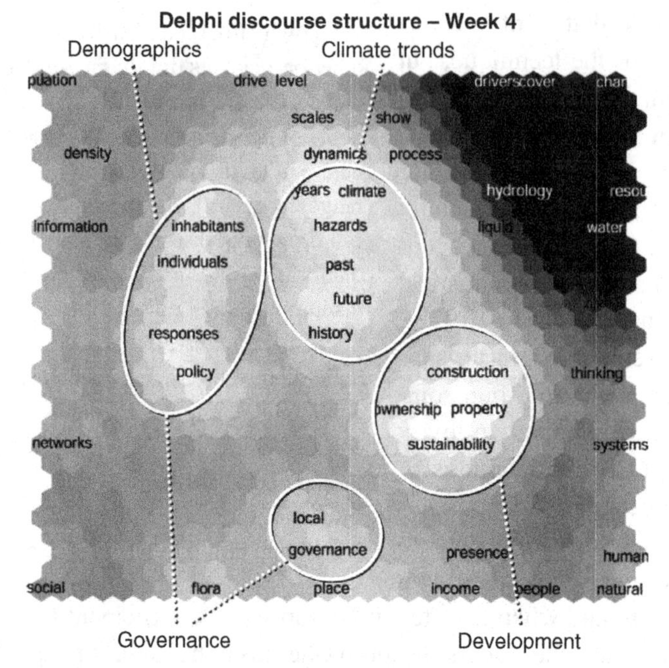

Figure 3.4. Self-organizing map (SOM)-based, two-dimensional representation of the multidimensional concept space that evolved through a Delphi exercise. The specific SOM representation shown reflects the concept space at week 4 in a discussion of human–environment interaction.

discourse. One addition was the application of computational text processing. This processing starts with simple term extraction, then moves to semantic abstraction (Mani 2001) and concept similarity measures (Lin 1995) to derive concept weights, and then processes these concept weights using self-organizing maps (SOMs) to produce a series of two-dimensional representations of the multidimensional concept space that evolves through a Delphi exercise (Figure 3.4).

The anonymous nature of the discussion helps make the final product representative of the group's choices, not those of a few opinionated or dominant members. Participants can also contribute to an activity whenever inspiration strikes. Still, we have found that the structured nature of e-Delphi lends itself to certain kinds of questions better than others. Those questions that involve identifying a few key themes, proposed outcomes, or policy decisions from many initial options often result in the most vigorous discussion. Appropriate problems might include selecting a set of core variables to monitor environmental change in local places or gathering and synthesizing public comment on land-use decisions. One of the shortcomings that is particularly apparent with asynchronous tools is that participants in a collaborative activity must feel that it is worthwhile to contribute. Questions that have been designed to satisfy the moderator's curiosity, but do not inspire the panel, result in lax participation. There must be an incentive to participate; such incentives include knowing that taking part in an activity will mean that one's opinions are included in an important decision, or that the result of the activity will be actionable information that can help guide research or policy-making. It is therefore important for the developers and administrators of collaborative systems to be mindful of appropriate use of the tools.

Data- and knowledge-sharing portals

Data integration is a common goal in any large-scale scientific endeavor in which measurement techniques and data formats vary from place to place and time to time. For scientists to make general statements about broad patterns, they must reconcile unlike measurements and formats and different conceptualizations of a problem. The development of national and international standards for data helps solve the problem,[4] but data integration alone is a partial solution. Scientists who monitor local environmental change must have ways to share their data, methods, and ideas so that they can build a picture of which characteristics are local, which are regional, and which are truly global.

Record-keeping and imagination-facilitating are key activities for all types of scientific research. Traditionally, these activities are mostly performed using paper notebooks manually. These notebooks usually contain project plans, raw data, experiment procedures, results, and so on. Paper notebooks have certain advantages – they are convenient, portable, and easy to use – and as a result have been widely used and have a long history. However, information in the paper notebook cannot be easily shared with other users, and the search and indexing capabilities are limited. In addition, subsequent analysis often requires transcription into digital form, a tedious and error-prone process. According to Lysakowski's study, on average, 14.3% of a highly trained scientist's time is devoted to transcription (Lysakowski and Doyle 1998; Myers *et al.* 2001).

Still, the traditional scientific notebook as a permanent, sequential archival record of research activities is an appealing metaphor for design of a digital counterpart that can support distributed, long-term encoding of and access to the heterogeneous information needed to support human–environment science. The concept of an electronic notebook (e-notebook) was initially proposed and implemented for laboratory science by researchers at the Environmental Molecular Science Laboratory (Edelson *et al.* 1996).

The HERO collaboratory included a Web portal that served as a modern-day scientific notebook, enabling environmental change scientists to share resources with their peers around the world – not just data and methods, but representations of conceptual models and theories, along with descriptions of how those resources were used to solve problems. Unlike paper notebooks, the portal supported

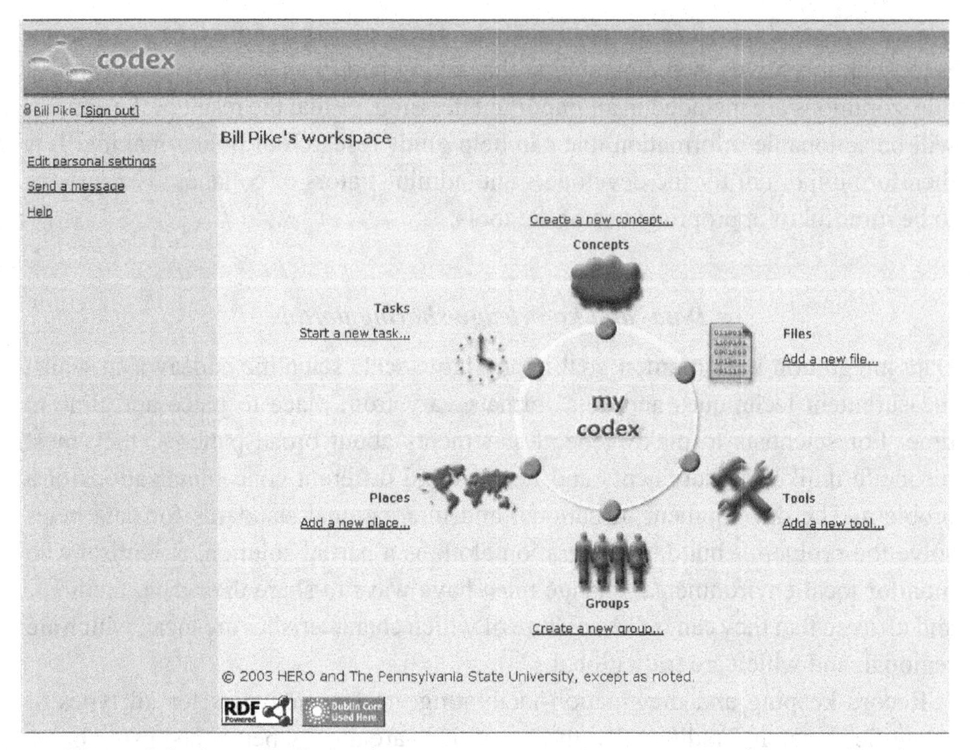

Figure 3.5. A user's Codex home page serves as an access point to a virtual workspace in which he or she stores and shares information. The individual workspace shown above can be nested within shared workspaces maintained by groups of collaborators.

searching, sharing, and reusing such resources involved in environmental change science.[5] The portal provided a personal workspace (see Figure 3.5) in which users could lay out descriptions of models or experiments, upload and download data to and from other users, and create concept maps that helped diagram their thinking and communicate it to others. The central feature of the portal was its support for distributed scientific resources: data, concept representations, and analysis tools stored on computers around the world. As long as these computers were accessible over the Internet, the portal provided a seamless interface that made them appear to a user as if they were stored on her computer. The portal's name, Codex, evoked manuscript notebooks – such as Leonardo da Vinci's – that document each step in the investigation process; to this traditional notebook we added supporting structure that made Codex a searchable, sharable, dynamic environment.

Design and implementation of Codex was driven by the scientific infrastructure research goals of the HERO and GEON projects and a firm belief that computational tools to support scientific exploration in a contested field such as human–environment interaction must do more than just enumerate a common frame of reference – they should also reflect how agreement is reached about the concepts and terms within the frame of reference (what Sowa [2000] calls ontological commitment). The experience of establishing this shared understanding of concepts is critical to working with and sharing group knowledge. Hence Codex attempted to contextualize the use of HERO resources by logging the situations that surrounded their use (including such aspects as who used them, where and when they used them, with what other resources they used them, and so forth). A key HERO objective was to develop and test protocols for data collection and analysis to support human–environment science. To support this objective, Codex managed the creation and management of personal and community conceptual understandings (ontologies), along with the aforementioned situations, used for scientific research.

Every resource that was available through Codex was stored as a "knowledge object."[6] A knowledge object can be anything that a researcher might use, such as a dataset, a description of a location, a hypothesis, or an analysis tool; it contains information about the connections between an object and other resources, such as other people, other places, or other tools.[7] Sitting on top of this architecture was a graphic interface that allowed users to draw diagrams (called concept maps) depicting how multiple objects connect, thus explaining broader associations among data, tools, concepts, and more (see Figure 3.6). An individual researcher could share

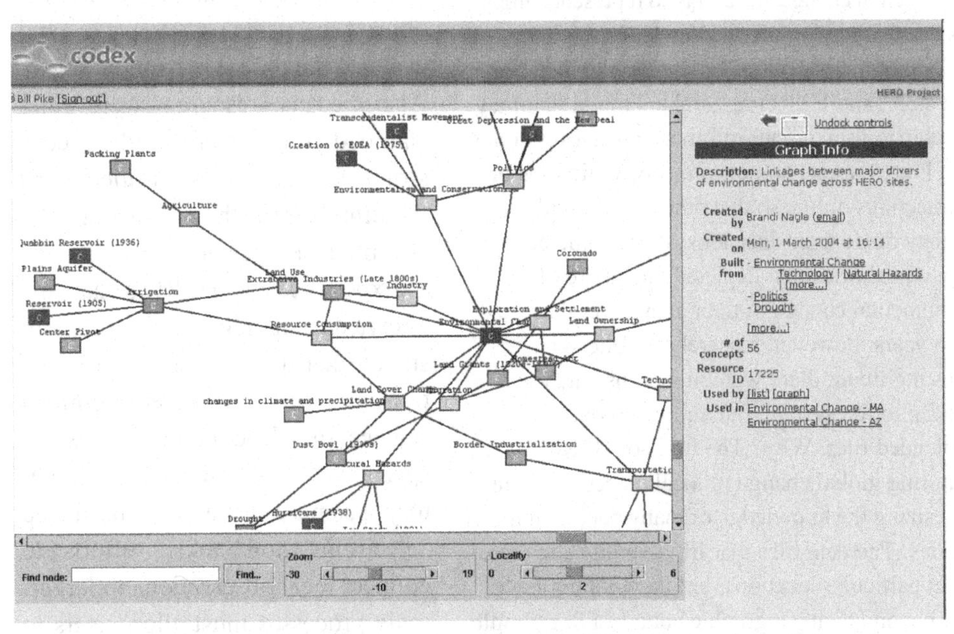

Figure 3.6. Codex uses concept maps to capture and display complex relationships among data, knowledge, and other types of information. These diagrams can be produced by the system in response to a user's query or can be sketched by users to describe and communicate their understanding quickly.

Case Study 3.3 Knowledge management practices and pitfalls

One of HERO's goals was to enable long-term monitoring by ensuring that researchers 100 years from now will be able to reconstruct our analyses of human–environment interactions. HERO began its efforts to implement a long-term data archiving solution with a tool called the Electronic Laboratory Notebook (ELN), developed by Pacific Northwest National Laboratory (Myers *et al.* 2001). ELN ran over the Web and had an interface like a traditional notebook, divided into chapters and pages. Each page held lists of notes that represented data files and their text descriptions. Beyond these annotations, there were limited ways to describe processes such as land-use change. There was no mechanism to show how notes stored in different areas of the notebook could be linked to create explanations of more complex phenomena. Given the fundamentally integrative nature of human–environment interactions and environmental change, the ability to represent complex ideas is crucial to long-term knowledge sharing.

ELN put no restrictions on the type of data that it could store, and HERO researchers found that they were able to think about information-sharing in new ways. For example, some teams put audio files of their meetings online so that others could understand how they made their decisions, while some built image repositories that depicted the nature of environmental change in their area. Over the course of two years, however, researchers noticed an interesting trend: there were numerous data files uploaded to the system, but almost no one downloaded files. Why? The basis of HERO – monitoring global change in local places – was in synthesizing the knowledge of many people in many locations. The data files that ELN could store were used in particular locations, but they did not have utility beyond their geographic bounds. For example, a land-use change map of central Pennsylvania is of limited use in southern Arizona; instead, it is more useful to know how the researchers constructed the

resources in his or her workspace with others, and groups of researchers could maintain shared workspaces representing the resources they all used. As a result, a user could have asked questions like "Who has used this data set, and how?" or "What are the different ways that people describe land-use change?", and Codex then returned not just individual data files, but linked sets of resources that showed how others tackled or explained environmental problems in their locations. One also could have asked the system to "compare conceptions of drought between researchers in the southwestern and northeastern United States." Codex could search through linked sets of knowledge objects in users' workspaces and evaluate the differences among them. Also, because resources were time-stamped, the system could reconstruct the evolution of ideas and illustrate how they came to be accepted or rejected.

Experience with HERO's data- and knowledge-sharing tools demonstrated that there are unique characteristics of human–environment science that lead to guidelines for specialized tools. While the laboratory sciences have also produced "electronic notebooks," efforts to connect local observations to larger-scale processes must allow users to connect resources in flexible ways.[8] These "e-notebooks" apply a conceptual model derived from traditional laboratory science; this model can ensure commensurability among researchers' results, but consequently it can be inflexible (see Case Study 3.3). Notions of change differ from person to person and place to place, and scientists need a structure that accommodates different perspectives. Flexibility means moving beyond flat lists of information in virtual notebook pages to dynamic, graphic depictions capable of reflecting the true complexity of processes. Moreover, for explanations to have an impact on stakeholders, researchers need to think about how to represent them in accessible ways. The graphic structure used by Codex was intended to provide a representation of researchers' thinking that would be readily understood by lay users. Codex was also designed with a long view; while laboratory scientists might need to preserve records for a few years, environmental change is a long-term process, the study of which crosses generations. As an aid to science work, Codex helped contemporary researchers organize and communicate their thinking; as an archiving medium, it provides future generations with insight into the observations, problems, and decisions of the past. Separating the knowledge that was stored in Codex from the tool itself helped insure that even when the tool was no longer used, the knowledge base would be accessible and readable by others.

map so others in southern Arizona can construct comparable ones. That is, it is not only the data that needed to be shared, but also the knowledge that went into (and came out of) the data. HERO researchers figuratively tore apart the old scientific notebook structure and built Codex with an eye toward creating explanations of "how" and "why," not just "what." Students working on the HERO project subsequently used Codex to share descriptions of human–environment interactions in their local regions; the graph structure of knowledge maps in Codex proved appealing to them since it could easily show complex connections that might otherwise be difficult to articulate. Students who interviewed local officials to understand environmental change in their area, for example, made concept maps of the resulting discussion to share with each other. The students could borrow relevant concepts produced by others to use in their own depictions. Difficulties in deploying Codex centered on the need to advance the sciences of tool development and environmental change simultaneously. Research into collaborative technologies was conducted in parallel with environmental research, resulting in one or the other of HERO's foci occasionally advancing without waiting for its complement to catch up. Keeping both halves of such a project tightly coupled requires clearly articulated goals.

Building networks that work

Because collaboratories consist of the people using the tools, not the tools themselves, the environmental change community needs to be concerned with two aspects of infrastructure development. First, how can advances in computational support for collaboration aid teams already working in the field? Second, what role can infrastructure play in increasing the involvement of other communities, including scientists in developing countries, students learning about their environment, and decision-makers coping with local and regional changes? After all, while the collaboratory concept is a popular one in the physical sciences and its users are often exclusively researchers (Kouzes *et al.* 1996; Hey and Trefethen 2003), environmental change includes social, economic, and political considerations, such that collaboratories must involve more than just scientists.[9]

With these questions in mind, the value of collaborative tools can be assessed against three critical roles of scientific infrastructure:

- *Fostering interaction among communities.* As noted in Chapter 1, most research and monitoring sites around the world that focus on environmental change and issues of sustainability function independently, collecting unique data in unique ways, thus guaranteeing that cross-site comparison is impossible now and in the future. Scientists working at these sites are sometimes aware of the work of colleagues at other research and monitoring sites through Web searches, published papers, and conferences, but often they are unaware of similar efforts. Likewise, students becoming familiar with the field can benefit from a mechanism that shows how their local studies can be integrated with, and are important to, larger efforts. It is therefore critical that an international network of researchers creates a consistent, verifiable, and comparable record of local environmental change over time and space. The HERO experience demonstrated that collaborative infrastructure does support the creation and application of this record by giving researchers a means to share ideas quickly (in online meetings), to negotiate measurement protocols (using e-Delphi discussions), and to store and access data, methods, and concept maps (via online portals). In addition, collaborative infrastructure ensures that contributions of transient members of the community – for instance, students who do important work as part of a course or a summer research project – will not be lost. Infrastructure to monitor local environmental change also offers a way to involve the developing world in research and policy decisions by giving researchers and decision-makers anywhere a direct pipeline to the global research community.
- *Leveraging the power of the Internet in novel ways.* Collaboration is not easy, and it is not only the technical challenges that pose difficulties. Simply providing the tools does not cause collaboration to happen, nor does it make collaboration work. Building an effective network requires a commitment from participants to share more than they traditionally would; technological aids to science can support participants by providing a return on their investment. For example, our experience shows that giving researchers the opportunity to work on problems in real time with desktop sharing creates excitement because it allows them to compare findings, ask questions, and share breakthroughs instantly. Students in particular take quickly to this technique; thus, the expense of the hardware is far outweighed by the pedagogical value of bringing students from across the country together to solve widespread problems that affect their local areas. Giving stakeholders the chance to contribute anonymously in e-Delphi discussions allows candid critiques of

scientific work that are sometimes impossible in traditional settings. Giving researchers one-stop access to data and knowledge from around the world in their desktop portals satisfies a fundamental curiosity and spurs discovery of larger patterns. In short, we have found that an effective network takes a broad approach to the task of collaboration. Together, these tools help provide continuous, ubiquitous, and informal access to the research process. Any one collaborative technique by itself is likely to be limited in success, as we have yet to discover a single tool that meets all three criteria; in combination, however, they allow collaborators to select the right tool for a problem, with each tool contributing different, yet complementary components of understanding.

- *Lowering barriers between private research and public good.* Ultimately, addressing local environmental change means making wise decisions about the future. The scientific community is not the arbiter of policy, and policy-makers rarely have access to the extent of scientific understanding. Effective infrastructure helps make each community's work more transparent. In the case of data- and knowledge-sharing portals, researchers must confront choices about what information to maintain in a private workspace and what to share. When individuals or teams expose their work in a globally accessible portal, it becomes more open to scrutiny and critical evaluation by scientists and non-scientists alike. Local decision-makers can see the progress of scientists' work as it happens, not years later when it finally reaches publication; the representations of environmental knowledge that the HERO portal maintained, when presented visually, were useful summaries to non-scientists. Most important, a researcher's (or a team's) knowledge will not exist in a vacuum – through the portal, that knowledge is connected to vast networks of other resources that reveal the depth of a community's understanding or the extent of its disagreement. By putting the practice of science, not just its products, on the Internet, communities can open the research process, making the transfer of knowledge into practice more efficient.

Next steps

HERO demonstrated that infrastructure to support remote collaboration can succeed, allowing environmental change scientists and stakeholders from diverse geographic settings to work together and learn from one another. Making collaborative tools available on the Internet attracted interest from other communities looking to make use of HERO's tools. Nevertheless, experience taught us that the best intentions do not overcome the substantial effort required for research and stakeholder communities to engage in effective collaboration.

Based on the three critical roles of scientific infrastructure described above, the following actions are advised for others interested in collaborative science:

- *Develop networks.* Use the power of collaboratory tools to collaborate and compare across sites. Use a variety of tools tailored to different kinds of work, and favor tools that support diverse audiences, not just domain experts.

- *Use tools*. Methods will never improve until people use them and recommend improvements. Collaborative tools will never become part of the culture until people get used to using them. Support a culture of sharing and integration, especially when sharing knowledge informally does not come easily to users.
- *Engage people*. Tools are only tools; they cannot make things happen. Only people can do that. Do not use collaborative systems for their own sake; use them where they fill a clear need, and constantly evaluate the increased efficiency or insight users receive in exchange for their efforts.

Even within the limited scope of the HERO prototype, its creators had to reinforce the long-term benefits of cooperation to maintain a vibrant network.[10] Researchers, decision-makers, and stakeholders are often comfortable in intellectual and geographic isolation, but as local environmental change accumulates to become regional and global in scope, they need to share solutions through the free communication of data and ideas. By using the growing interconnectivity of the world's individuals and communities to involve new users and user communities over time, HERO collaborators hope to create an infrastructure that serves as a forum for a global network of local places addressing environmental change.

Notes

1. Zare was chair of the National Science Board at the time this chapter was written.
2. There are many commercial videoconferencing products available. All of them require conference participants to connect to a centralized server that brokers communication between them (direct point-to-point communication is only possible when there are two participants). Many of these products are subscription or pay-per-use services in which participants use their own camera to connect to the service provider, which in turn connects them to other participants. Usually, the cost of such meetings increases with each additional participant. Another option is for a group of collaborators to manage their own videoconferencing server, with specialized hardware or software purchased from a third-party vendor. This approach has higher front-end costs but has no per-conference cost. A group must have the human and financial resources to manage a videoconferencing server, but can also provide this service to other groups that have fewer resources. HERO tested several videoconferencing servers, both those sold as hardware products and as software packages for installation on an existing server. The former (such as those sold by Polycom) often required that participants use a special camera; inexpensive "webcams" would not suffice. The latter (such as Lotus Sametime and CUSeeMe) were often more flexible. HERO selected Sametime as its real-time collaboration tool; local human–environmental research groups from around the world therefore could have used HERO's videoconferencing server in lieu of expensive commercial services, lowering the financial barriers to real-time communication.
3. Evidence suggests that videoconferences enhance the quality of interpersonal interaction over more traditional telephone conversations, as the impact of being able to see one's collaborators and read their body language provides an experience much like face-to-face meetings (Daly-Jones *et al.* 1998).
4. See Chapter 1 for a brief description of organizations promulgating these standards.
5. Efforts to organize the resources involved in scientific investigation are increasingly common in a variety of fields, notably bioinformatics. The myGrid project (www.mygrid.org.uk), for example, is building an "electronic workbench" for biological research that helps integrate data sources and

analysis tools. In the USA, the NSF-funded GEON (www.geongrid.org) and SEEK (seek.ecoinformatics.org) projects are developing data and knowledge sharing tools for the Earth and ecological sciences, respectively. Our portal work was among the first that was designed to accommodate the unique interdisciplinary nature of researchers studying environmental change in general, and human–environment interaction in particular. Whereas many other efforts in the physical and life sciences focus largely on managing quantitative data, tools to support environmental change science require equal attention to managing qualitative information.

6. The Codex portal represented a translation of the ideals of knowledge management, an initiative popular in industry, to the practice of science. Knowledge management is an effort to enable members of an organization to access and apply the wisdom and experience of their peers and to preserve organizational memory for future generations. See also Fischer and Ostwald (2001).
7. Every object accessed through Codex was described in a special language called OWL (Web Ontology Language: http://www.w3.org/2001/sw/WebOnt) which is designed for sharing knowledge over the "Semantic Web." This language allowed data and knowledge stored in our portal to be used in any other tool based on the same standard.
8. Tools that help represent the relationships among scientific concepts often rely on fixed notations for relationship types, which ensure commensurability between different users, but can impede individual expression. Examples include ScholOnto (Buckingham Shum *et al.* 2000) and Belvedere (Suthers 1999). The alternative afforded in Codex was to allow communities to use both standard and custom vocabularies; through the use of a notation that preserves semantic relationships, even custom vocabularies could be related back to the standardized terms from which they emerged.
9. The technological mechanisms to support collaboratories are part of nascent efforts to develop scientific cyberinfrastructure. One of the goals of cyberinfrastructure is to build online scientific workbenches by connecting distributed high-performance computing and data-storage resources. Such workbenches would be virtual toolkits where researchers can find, analyze, and share data. A National Science Foundation (2003) panel argued that current technological capabilities – in terms of computational power, storage capacity, and data transfer speed – have the potential to revolutionize the way science is performed, not just by enabling individual research centers to achieve results, but by making it possible to transcend geographic boundaries in the pursuit of knowledge. Of six critical applications for cyberinfrastructure that the panel noted, fully half involve aspects of human–environment interaction: understanding global climate change, protecting our natural environment, and predicting and protecting against natural and human disasters.
10. While the long-term viability of any technological endeavor is never assured, we anticipated that our tools and techniques would continue to be used, in changing ways, over time. Codex, for example, was built on widely supported open standards for knowledge sharing. When this knowledge representation format becomes obsolete, its human-readability and broad user base makes it likely that a future researcher could translate information stored in Codex into a new format. We hope the richness of the record stored within Codex makes this an attractive proposition for future scientists. Similarly, the Delphi decision-making process has been used for nearly 50 years and has successfully made the transition from paper to Web. The technique is not technology-dependent, allowing generations to refine implementations as technology advances. Researchers within our project successfully transitioned from using one Web conferencing tool to another, and from the ELN to Codex, as we found implementations better suited to our needs. These experiences make us confident that research communities can embrace changing techniques as long as the underlying motivations remain the same – and we believe that they have remained the same since long before the days of Leonardo's notebooks!

References

Association of American Geographers Global Change in Local Places Research Team, (eds.,) 2003. *Global Change in Local Places: Estimating, Understanding, and Reducing Greenhouse Gases*. Cambridge: Cambridge University Press.

Berners-Lee, T., J. Hendler, and O. Lassila, 2001. The semantic web. *Scientific American* **284**(5): 34–43.

Brewer, I., A. M. MacEachren, H. Abdo, J. Gundrum, and G. Otto, 2000. Collaborative geographic visualization: enabling shared understanding of environmental processes. In *Proceedings IEEE Symposium on Information Visualization (InfoVis 2000)*, October 9–10, 2000, p. 137.

Buckingham Shum, S., E. Motta, and J. Domingue, 2000. ScholOnto: an ontology-based digital library server for research documents and discourse. *International Journal on Digital Libraries* **3**(3): 237–248.

Cerf, V., 1993. *National Collaboratories: Applying Information Technology for Scientific Research*. Washington, D.C.: National Academy Press.

Craver, J. M. and R. S. Gold, 2002. Research collaboratories: their potential for health behavior researchers. *American Journal of Health Behavior* **26**(6): 504–509.

Dalkey, N. C., 1969. *The Delphi Method: An Experimental Study on Group Opinion*. Santa Monica, CA: RAND Corporation.

Daly-Jones, O., A. Monk, and L. Watts, 1998. Some advantages of video conferencing over high-quality audio conferencing: fluency and awareness of attentional focus. *International Journal of Human–Computer Studies* **49**(1): 21–58.

Edelson, D., R. Pea, and L. M. Gomez, 1996. The Collaboratory Notebook. *Communications of the Association for Computing Machinery* **39**(4): 32–33.

Finholt, T., 2002. Collaboratories. *Annual Review of Information Science and Technology* **36**: 73–108.

Fischer, G., and J. Ostwald, 2001. Knowledge management: problems, promises, realities, and challenges. *IEEE Intelligent Systems* **16**(1): 60–72.

Henline, P., 1998. Eight collaboratory summaries. *Interactions* **5**(3): 66–72.

Hey, T., and A. Trefethen, 2003. e-Science and its implications. *Philosophical Transactions of the Royal Society of London A* **361**: 1809–1825.

Kates, R., and T. Wilbanks, 2003. Making the global local: responding to climate change concerns from the ground up. *Environment* **45**(3): 12–23.

Kaur, S., V. Mann, V. Matossian, R. Muralidhar, and M. Parashar, 2001. Engineering a distributed computational collaboratory. *Proceedings of the 34th Annual Hawaii International Conference on System Sciences*, Vol. 9, p. 9026.

Keahey, K., T. Fredian, Q. Peng, D. P. Schissel, M. Thompson, I. Foster, M. Greenwald, and D. McCune, 2002. Computational grids in action: the national fusion collaboratory. *Future Generation Computer Systems* **18**(8): 1005–1015.

Kouzes, R., J. D. Myers, and W. A. Wulf, 1996. Collaboratories: doing science on the internet. *Computer* **29**(8): 40–46.

Kuhlman, K. M., A. Soffer, and T. W. Foresman, 1997. Development of a three-tier metadata documentation scheme: examining Level 1 as an Internet accessible metadata input and search tool. *Proceedings of the 2nd IEEE Metadata Conference*, Piscataway, NJ: IEEE.

Lin, C.-Y., 1995. Topic identification by concept generalization. *Proceedings of the 33rd Conference of the Association of Computational Linguistics*, Boston, MA.

Linstone, H., and M. Turoff, 1975. *The Delphi Method*. Reading, MA: Addison-Wesley.

Lysakowski, R., and L. Doyle, 1998. Electronic lab notebooks: paving the way of the future of R&D. *Records Management Quarterly*: 23–28.

Mani, I., 2001. *Automatic Summarization*. Amsterdam: John Benjamins.

Myers, J., E. Mendoza, and B. Hoopes, 2001. A collaborative electronic notebook. *Proceedings of the IASTED International Conference on Internet and Multimedia Systems and Applications*, Honolulu, HI, August 13–16, 2001, pp. 334–338.

Nagels, J. W., R. J. Davies-Colley, and D. G. Smith, 2001. A water quality index for contact recreation in New Zealand. *Water Science and Technology* **43**(5): 285–292.

National Research Council (NRC), 2003. *NEON: Addressing the Nation's Environmental Challenges*. Washington, D.C.: National Academies Press.

National Science Foundation (NSF), 2003. *Revolutionizing Science and Engineering through Cyberinfrastructure*, Report of the National Science Foundation Blue Ribbon Advisory Panel on Cyberinfrastructure. Accessed at www.communitytechnology.org/nsf_ci_report/.

Olson, G. M., and J. S. Olson, 2000. Distance matters. *Human–Computer Interaction* **15**: 139–178.

Olson, G. M., D. E. Atkins, R. Clauer, and T. A. Finholt, 1998. The upper atmosphere research collaboratory. *Interactions* **5**(3): 48–55.

Olson, G. M., S. Teasley, M. J. Bietz, and D. L. Cogburn, 2002. Collaboratories to support distributed science: the example of international HIV/AIDS research. *Proceedings of the 2002 Annual Research Conference of the South African Institute of Computer Scientists and Information Technologists on Enablement through Technology*, Pretoria, South Africa.

Pike, W., and M. Gahegan, 2003. Constructing semantically scalable cognitive spaces. In *Lecture Notes in Computer Science* 2825/2003, pp. 332–348. Berlin: Springer.

Russell, M., G. Allen, G. Daues, I. Foster , E. Seidel, J. Novotny, J. Shalf, and G. von Laszewski, 2001. The astrophysics simulation collaboratory: a science portal enabling community software development. *Proceedings of the 10th IEEE International Symposium on High Performance Distributed Computing*, San Francisco, CA.

Schissel, D. P., 2002. An advanced collaborative environment to enhance magnetic fusion research. *Workshop on Advanced Collaborative Environments*, Edinburgh, UK.

Schissel, D. P., A. Finkelstein, I. T. Foster, T. W. Fredian, M. J. Greenwald, C. D. Hansen, C. R. Johnson, K. Keahey, S. A. Klasky, K. Li, D. C. McCune, Q. Peng, R. Stevens, and M. R. Thompson, 2002. Data management, code deployment, and scientific visualization to enhance scientific discovery in fusion research through advanced computing. *Fusion Engineering and Design* **60**(3): 481–486.

Sonnenwald, D. H., and B. Li, 2003. Scientific collaboratories in higher education: exploring learning style preferences and perceptions of technology. *British Journal of Educational Technology* **34**(4): 419–431.

Sonnenwald, D. H., M. C. Whitton, K. L. Maglaughlin, 2003. Evaluating a scientific collaboratory: results of a controlled experiment. *ACM Transactions on Computer–Human Interaction* **10**(2): 150–176.

Sowa, J. F., 2000. *Knowledge Representation: Logical, Philosophical, and Computational Foundations*. Pacific Grove, CA: Brooks/Cole.

Suthers, D., 1999. Representational support for collaborative inquiry. *Proceedings of the 32nd Hawaii International Conference on System Sciences*, p. 1076.

Tapio, P., 2002. Climate and traffic: prospects for Finland. *Global Environmental Change: Human and Policy Dimensions* **12**(1): 53–68.

Taylor, J. G., and S. D. Ryder, 2003. Use of the Delphi method in resolving complex water resources issues. *Journal of the American Water Resources Association* **39**(1): 183–189.

Turoff, M., 1971. Delphi and its potential impact on information systems. *Proceedings AFIPS Conference*, vol. 39.

Turoff, M., 1972. Delphi conferencing: computer-based conferencing with anonymity. *Technological Forecasting and Social Change* **3**: 159–204.

Turoff, M., and S. Hiltz, 1996. Computer based Delphi processes. In *Gazing into the Oracle: The Delphi Method and its Application to Social Policy and Public Health*, eds. M. Adler and E. Ziglio, pp. 3–33. London: Kingsley.

Zare, R. N., 1997. Knowledge and distributed intelligence. *Science* **275**: 1047.

4

Representing and reasoning with conceptual understanding

OLA AHLQVIST AND CHAOQING YU

Introduction

One of the primary goals for HERO was to provide a knowledge management system for interdisciplinary research that also provides a link between human understanding and formal systems, for example databases, analyses, and models. Chapter 2 elaborated extensively on how concepts that people create and use in their attempts to understand and manage Earth's dynamic systems are defined differently depending on place and situation. It was specifically pointed out that it is of particular importance for multidisciplinary research such as HERO to articulate how concepts and understanding change with context. While Chapter 3 demonstrated progress made in developing support for the process of collaboratory research this chapter addresses representational issues involved in linking human understanding with formal systems. We present two ways of modeling knowledge about both the conceptual understanding of human–environment interaction and the process of decision-making.

A parameterized representation of uncertain conceptual spaces

The collaboratory Web portal (Chapter 2) embodies the idea of a customizable window onto distributed resources and ways to make these accessible to a group of users. For a portal to be able to filter and customize the content to a specific user community, one of the critical components to any such solution is a metadata structure that describes and represents available resources. The goal is to enable users to exchange methods, data, ideas, and results. Most results presented in this book were achieved by negotiating a common understanding, adhering to a shared vocabulary, and using a common set of methods. Examples are the common classification system for the land-use and land-cover change analysis (Chapter 7) and the shared analysis procedures for the vulnerability studies in Part III (Chapters 5, 8, and 9). This commonality was only possible through close collaboration among

Sustainable Communities on a Sustainable Planet: The Human–Environment Regional Observatory Project, eds. Brent Yarnal, Colin Polsky, and James O'Brien. Published by Cambridge University Press.

participating researchers, which is something that we cannot normally expect when we incorporate methods or data from different times and places.

In general, it is difficult to negotiate common representations because different people model the same concepts in diverse ways or use identical terms for dissimilar concepts (known as semantic heterogeneity). These semantic differences arise from alternate perspectives, knowledge, and backgrounds; they arise because data were collected by different agencies or by the same agency at different times; and they arise when there are different organizations, professions, or even nationalities involved in the negotiation process. Current efforts to provide mechanisms for improving information exchange create meta-languages to leverage the potential of the World Wide Web communication infrastructure. Despite such attempts, there are currently no established solutions for realizing such semantic negotiations. Moreover, most attempts by geographic information scientists (see Bishr 1997; Gahegan 1999; Mennis 2003; Rodriguez and Egenhofer 2004) have used object-oriented modeling techniques in which classes (types of objects) define what its objects look like, such as the composition of properties and class-subclass relations.

There is a growing realization that any chosen technique must be able to handle the vagaries and ambiguities typically present in such conceptualizations. As an example, here we will look at how elements of a definition of vulnerability can be represented formally to preserve some of the inherent vagueness. We will then incorporate this uncertain representation in a general representation for geographic categories.

Example 1. Representation of vulnerability parameters: how many old people are there in a census unit?

Many of the ideas that go into a vulnerability assessment are inherently vague. To many people, it would make perfect sense to say that an old water system together with a poor and aging population would contribute to the vulnerability of a place. If we want to evaluate vulnerability quantitatively, however, we find ourselves interpreting these linguistic variables into measurable terms such as age of water system and percentage of population below the poverty level and above a certain age. To preserve and communicate our thinking and understanding of vulnerability effectively, we need a way to represent vague categories.

Let us look at how we can measure the number of old people in an area. We could say that all people above the age of 64 years are old and look for high percentages of old people as one indicator of vulnerability. This strategy would work reasonably well, but most people would agree that one does not turn "old" upon celebration of the 65th birthday.

The concept of "old" is a typical fuzzy linguistic variable. Instead of following a crisp, yet artificial definition that anyone older than 64 years is "old," we can define "old" as a continuous function of age. By doing so, we use the idea of membership values to indicate the degree to which a certain age is considered "old."

We could, for example, decide that people under 60 years are "not old" and people over 70 years are "old," and ages between these two are "old to some degree." This way of formally representing vagueness as fuzzy sets was introduced by Zadeh (1965) and has found many applications in fields such as decision-making, psychology, and engineering. An excellent review of fuzzy set theory and its application to geographic information can be found in Robinson (2003).

Let us now assume that we have settled on a theoretical function of "old" as the slanted line in Figure 4.1. Age is normally given as integers, which means that we will not find any instances that lie between whole years, for example, 61 and 62. Following this graded definition of "old" on the figure, a 61-year-old person is 0.2 old even if it is his or her birthday tomorrow, which would turn that person into 0.3 old. So, in the face of age recordings with limited resolution (simply whole years) the realized membership function will be an interval function (Figure 4.2).

Fuzzy set theory does not give a completely correct representation of the thinking and available data. Pawlak (1991) introduced *rough set theory* as a general framework for treating effects of limited resolution in situations like the one presented in Figure 4.1. The next section will explain how fuzzy and rough set theories can be

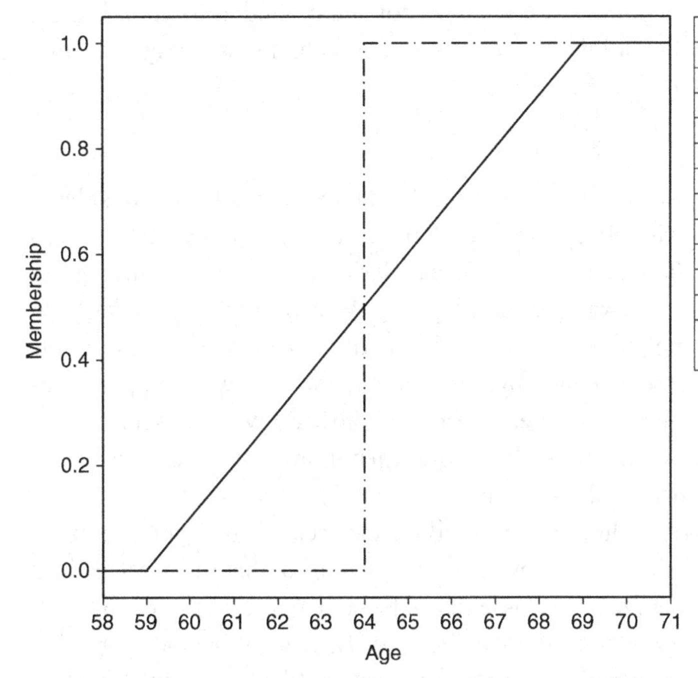

Age	Membership
<59	0
59	0
60	0.1
61	0.2
62	0.3
63	0.4
64	0.5
65	0.6
66	0.7
67	0.8
68	0.9
69	1.0
>69	1.0

Figure 4.1. Fuzzy membership function representing a vague notion of the concept "old."

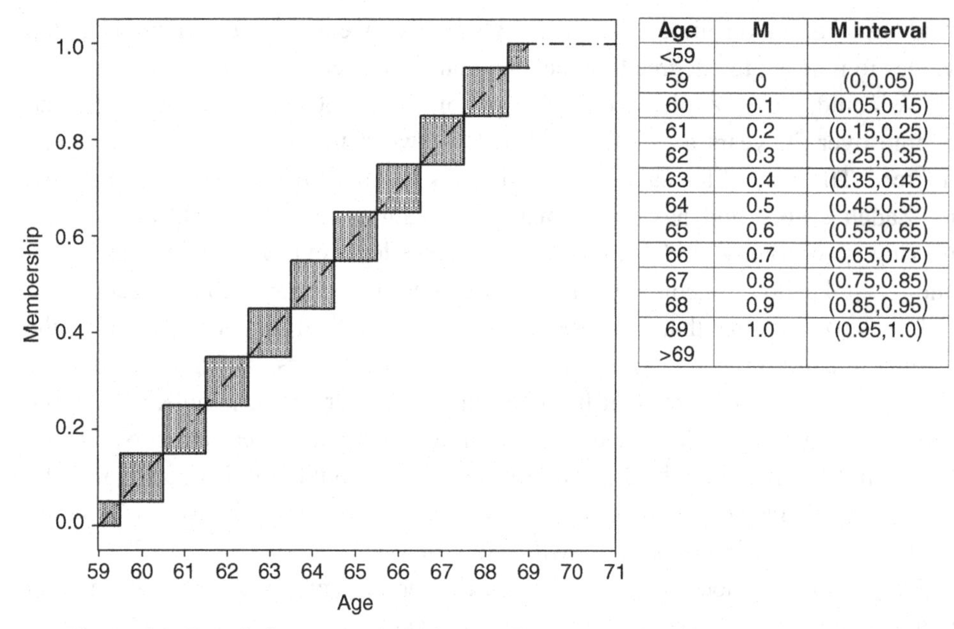

Age	M	M interval
<59		
59	0	(0,0.05)
60	0.1	(0.05,0.15)
61	0.2	(0.15,0.25)
62	0.3	(0.25,0.35)
63	0.4	(0.35,0.45)
64	0.5	(0.45,0.55)
65	0.6	(0.55,0.65)
66	0.7	(0.65,0.75)
67	0.8	(0.75,0.85)
68	0.9	(0.85,0.95)
69	1.0	(0.95,1.0)
>69		

Figure 4.2. Rough fuzzy membership functions approximating the concept "old" using a granularity of whole years.

further generalized into a joint representation for vague and resolution-limited information, and how that joint representation can help in handling semantic uncertainties of vulnerability variables.

Rough fuzzy "old"

Following Dubois and Prade (1992), let us call F_{old} the fuzzy set of ages considered as "old" for people, which is a fuzzy subset of A, the set of all ages. If we take the US Census Bureau Census 2000 data, it imposes granularity on A because some classes consist of a range of ages, for example, 25–29 years. We can call this granularity an equivalence relation R, such that X/R will be the set of equivalence classes (in this case the Census 2000 classes) created by this relation, and $[x]R$ will be an equivalence class (for instance, the 25–29 years class). So how do we represent F_{old} by means of X/R? That is, how can we use the theoretical notion of "old" represented by the graded function when we only have access to pre-classified data?

Because of the limited resolution, we introduce the idea of an upper and lower approximation of F_{old}. The upper and lower approximations of that fuzzy set are also fuzzy sets of X/R (all the equivalence classes each denoted X_i, e.g., X_1 = [62–64 yrs]) and a pair of membership functions defines them: $(\underline{\mu}, \bar{\mu})$. Thus, if we look for "old" people in the census data, we get a rough fuzzy set, $(\underline{old}, \overline{old})$, defined by two membership functions as in Figure 4.3. This means for example that each instance

Table 4.1 *Example data on population age for one hypothetical 2000 census block together with membership degrees to the rough fuzzy set "old"*

2000, Block 1	Count	$(\underline{old}, \overline{old})$
55 to 59 years	89	(0 , 0.05)
60 and 61 years	25	(0.05 , 0.25)
62 to 64 years	21	(0.25 , 0.55)
65 and 66 years	18	(0.55 , 0.75)
67 to 69 years	36	(0.75 , 1)
70 to 74 years	51	(1 , 1)
75 to 79 years	36	(1 , 1)
80 to 84 years	23	(1 , 1)
85 years and over	15	(1 , 1)

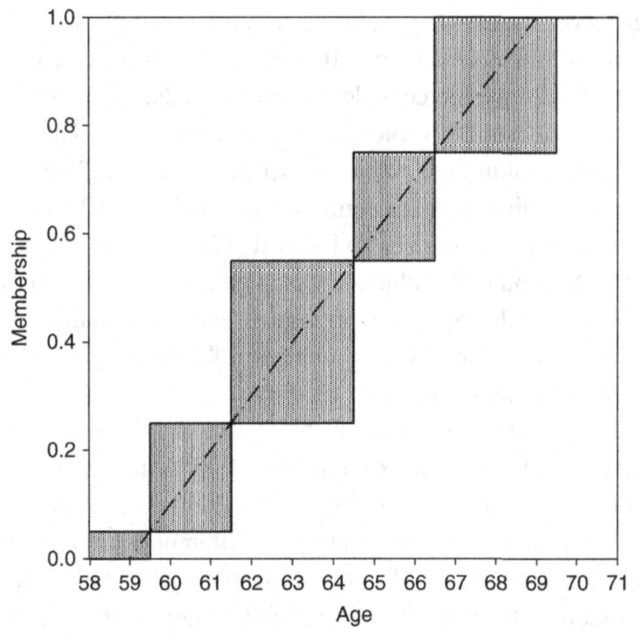

Figure 4.3. Rough fuzzy membership functions approximating the concept "old" for census data.

of the [62–64 yrs] group is a member of the rough fuzzy set "old" with membership degrees (0.25,0.55), and each instance of the [67–69 yrs] group is a member of the rough fuzzy set "old" with membership degrees (0.75,1.0).

Consider the following data for one census block (Table 4.1). With the census class resolution, there are 125 (15+23+36+51) people who are definitely "old,"

Table 4.2 *Example data on population age for one hypothetical 2000 census block together with membership degrees and approximations of the number of people belonging to the rough fuzzy set "old"*

2000, Block 1	Count	$(\underline{old}, \overline{old})$	$\lvert(\underline{old}, \overline{old})\rvert$
55 to 59 years	89	(0 , 0.05)	(0,4.45)
60 and 61 years	25	(0.05 , 0.25)	(1.25,6.25)
62 to 64 years	21	(0.25 , 0.55)	(5.25,11.55)
65 and 66 years	18	(0.55 , 0.75)	(9.9,13.5)
67 to 69 years	36	(0.75 , 1)	(27,36)
70 to 74 years	51	(1 , 1)	51
75 to 79 years	36	(1 , 1)	36
80 to 84 years	23	(1 , 1)	23
85 years and over	15	(1 , 1)	15
Sum	314		(168.4,196.75)

36 people considered (somewhat) "old" to a degree of 0.75–1.0, 18 people considered (somewhat) "old" to a degree of 0.55–0.75, 21 people considered "old" to a degree of 0.25–0.55, 25 people considered "old" to a degree of 0.05–0.25, and 89 people considered (only slightly) "old" to a degree of 0.0–0.05. The total population in the block is 1513, making the proportion of definitely "old" people 125/1513 = 8.26%, and the proportion "old" to some degree 314/1513 = 20.75%. In addition, we could express that information as an interval (125, 314) or a percentage interval (8.26%, 20.75%). We could also obtain a weighted approximation of the amount of "old" people based on the membership values using cardinalities from the sets (Table 4.2), giving us an interval for the number of "old" people (168.4, 196.75) or the percentage of "old" people (11.1%, 13.0%).

The methodology gives us the possibility to compare these data with older data that use different class limits. For example, the 1990 Census had slightly different categories that did not exactly match the Census 2000 categories (Table 4.3). The procedure described above would produce a different interval estimate for the number of "old" people (106.5, 143.4) or percentage of "old" people (9.66%, 13.0%) in the total population (1103). The slightly larger interval range results from the coarser granularity of the categories. The total uncertainty in the 65–69 year interval is 2.2 percentage units, whereas in the Census 2000 data, the same interval is only 0.8 percentage units.

In summary, the generic rough fuzzy $(\underline{old}, \overline{old})$ representation can accommodate crisp, rough, and fuzzy sets, representing vague linguistic terms even in situations where resolution in the data is limited. The representation also enables comparison across incompatible data specifications. Further details on the rough fuzzy formalism can be found in Ahlqvist *et al.* (2003).

Table 4.3 *Example data on population age for one hypothetical 1990 census block together with membership degrees and approximations of the number of people belonging to the rough fuzzy set "old"*

1990, Block 1	Count	$(\underline{old}, \overline{old})$	$\lvert(\underline{old}, \overline{old})\rvert$
55 to 59 years	40	(0 , 0.05)	(0,2)
60 and 61 years	26	(0.05 , 0.25)	(1.3,6.5)
62 to 64 years	18	(0.25 , 0.55)	(4.5,9.9)
65 to 69 years	54	(0.55 , 1)	(29.7,54)
70 to 74 years	49	(1 , 1)	49
75 to 79 years	14	(1 , 1)	14
80 to 84 years	6	(1 , 1)	6
85 years and over	2	(1 , 1)	2
Sum	209		(106.5,143.4)

Example 2. Interoperating different land-use/land-cover classes

The concept of vulnerability is not as simple as deciding how many old people live in an area. Vulnerability and many other concepts used by human–environment researchers are complex combinations of many properties. Gärdenfors (2000) argues from a psychological perspective that concepts are made up of a collection of defining attribute domains. For example, the concept of vulnerability has numerous domains, including population density, educational attainment, poverty rate, and much more. In each of these attribute domains, a concept property defines a point or fuzzy region, such as the interval or intervals of poverty rate values that contribute to the concept of vulnerability. Moreover, for any concept definition, each property of that concept is assigned a certain importance, or *salience*, in relation to other properties of the concept. Salience enables researchers to declare that some properties are more important than others for defining a concept. A framework termed *uncertain conceptual spaces*, which uses the same type of fuzzy- and rough-set-based constructs described in the previous example, was implemented to provide a way to represent the many vague concepts that geographers and other scientists use (Ahlqvist 2004).

One of the many challenges facing the HERO team was to get some baseline knowledge of local land-use and land-cover change that could be compared among the four sites. Comparing land-cover classifications across sites or through time is often problematic because of differences among the classification systems used. The solution presented in Chapter 7 is to standardize the classification among all four sites by using the same Anderson Level 1 land-cover classes (Anderson *et al.* 1976). Even when classes are homogenized, change analysis can be cumbersome due to the many possible change categories. For a study that has 15 classes, the number of possible change categories is 15*15 = 225, so it can be a daunting task to identify the important changes. In the following section, we will demonstrate how to represent land-cover classes using the uncertain conceptual spaces framework noted above. This framework makes it possible not only to represent the vagueness of land-use and land-cover categories explicitly, but also to compare categories among incompatible classification systems, either over time or across sites. We also argue that other concepts, such as vulnerability and its defining attributes, can be represented and analyzed across time and space using the same approach.

Data and methods

In the same way that we can define "old" as a fuzzy set, we can define land-cover classes using fuzzy attribute characteristics. For example, category definitions of classes that relate to developed land are partly based on the amount of impervious surface cover. Figure 4.4 and the description below illustrate how different ranges of impervious surface cover can help to differentiate between Anderson *et al.* (1976) categories "Low Intensity Residential" and "High Intensity Residential" areas:

> 21. Low Intensity Residential: Includes areas with a mixture of constructed materials and vegetation. Constructed materials account for 30 to 80 percent of the cover. Vegetation may account for 20 to 70 percent of the cover. These areas most commonly include single-family housing units. Population densities will be lower than in high intensity residential areas.
> 22. High Intensity Residential: Includes heavily built up urban centers where people reside in high numbers. Examples include apartment complexes and row houses. Vegetation accounts for less than 20 percent of the cover. Constructed materials account for 80 to 100 percent of the cover.

In this example, a residential area with more than 80% impervious surface cover would be categorized as "High Intensity Residential." If impervious surface cover is below this percentage, the membership in the "High Intensity Residential" fuzzy-set-would be lower than membership in the "Low Intensity Residential" class. In this way, a fuzzy-set-based representation acknowledges the graded character of these class definitions and enables a quantitative evaluation of the class similarities

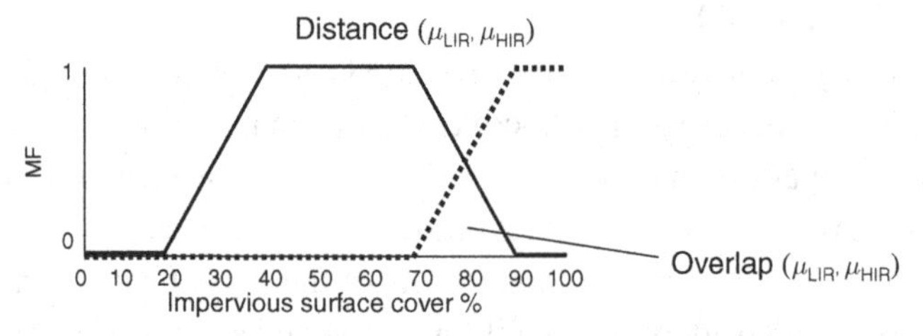

Figure 4.4. Schematic illustration of two fuzzy membership functions describing the amount of impervious surface cover for two land-cover classes, "Low Intensity Residential" and "High Intensity Residential."

using measures of semantic overlap and distance (Ahlqvist 2004), as indicated in Figure 4.4. Moreover, most classes require several characterizing attributes to provide a useful description. The descriptions above explicitly mention vegetation cover and a few qualitative attributes. All can be defined as rough fuzzy sets to build a parameterized definition of each category.

The above approach was used to develop parameterized concept definitions for all land-cover classes in two land-cover datasets covering Centre County, Pennsylvania: the 1992 National Land Cover Data (USGS 2005) (Figure 4.5a) and the 2000 Pennsylvania Land Cover (RESAC 2004) (Figure 4.5b). Both datasets use similar variations of the Anderson land-cover classification system (Anderson *et al.* 1976). Similar to the example given above, the parameterized definitions were developed from textual descriptions of the class definitions provided by the data producers.

Analogous with the semantic distance and overlap measures outlined in Figure 4.4, the parameterized class definitions can be used to compare all land-cover classes by calculating a compound semantic distance and overlap metrics based on all attributes of a definition. The full comparison, with each class compared to every other class, can be summarized in two matrices; one for the semantic distance measures and one for the overlap measures.

(a)

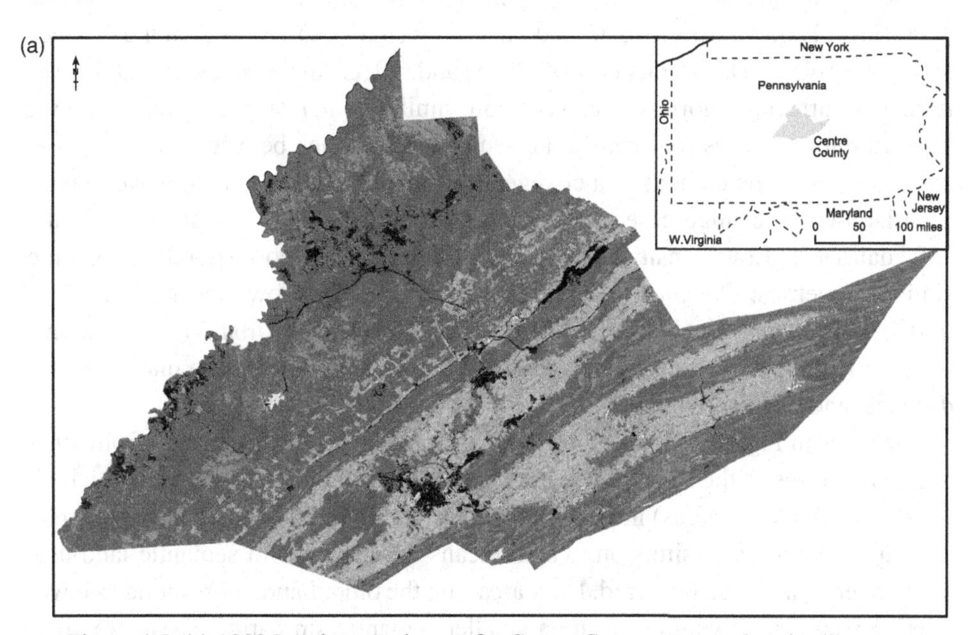

Figure 4.5. (a) 1992 Land-cover dataset for Centre County, Pennsylvania. (Source: USGS 2005.) (b) 2000 Land-cover dataset for Centre County, Pennsylvania. (Source: RESAC 2003.)

(b)

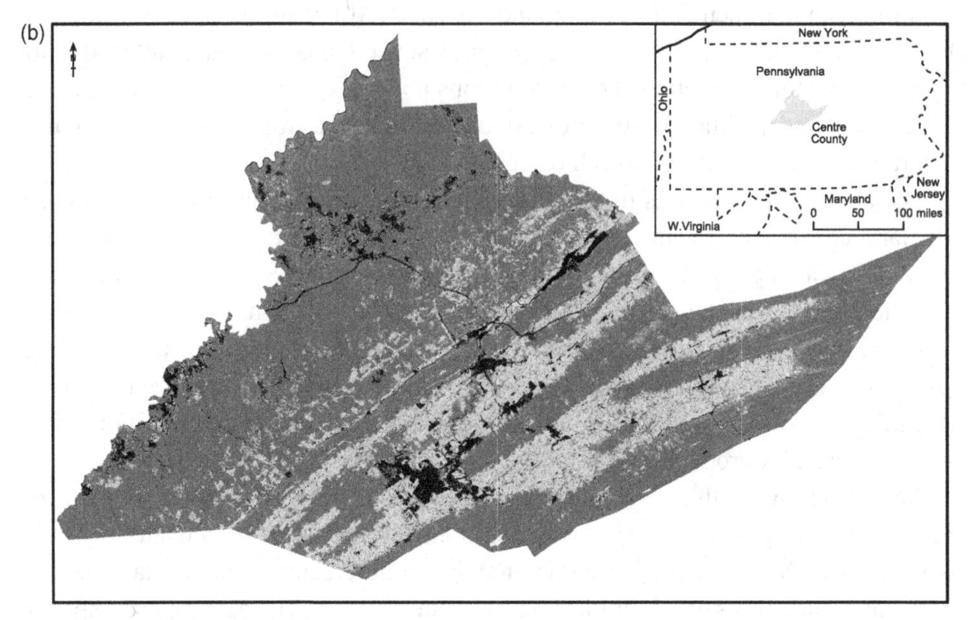

Figure 4.5. (cont.)

An extract of the full matrices is provided in Table 4.4; see also Figure 4.6. Here we see that the distance between the 1992 class "Low Intensity Residential" and the 2000 class "High Density Urban" is 0.22. This is significantly less than the distance to the 2000 class "Deciduous Forest" (0.73) and reflects the perception that the two residential/urban categories are closer (more similar) than residential and forest. We now use these metrics to quantify the semantic difference between any two land-cover classes in the context of a change assessment. More specifically, we replace each land-cover change case, e.g., change from "Deciduous Forest Land" in the 1992 dataset to "Low Density Urban" in the 2000 data, with corresponding semantic similarity metrics, *Distance* ("Deciduous Forest Land", "Low Density Urban") = 0.51, and *Overlap* ("Deciduous Forest Land", "Low Density Urban") = 0.89. In this way, we replace the nominal change map with two semantic change maps: one for distance, and one for overlap.

The maps in Figure 4.7 and Figure 4.8 illustrate the landscape change from 1992 to 2000 in terms of the semantic overlap and distance metrics. In Figure 4.7, a high overlap value (white areas) indicates that the situation in 2000 had a large semantic overlap with the 1992 situation, which means that little or no semantic land-use/land-cover change had occurred. Dark areas, on the other hand, correspond to lower overlap values and would indicate a smaller semantic similarity. These are areas where some kind of change had occurred, either in the category definition or in the form of an actual landscape change. A similar interpretation is valid for Figure 4.8

Table 4.4 *Extract from the semantic distance and overlap matrices derived from parameterized land-cover category definitions*

Distance		2000 Low Density Urban	High Density Urban	Deciduous Forest	Coniferous Forest
1992	Low Intensity Residential	0	0.22	0.73	0.73
	High Intensity Residential	0.22	0	0.81	0.81
	Deciduous Forest	0.51	0.54	0	0.43
	Evergreen Forest	0.51	0.54	0.43	0

Overlap		2000 Low Density Urban	High Density Urban	Deciduous Forest	Coniferous Forest
1992	Low Intensity Residential	0.70	0.79	0.5	0.5
	High Intensity Residential	0.70	0.70	0.16	0.16
	Deciduous Forest	0.89	0.83	1	0.77
	Evergreen Forest	0.89	0.83	0.77	1

where semantic distance values are displayed. Dark areas correspond to large distance (big change) and white areas correspond to small distances (little or no change). The overlap and distance metrics convey different aspects of semantic similarity; by looking at different combinations of these values we can distinguish four general change situations (Table 4.5).

In the maps described above (Figures 4.7 and 4.8), the overlap and distance values appear to co-vary in most places. This co-variance will mean that most places either have large overlap and small distance values, which corresponds to little or no change (very similar classes), or have small overlap and large distance, which corresponds to big change (very different classes). There are, however, two other possibilities: a small distance and small overlap value would indicate that the 2000 situation corresponds to a similar, but still not the same class compared to the 1992 situation. A change from "Low Intensity Residential" to "High Density Urban" would be an example of this situation. Generally, we can associate this type of change with gradual processes, such as infilling of residential areas and vegetation growth transitions. Another possibility can be illustrated with a place that in 1992 was classified as "Urban/Recreational Grass." This class did not exist in the 2000 dataset and typically would have been classified as "Low Density Urban." The two classes could be looked at as class and subclass since "Recreational Grasses" could be part of a "Low Density Urban" land cover. These two cases of semantic distance

4. FOREST LAND
Forest Lands have a tree-crown areal density (crown closure percentage) of 10 percent or more, are stocked with trees capable of producing timber or other wood products, and exert an influence on the climate or water regime. [...]

41. DECIDUOUS FOREST LAND
Deciduous Forest Land includes all forested areas having a predominance of trees that lose their leaves at the end of the frost-free season or at the beginning of a dry season. [...]

Source: Anderson et al. (1976)

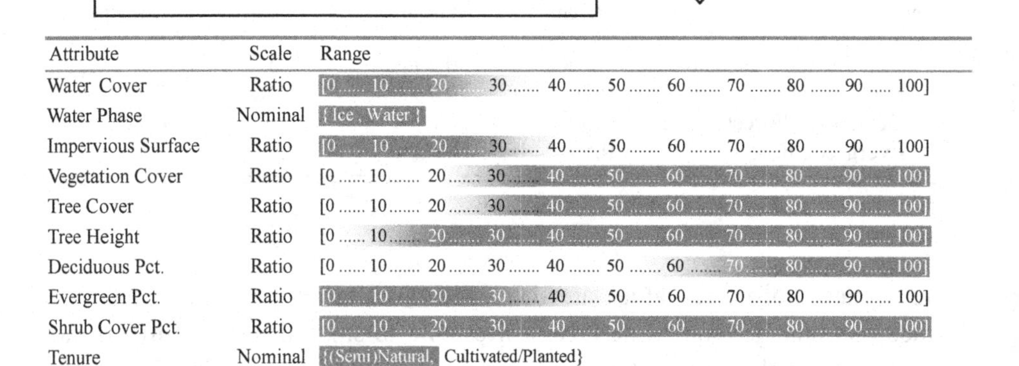

Attribute	Scale	Range
Water Cover	Ratio	[0 ...... 10 .... 20 .... 30 .... 40 .... 50 .... 60 .... 70 .... 80 .... 90 ..... 100]
Water Phase	Nominal	{Ice, Water}
Impervious Surface	Ratio	[0 ...... 10 .... 20 .... 30 .... 40 .... 50 .... 60 .... 70 .... 80 .... 90 ..... 100]
Vegetation Cover	Ratio	[0 ...... 10 .... 20 .... 30 .... 40 .... 50 .... 60 .... 70 .... 80 .... 90 ..... 100]
Tree Cover	Ratio	[0 ...... 10 .... 20 .... 30 .... 40 .... 50 .... 60 .... 70 .... 80 .... 90 ..... 100]
Tree Height	Ratio	[0 ...... 10 .... 20 .... 30 .... 40 .... 50 .... 60 .... 70 .... 80 .... 90 ..... 100]
Deciduous Pct.	Ratio	[0 ...... 10 .... 20 .... 30 .... 40 .... 50 .... 60 .... 70 .... 80 .... 90 ..... 100]
Evergreen Pct.	Ratio	[0 ...... 10 .... 20 .... 30 .... 40 .... 50 .... 60 .... 70 .... 80 .... 90 ..... 100]
Shrub Cover Pct.	Ratio	[0 ...... 10 .... 20 .... 30 .... 40 .... 50 .... 60 .... 70 .... 80 .... 90 ..... 100]
Tenure	Nominal	{(Semi)Natural, Cultivated/Planted}

Figure 4.6. Schematic illustration of the construction of parameterized concept definitions based on textual land-over class descriptions.

and overlap metrics are examples of changes that are not dramatic; it could be argued that it is more important to highlight larger landscape changes, that is, the situation where there is a low overlap and large distance. A summary metric, here called overall semantic change, can be constructed by taking distance × (1 – overlap). This will cause cases II and III to be cancelled out and only the large changes, such as case IV, to be retained (see Table 4.5).

The map in Figure 4.9 shows where the largest land-cover changes in the landscape have occurred according to the overall semantic change metric. We can note that most of the changes occur near the population centers of State College and Bellefonte. Around those areas, the biggest changes are concentrated along the major roads leading through and around the town centers. Also notable is that the change metric seems to pick up a pattern of a non-changing (white) town center and a surrounding 'band' with larger changes (darker shades), which is associated with suburban and exurban development.

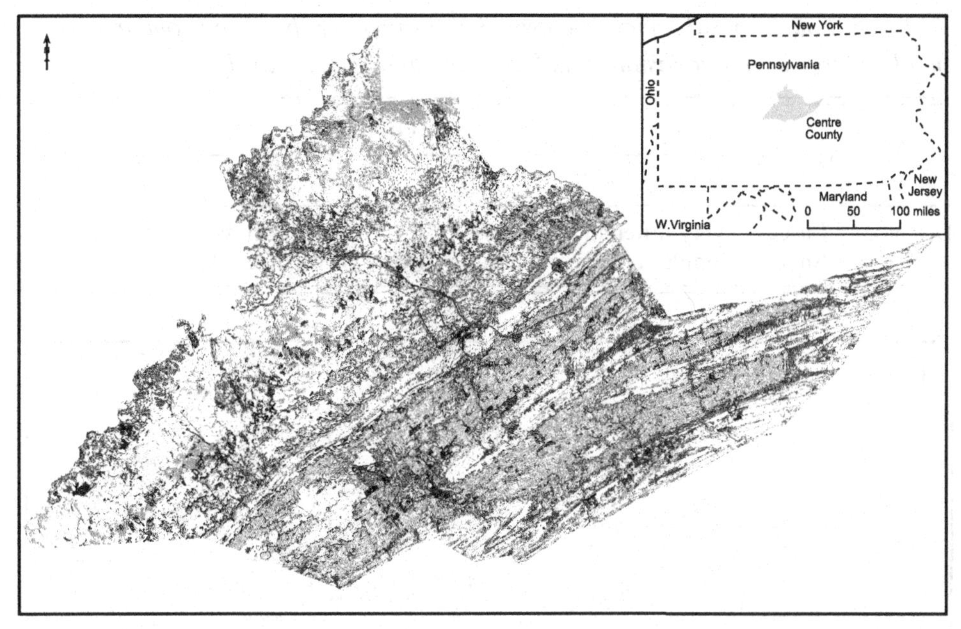

Figure 4.7. 1992–2000 land-use/land-cover change displayed as semantic overlap in which dark areas correspond to low overlap (big change) and white areas correspond to large overlap (little or no change).

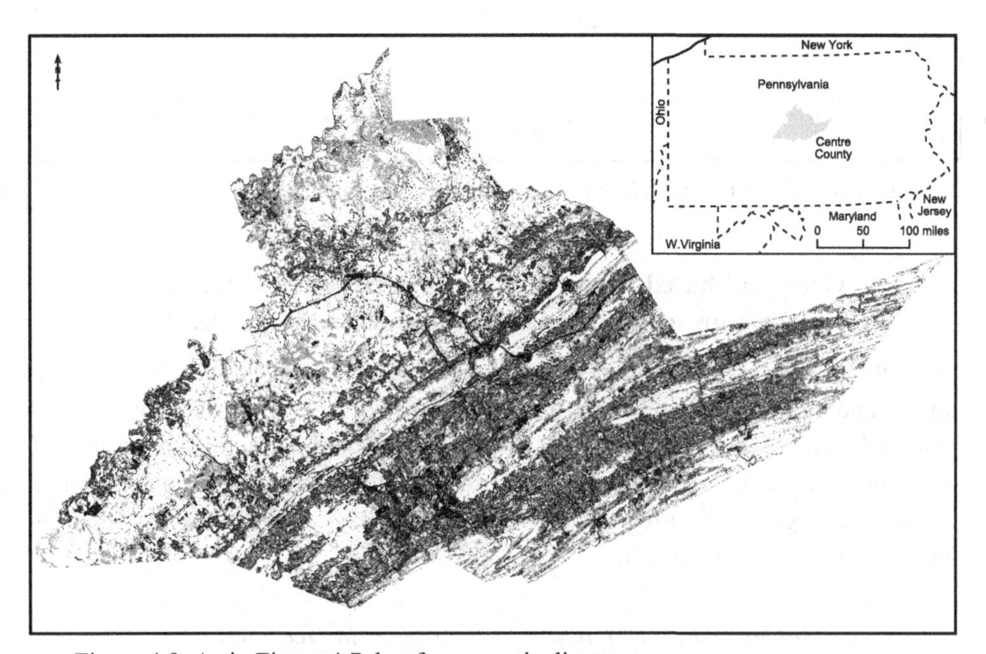

Figure 4.8. As in Figure 4.7, but for semantic distance.

Table 4.5 *Four general semantic class relationship cases based on combination of small or large semantic distance and overlap values, respectively*

		Overlap	
		Small	Large
Distance	Large	Very different classes (I)	Class/subclass relationship (II)
	Small	Similar but disjoint classes (III)	Very similar classes (IV)

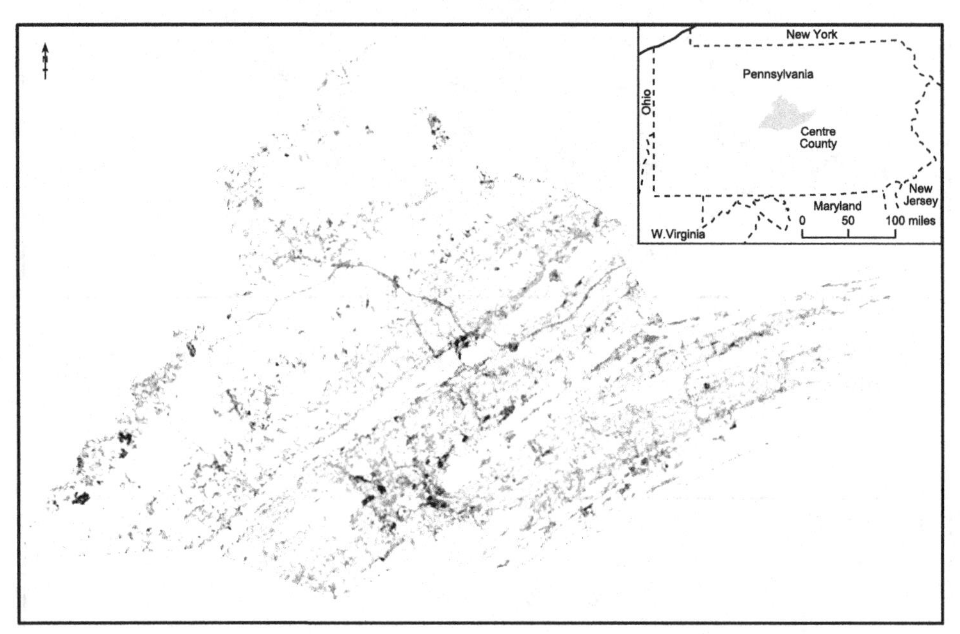

Figure 4.9. As in Figure 4.7, but for overall semantic change.

GeoAgent-based Knowledge System (GeoAgentKS): a tool for representing human–environment interactions

In addition to representing observed changes, the HERO team was also interested in understanding the dynamic interactions among the social and physical components that underlie observations. As one way of understanding human–environment interactions, Yu (2005) developed a Java-based tool, the GeoAgent-based Knowledge System (GeoAgentKS), to represent diverse kinds of knowledge about how humans interact with their environment.

Integrated representation technologies in GeoAgentKS

Representing human–environment interactions needs to address the complex interactions of both social processes and physical processes. To achieve such representation in computer systems, it is necessary to store and represent not only the observed data, but also human knowledge and knowledge-driven human actions. This representation requires an integration of multiple data and knowledge-representation technologies. The techniques used in GeoAgentKS include graph-based concept maps, expert systems, agent-based technologies, mathematical models, and geospatial databases. This subsection focuses on discussing the usefulness of these technologies in representing human–environment relations.

Concept maps

To represent human–environment relations, the techniques must be able to show how social and natural elements interrelate in a given system. Graph-based knowledge representation technologies, such as semantic networks (Collins and Quillian 1969), frames (Minsky 1975), concept maps (Novak and Gowin 1984), and conceptual graphs (Sowa 1984), are widely used to show relationships among concepts. We adopted the concept-map technology for GeoAgentKS because it can provide vivid knowledge visualization, including time representation, for fast learning and knowledge sharing. According to Novak and Gowin (1984), "concepts" in concept maps are perceived regularities in events, objects, and records of events or objects, all designated by labels (http://cmap.coginst.uwf.edu/info/printer.html). Although concept maps and other graph-based technologies can represent relationships well, they have limited capabilities of performing automated reasoning and knowledge-driven actions. Usually, they are incorporated with rule-based systems, such as expert systems, to achieve more advanced knowledge representation.

Expert systems

Expert systems are suitable for representing qualitative knowledge and knowledge-driven actions (Giarrantano 1998). A rule-based expert system comprises two parts: the knowledge base for representing knowledge within a given domain (i.e., the rules), and the inference engine, which evaluates conclusions using the knowledge base via a logical reasoning process. Domain knowledge in the knowledge base is usually expressed as qualitative IF…THEN… rules:

IF <conditions> THEN <conclusions>

The "conclusions" part is also called an "action," so that the rules are often described as condition-action rules. To implement the general knowledge represented in the rules, case-specific "facts" are needed to meet the conditions in the "IF" part of the rules. "Facts" are unconditional information assumed to be true at the time they are used. The inference engine has at least has three functions: collecting rules whose conditions in the "IF" parts can match the available "facts," performing actions in the "THEN" parts in rules that are used, and resolving conflicts to ensure that only one rule will be used if conditions are matched for multiple rules.

Geographic agents (GeoAgents)

Conventional implementations of expert systems mean that they have limited ability to represent spatially distributed and heterogeneous knowledge-driven human–environment interactions. To solve this problem, we used agent-based technologies to represent the interactions of spatially distributed humans or human activities. In general, an agent is a software entity that can perform goal-driven actions (Luck and d'Inverno 2001) and can interact with other agents to accomplish tasks cooperatively.

GeoAgents are agents having geographic characteristics. Namely, they are spatial (they have a location in geographic space), dynamic, and scale dependent. Moreover, they have the capability of being "aware" of their natural and social environment and can respond to the environmental changes in that space. As a result, linking the behavioral rules of individual GeoAgents with corresponding geospatial databases is essential.

GeoAgents are dynamic for three reasons. First, GeoAgents often have lifespans, which can include birth and death. Second, they can respond to environmental conditions and perform actions. Third, their action rules are subject to change over time and space as the environment and the states of other GeoAgents change. GeoAgents are scale-dependent because they can interact not only with other GeoAgents at the same geographic scale, but also with GeoAgents at larger or smaller scales. In the current research, each agent can have an independent knowledge base to store its behavioral rules and a database to store the spatial information of its environment. Due to their independence, agents can be used to represent different levels of social components to address the scale issue. For example, when an agent represents a state-level government agency, its knowledge base can store the laws, regulations, or plans related to this agency, and its database can store the statewide environmental conditions. Similarly, if an agent represents a local institution, its geospatial database stores the corresponding local environmental conditions.

To achieve dynamic interactions among multiple GeoAgents, this research adopted the Agent Communication Language (ACL) for inter-agent communication. ACL, developed by the Foundation for Intelligent Physical Agents (www.fipa.org), provides a protocol for sending and receiving messages, has an ontology to provide vocabulary, and has the capability for an agent to express its intentions. In each message, it contains the information of the sender, the receiver, the action (e.g., *A* requests *B* to perform action *X*), and the content (or message) to be interpreted by the receiver. The communication-based interactions between GeoAgents can be used to represent social processes.

Models

Scientists from many disciplines have developed models (mostly in mathematical form) to explore diverse scientific understandings of relationships among sundry variables or parameters. Researchers use models to perform data analysis, simulation, and prediction, and to support decision-making. Models usually have direct connections with data, so GeoAgents can use them to retrieve and interpret relevant data and to respond to environmental change presented by a model. Thus, it is possible to represent dynamic human–environment interactions by integrating GeoAgents with models and databases.

Databases

Geospatial (or GIS) databases provide geographic information to the user. No single technology, however, can fully address the complexity needed to represent the complexity of human–environment relations. Indeed, many different representations must be implemented to solve such complex problems (Minsky 1991). In this research, GeoAgentKS integrates the concept maps, expert systems, agent-based technologies, mathematical models, and geospatial databases discussed here to represent complex geographic processes. The architecture and implementation of GeoAgentKS are not discussed here; see Yu (2004, 2005) for more information. The reminder of this section presents a case study to demonstrate the use of GeoAgentKS in representing human–environment interactions relevant to community water systems (CWSs) in Centre County, Pennsylvania.

Overview of the case study

The United States Environmental Protection Agency (EPA) defines CWSs as water systems that serve at least 15 connections or 25 people on a year-round basis. There are approximately 54,000 CWSs in the United States drawing on surface and groundwater supplies to serve roughly 268 million people (EPA 2003). Whether serving large cities or small communities, CWSs are key institutions for providing a sustainable, safe water supply to the general population and for protecting public health and community well-being. The natural environment, including such factors as geology, land cover, climate, and extreme weather events, has a large impact on water quality in streams and aquifers, which are critical sources of CWS water. The quality of water delivered by CWSs is further influenced by public policy, regulations, financial constraints, and other socially based factors. Each CWS has a unique context, which includes its physical infrastructure and interactions with the social and natural environment.

This research used two methods to capture and represent knowledge of CWS interactions with their physical and socioeconomic environment: (1) interpretation

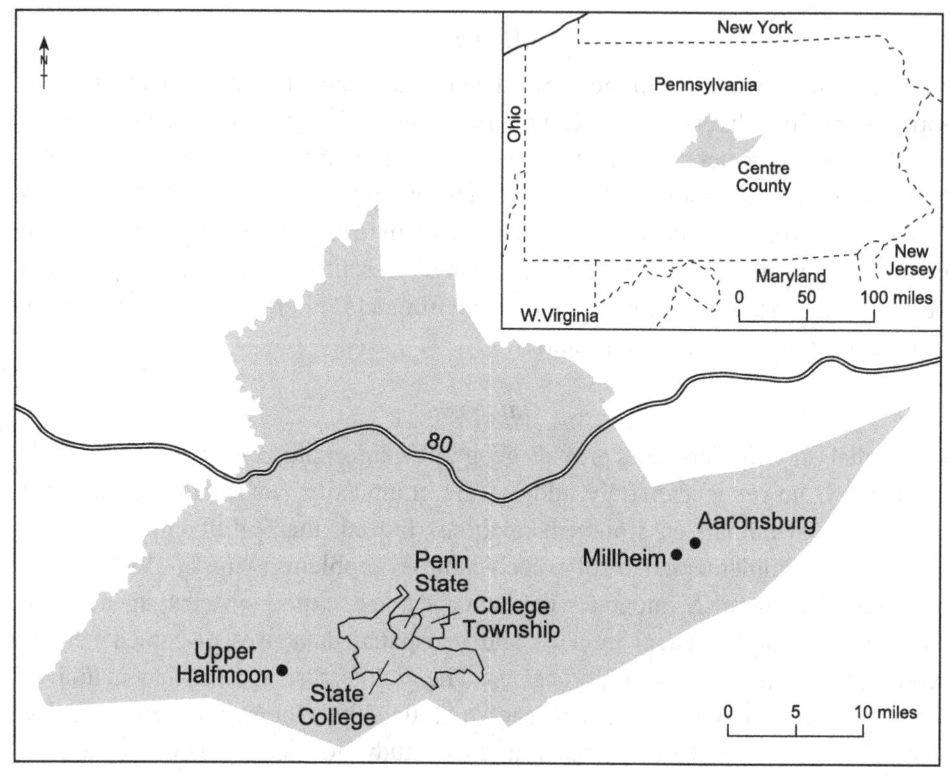

Figure 4.10. The CWSs involved in the case study.

of text documents (e.g., laws, regulations, and plans), and (2) interviews with experts (e.g., local CWS managers). The documents formalized the GeoAgents' behavioral rules, concept maps, and models. We captured undocumented knowledge, such as very recent changes related to the CWSs, via interviews with experts. During the interviews, we used GeoAgentKS to capture the experts' knowledge as concept maps.

Six CWSs were involved in this research: the Aaronsburg Waterpipes Corporation, the College Township Water Authority, the Millheim Borough Water System, the Penn State University Water System, the State College Borough Water Authority, and the Upper Halfmoon Water Association (Figure 4.10). The following example presents a concept map derived from GeoAgent-based representations of human–environment interactions (i.e., text), whereas the succeeding example shows a concept map developed from an interview.

Example 3. Using GeoAgentKS to represent dynamic human–environment interactions

This subsection provides an example that integrates concept maps with GeoAgents, thus allowing users to see not only the relations among the various natural and human elements of the context, but also the spatial information and possible dynamic responses (e.g., direct suggestions to the user) to the environmental changes in GeoAgents. The example covers how the College Township CWS responds to a contamination event (e.g., a chemical spill). The knowledge encoded in this example came directly from the documented emergency plan obtained from this water system.

Figure 4.11 shows part of the concept map of the College Township CWS (labeled "CollegeTownship_CWS"). The three rectangular concept nodes represent three GeoAgents, which are DEP_PA (Pennsylvania Department of Environment Protection), Centre_Daily_Times (the local newspaper), and CollegeTownship_CWS. These concept nodes are internally linked with these three GeoAgents being displayed below the concept map while they run. Each GeoAgent has a knowledge base (i.e., an expert system) to store its behavioral rules and a database to store its environmental data (e.g., geology, land cover, and well depth). The concept nodes are also directly linked to geographic features in the database to allow the user or the GeoAgents to retrieve the spatial information from the database via the concept map. The GeoAgents keep checking the conditions of the relevant concept nodes; the user decides for the GeoAgents which nodes need to be checked. For example, the condition (or FACT) of the node "power" can be "outage" or "normal," and the node "contamination" can be "identified" or "unidentified." These conditions are be derived either from the database or from the user's direct inputs. According to their internal rules, GeoAgents can take actions to respond to the environmental changes.

The GeoAgents' behavioral rules in this example are derived from documented emergency plans, laws, and regulations. According to its emergency plan, the College Township CWS considers its environment to be a circle within three miles of the primary water source, Spring Creek Park Well. If the distance between a possible contamination event and the water source is less than three miles, the CollegeTownship_CWS GeoAgent must perform a set of actions (e.g., isolate the contaminated water source) to address the contamination event; otherwise, it does not have to respond. As shown in Figure 4.11, the CollegeTownship_CWS GeoAgent checks the concept node "contamination," which is present in the database and highlighted on the map. The GeoAgent calculates the Euclidean distance between the contamination point and the water source to decide if the distance is greater than three miles; in this case, the contamination is less than three miles. The GeoAgent of CollegeTownship_CWS recommends the system to isolate the contaminated sources and flush the contaminants. At the same time, it sends a message to the GeoAgent of DEP_PA to report this contamination and another message to the GeoAgent of Centre_Daily_Times to warn the public. Both of these GeoAgents receive the messages; both can take further responses

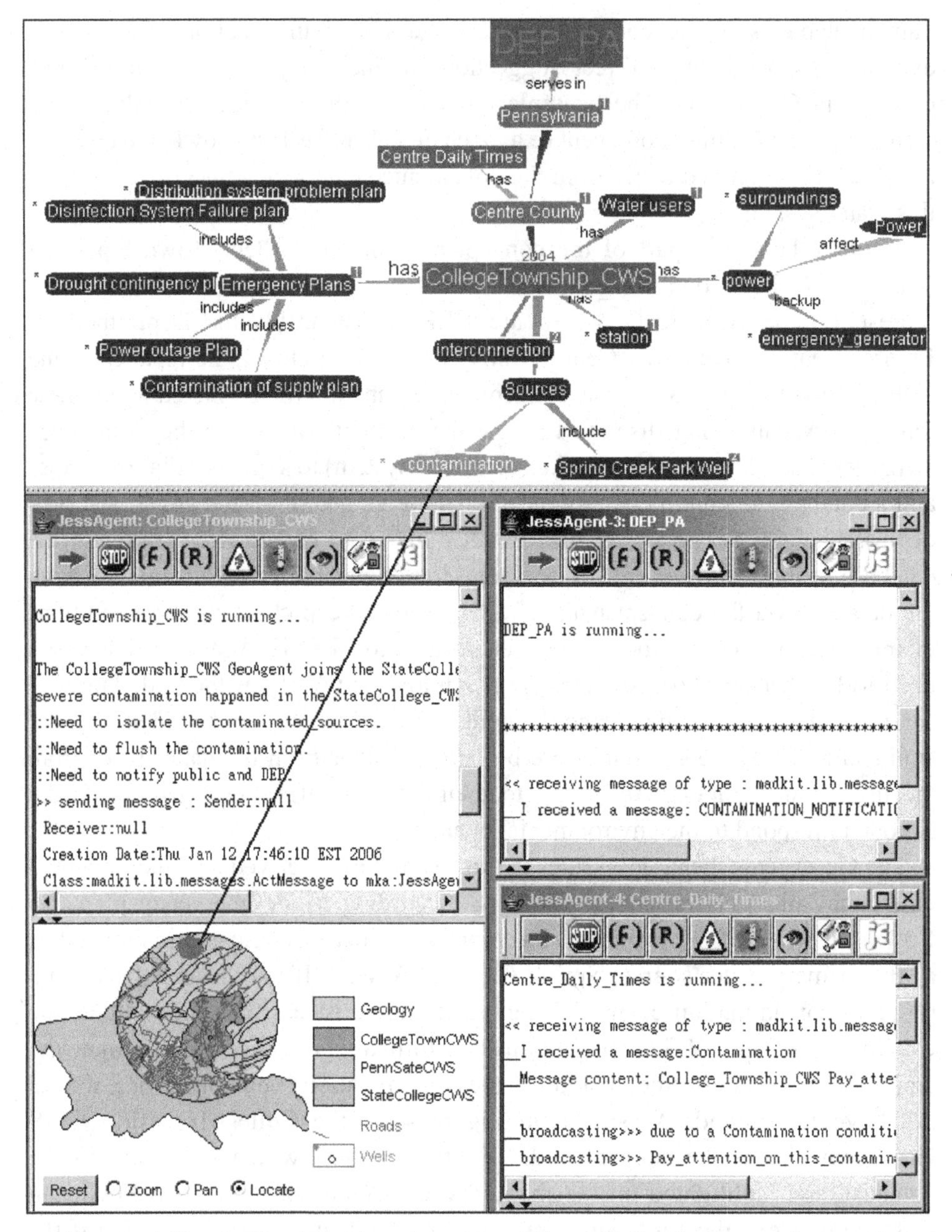

Figure 4.11. A GeoAgent-based representation of the human–environment interactions observed during a contamination event.

according to their own sets of behavioral rules. For example, once the Centre_Daily_Times receives the message, it broadcasts this contamination event to the public and requests that water users find alternate water sources of drinking water.

This example shows that the integration of the concept maps and GeoAgents in GeoAgentKS promotes the representation of very specific human–environment interactions. Yu (2005) presents more complex examples about dealing with power outages and incorporating GeoAgents in drought models to represent the scale-dependent and dynamic human–environment interactions in water management.

Example 4. Using GeoAgentKS to capture and represent experts' knowledge via interviews

Part of the concept map constructed by the manager of the Millheim Borough Water System (i.e., labeled as "Millheim_CWS" in the concept map) is illustrated in Figure 4.12. In this concept map, the water manager identified the basic social and natural elements of the water system and described how this system responded to a major snow event in 1995. The concept map shows that the Millheim water system started in 1995 as a public water supplier. It uses surface water from Phillips Creek and Elk Creek as its water sources. It connects to 380 households and serves 750 water users. Currently, this system has a full-time water operator, a backup operator, and online support. The primary infrastructure of this water system include a reservoir, a filtration plant, a pump station, a storage tank, seven miles of small pipes, and an online monitoring system. According to the water manager's

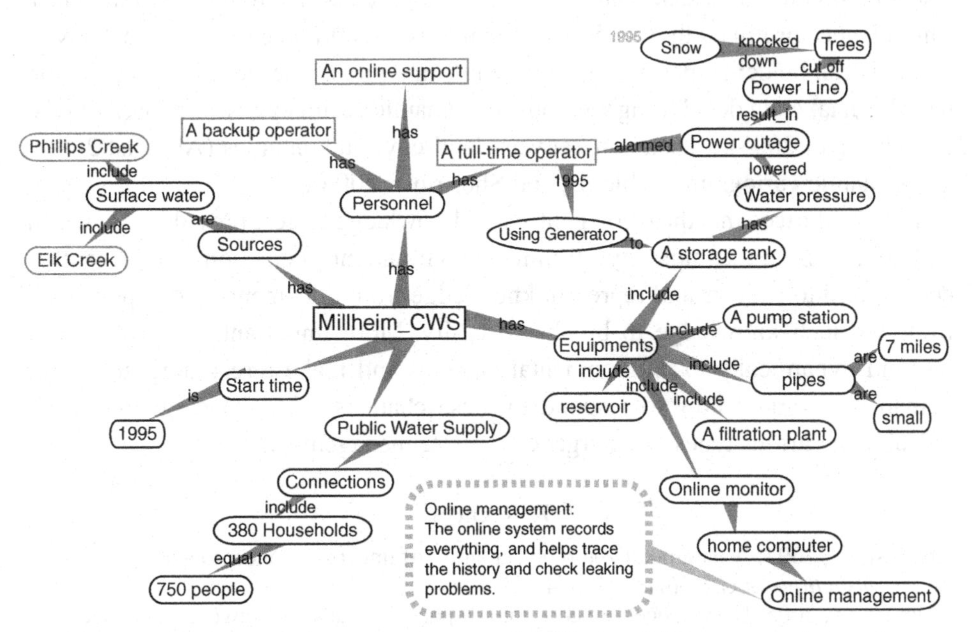

Figure 4.12. Part of the concept map of Millheim CWS created during the interview with the system's water manager on May 31, 2004.

comments on the concept node "online management", the online system automatically records all necessary information and helps the CWS operators trace system history and check leaks.

The concept map also shows that a severe snowstorm knocked down the trees and cut power in 1995, thus lowering the water pressure and triggering an alarm with the full-time water operator. The operator then used the emergency generator to recover the power supply and to pump water into the storage tank.

Conclusions

This chapter has built on the theory introduced in Chapter 2 about consensus building and utilized the HERO collaboratory discussed in Chapter 3 to explore attempts to link human understanding and formal systems, such as databases, analyses, and models using rough fuzzy sets and GeoAgents. We used these approaches to represent and reason with different conceptual understandings to achieve a common conceptual understanding.

In particular, we have demonstrated how the rough fuzzy formalism is able to represent geographic categories that have a graded character and often come as resolution-limited information. We showed that is possible to develop a representational framework based on uncertain conceptual spaces by treating such uncertainties explicitly. This treatment facilitated an assessment of graded change in categorical land cover data that go beyond standard change/no-change outcomes. This advance promises that additional quantitative geographic information analysis methods can be developed for qualitative information. Some steps in this direction have been taken in developing methods for semantic accuracy assessment (Ahlqvist and Gahegan 2005), translation between land-cover taxonomies (Ahlqvist 2005), and semantic variograms (Ahlqvist and Shortridge 2006).

We also showed that the GeoAgent-based knowledge system provides a powerful way to address the complexity of human–environment interactions. GeoAgentKS can be used to capture and represent knowledge from documents and experts, and it allows such knowledge to be shared easily. More important, GeoAgents can respond dynamically to environmental changes and make direct suggestions for human users about how to respond to these changes. Thus, GeoAgents can be valuable in various types of emergency/disaster management.

References

Ahlqvist, O., 2004. A parameterized representation of uncertain conceptual spaces, *Transactions in GIS* **8**(4): 493–514.

Ahlqvist, O., 2005. Using uncertain conceptual spaces to translate between land cover categories. *International Journal of Geographical Information Science* **19**(7): 831–857.

Ahlqvist, O., and M. Gahegan, 2005. Probing the relationship between classification error and class similarity. *Photogrammetric Engineering and Remote Sensing* **71**(12): 1365–1373.

Ahlqvist, O., and A. Shortridge, 2006. Characterizing land cover structure with semantic variograms. In *Progress in Spatial Data Handling*, eds. A. Reidl, W. Kainz, and G. A. Elmes, pp. 401–415. Berlin: Springer.

Ahlqvist, O., J. Keukelaar, and K. Oukbir, 2003. Rough and fuzzy geographical data integration. *International Journal of Geographic Information Science* **17**(3): 223–234.

Anderson, J. R., E. E. Hardy, J. T. Roach, and R. E. Witmer, 1976. *A Land Use and Land Cover Classification System for Use with Remote Sensor Data*, Professional Paper No. 964. Washington, D. C.: United States Geological Survey.

Bishr, Y., 1997. *Semantic Aspects of Interoperable GIS*. ITC Ph.D. Publication Series No. 56. Enschede, The Netherlands: International Institute for Aerospace Survey and Earth Sciences.

Collins, A. M., and M. R. Quillian, 1969. Retrieval time from semantic memory. *Journal of Verbal Learning and Verbal Behavior* **8**: 240–248.

Dubois, D., and H. Prade, 1992. Putting rough sets and fuzzy sets together. In *Intelligent Decision Support: Handbook of Applications and Advances of the Rough Sets Theory*, ed. R. Slowinski, pp. 203–232. Dordrecht: Kluwer.

Gahegan, M. N., 1999. Characterizing the semantic content of geographic data, models, and systems. In *Interoperating Geographic Information Systems*, eds. M. F. Goodchild, M. J. Egenhofer, R. Fegeas, and C A Kottman, pp. 71–83. Boston, MA: Kluwer.

Gärdenfors, P., 2000. *Conceptual Spaces: The Geometry of Thought*. Cambridge, MA: MIT Press.

Giarrantano, J. C., 1998. Otter: an automated deduction system. Accessed at www-unix.mcs.anl.gov/AR/otter.

Luck, M., and M. d'Inverno, 2001. Autonomy: a nice idea in theory. In *Intelligent Agents VII: Agent Theories, Architectures and Languages*, eds. C. Castelfranchi and Y. Lesperance, pp. 351–353. Berlin: Springer.

Mennis, J. L., 2003. Derivation and implementation of a semantic GIS data model informed by principles of cognition. *Computers, Environment and Urban Sytems* **27**: 455–479.

Minsky, M., 1975. A framework for representing knowledge. In *The Psychology of Computer Vision*, ed. P. H. Winston, pp. 211–277. New York: McGraw-Hill.

Minsky, M., 1991. Logical versus analogical or symbolic versus connectionist or neat versus scruffy. *AI Magazine* **12**(2): 34–51.

Novak, J. D., and D. B. Gowin, 1984. *Learning How to Learn*. New York: Cambridge University Press.

Pawlak, Z., 1991. *Rough Sets: Theoretical Aspects of Reasoning about Data*. Dordrecht: Kluwer.

Regional Earth Science Applications Center (RESAC), 2004. RESAC Chesapeake Bay Watershed Land Cover – 2000, Version 1.05. Accessed at www.geog.umd.edu/resac/.

Robinson, V. B., 2003. A perspective on the fundamentals of fuzzy sets and their use in geographic information systems. *Transactions in GIS* **7**(1): 3–30.

Rodriguez, M. A. and M. Egenhofer, 2004. Comparing geospatial entity classes. *International Journal of Geographic Information Science* **18**: 229–256.

Sowa, J. F., 1984. *Conceptual Structures: Information Processing in Mind and Machine*. Reading, MA: Addison-Wesley.

United States Census Bureau, 2000. Accessed at http://factfinder.census.gov.

United States Geological Survey (USGS), 2005. National Land Cover Dataset 1992. Accessed at http://landcover.usgs.gov/natllandcover.asp.

United States Environmental Protection Agency (EPA), 2003. *Water on Tap: What You Need to Know*, EPA 816-K-03-007. Washington, D.C.: Environmental Protection Agency, Office of Water.

Yu, C., 2004. GeoAgent-based knowledge acquisition, representation, and validation. In *GIScience 2004*, eds. M. J. Egenhofer, C. Freksa, and M. Harvey, pp. 338–341. Adelphi, MD: University of Maryland.

Yu, C., 2005. GeoAgent-based knowledge systems. Unpublished dissertation. University Park, PA: The Pennsylvania State University.

Zadeh, L. A., 1965. Fuzzy sets. *Information and Control* **8**: 338–353.

Part III

5

Establishing vulnerability observatory networks to coordinate the collection and analysis of comparable data

COLIN POLSKY, ROB NEFF, AND BRENT YARNAL

Introduction

Vulnerability has emerged in recent years as one of the central organizing concepts for research on global environmental change (e.g., Downing 2000; O'Brien and Leichenko 2000; Turner *et al.* 2003; Schröter *et al.* 2005; Parry *et al.*, 2007).[1] This concept is appealing because it is inclusive. From this perspective, humans and the natural environment are not independent systems, homogeneous and unable to adapt to threats, be they anticipated, realized, or perceived but not realized. Instead, human and natural systems are viewed as intimately coupled, and differentially *exposed*, *sensitive,* and *adaptable* to threats. This logic, followed to its natural conclusion, means that adopting a "vulnerability" perspective demands a thorough investigation of biophysical, cognitive, and social dimensions of human–environment interactions. Strictly speaking, to conduct a vulnerability assessment means that no element of the human–environment system may be simplified away or considered a mere boundary condition.

This conceptual inclusiveness complicates the analytical task (compared to the simpler impacts-only approach), which partially explains why there are few, if any, studies that deeply engage this vast set of intellectual dimensions. This inclusiveness also raises important methodological questions. Consider two vulnerability assessments that examine local-scale vulnerabilities associated with hydroclimatic variability. Mustafa (1998) examines flood-related vulnerabilities in five Pakistani farming communities; Hill and Polsky (2005) assess drought-related vulnerabilities in ten non-farming Massachusetts (USA) towns. Can the vulnerability indicators produced by these assessments be easily compared such that potential common findings on how exposure, sensitivity, and adaptive capacity contribute to local vulnerabilities may be identified? We argue that the answer to this question for much of the vulnerability literature is no because many assessments use dissimilar measurements (often of necessity). We also argue that a suitable graphical

Sustainable Communities on a Sustainable Planet: The Human–Environment Regional Observatory Project, eds. Brent Yarnal, Colin Polsky, and James O'Brien. Published by Cambridge University Press.

organization of the vulnerability assessment's findings – using the Vulnerability Scoping Diagram (VSD) presented below (p. 89) – may provide the basis for the inter-assessment comparisons that are a precondition for advancing vulnerability science (Cutter 2001, 2003). Meeting this comparison challenge is important. If the vulnerability perspective is to represent not only an appealing conceptual framework but also a meaningful catalyst for empirical research, then researchers must be able to identify common lessons from multiple, independent vulnerability assessments – even if the assessments use dissimilar measurements.

The above observations motivate the overarching goal of this chapter: to facilitate the construction of comparable global change vulnerability assessments. There are three specific objectives: (1) to argue for the need for vulnerability researchers to adopt a common and replicable schema – the VSD – for representing and organizing the findings of a given vulnerability assessment, (2) to present an example of such a framework, illustrated by our recent vulnerability research with the HERO program, and (3) to situate this *specific* methodological innovation within a recently proposed *general* vulnerability methodological protocol, namely the "Eight Steps" protocol described by Schröter *et al.* (2005), to highlight the limited but important role of the VSD within a broader vulnerability research project. In Chapters 8, 9, and 10, we expand on the methodological needs associated with conducting vulnerability assessments.

Global change vulnerability assessments: concepts, methods, and approaches

Vulnerability assessment concepts

One of the principal developments in the recent global change literature is a basic shift in the conceptualization of the problem under study. There has been a movement to favor vulnerability assessment over the more familiar "impacts" approach (e.g., Liverman 1990; Downing 1991; NRC 1999; Kelly and Adger 2000; Kasperson 2001; Parry 2001; Turner *et al.*, 2003; Schröter *et al.* 2005). In this literature, vulnerability is a function of three main dimensions: *exposure* to specific social and/or environmental stresses, associated *sensitivities*, and related *adaptive capacities*.

Using this conceptualization, to be vulnerable to the effects of stresses associated with global change, human–environment systems not only must be exposed and sensitive to the effects, but also must have limited ability to adapt. Conversely, systems are less vulnerable – perhaps even sustainable, i.e., able to persist in the long-term in the face of threatening stresses – if they are less exposed, less sensitive, or possess strong adaptive capacity (Smit *et al.* 1999; Finan *et al.* 2002).

In vulnerable systems, some form of anticipatory action would be justifiable to mitigate the ecological, social, and economic damages anticipated from global change. In relatively sustainable systems, there would be less reason for concern and pre-emptive action. Vulnerability assessments are therefore a necessary part of sustainability science; that is, basic research intended to inform ways to protect social and ecological resources for present and future generations (Kates *et al.* 2001; Clark and Dickson 2003).

Vulnerability assessment approaches

Researchers sympathetic to the vulnerability perspective typically blend methods from more than one scholarly tradition in an attempt to give added weight to one of the vulnerability dimensions (exposure, sensitivity, or adaptive capacity) previously downplayed or assumed away (e.g., Kelly and Adger 2000; Downing *et al.* 2001; Polsky 2004). Thus expertise in vulnerability research methods is predicated on expertise with pre-existing research methods from a variety of research domains. Yet expertise with many distinct methods does not necessarily ensure expertise in cobbling the methods together (Young *et al.* 2006). Indeed, methods are not well established for conducting a vulnerability assessment that synthesizes across such an array of substantive domains. Thus the novelty of global change vulnerability assessments lies not so much in the development of new conceptual domains as in the methodological integration across existing research traditions (Polsky and Cash 2005; Schröter *et al.* 2005).

This need to integrate across research traditions calls for an all-embracing methodological approach. Schröter *et al.* (2005), drawing primarily from work by Kates *et al.* (1985), Carter *et al.* (1994), Klein *et al.* (1999), Smit *et al.* (1999), Downing *et al.* (2001), and Turner *et al.* (2003), propose that researchers will capture the vulnerability perspective if they adopt an overarching approach comprising eight general steps (Figure 5.1). This eight-step approach tries to describe, in *general* terms, the full array of analytical activities needed to characterize vulnerability in all its complexities. The ordering of these eight steps is: (1) define the study area together with stakeholders, (2) get to know the place over time, (3) hypothesize who is vulnerable to what, (4) develop a causal model of vulnerability, (5) find indicators for the elements of vulnerability, (6) operationalize model(s) of vulnerability, (7) project future vulnerability, and (8) communicate vulnerability creatively.

These steps are ordered because there is a natural flow to the analytical activities. However, over the course of an entire vulnerability assessment, researchers will not necessarily follow this order strictly. Some steps will be performed in parallel, and most steps will be performed iteratively. For example, researchers will likely revisit

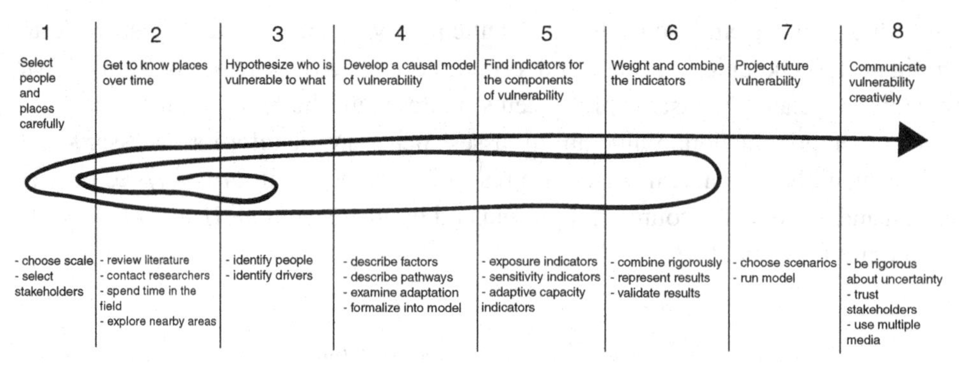

Figure 5.1. Assessing vulnerabilities to the effects of global change: an eight-step approach. (Modeled after Schröter *et al.* 2005.)

and revise Step 4 (in which they develop a causal model) based on findings from Steps 6 to 8 (in which they operationalize the vulnerability model, project its future status, and creatively communicate the findings with stakeholders). Step 8 warrants particular attention here. Although the step might appear relatively unimportant because of its placement in the order, this step is no less important to the production of findings than the other steps. The process of communicating results with – not to – stakeholders is crucial to the overall vulnerability assessment for the ethical purpose of sharing insights gained about vulnerability with the people who are the subject of the vulnerability assessment.

Vulnerability assessment comparisons

The Eight Steps enumerated by Schröter *et al.* (2005) represent an iterative, flexible, and comprehensive *general* analytical approach, the particular application of which may be tailored to the demands of a *specific* research project consistent with the vulnerability conceptualization emerging in the literature. However, the Eight Steps protocol does not encourage the production of assessment results that are comparable with other assessment results. Of course, when data collection and analysis methods differ between vulnerability assessments, comparing results will be difficult. Yet it is crucial for the global change research community to be able to compare results from independent vulnerability assessments (Cutter 2003; Adger 2006). If vulnerability researchers cannot systematically compare lessons learned across independent vulnerability assessments, then they can have little hope of making generalizations. Some form of generalization is necessary to allow people who lack the time or other resources to conduct their own vulnerability assessment to take anticipatory, mitigative actions by referencing results from systematic syntheses of existing vulnerability assessments.

It is natural to inquire how, in theory, one might conduct such comparisons. Perhaps the simplest, and most common, form of comparing independent studies that were not necessarily designed to permit direct comparisons is the "literature review" technique. Literature reviews will always play an important role in scholarly research, but they are not as systematic as they are familiar and simple to use. A systematic approach is needed for topics such as vulnerability assessments in which the universe of potential comparisons is large. The literature on "meta-analysis" (e.g., Wolf 1986) provides a methodological basis for systematically comparing independent studies in which the measurements are quantitative – even if the studies were not designed with the goal of comparison in mind. In the global change literature, variants of this approach have been used by, for example, Geist and Lambin (2002), who compared 152 independent studies of deforestation, and Geist (2005), who compared 132 studies of desertification. There is even an emerging literature for systematically comparing independent studies on the basis of qualitative information, such as Ragin's (1987) "qualitative comparison analysis" method. Rudel (2005) used this method to draw commonalities among 278 studies of tropical deforestation.

As helpful as these methods are for the cases of deforestation and desertification, comparing results from independent vulnerability assessments presents a challenge that even the emerging meta-analytical methods may not be able to overcome. There is little stability in the underlying concepts being operationalized in vulnerability research (Rudel 2008). For example, studies of tropical deforestation typically share a similar, if not identical, operationalization of the dependent variable: felled trees. The measurements may differ in terms of tree type (e.g., hardwood versus softwood), frequency (e.g., annual versus decadal), or observation type (e.g., satellite imagery versus visual confirmation), but there is a fundamental conceptual commonality: felled trees. By contrast, the dependent variable in vulnerability assessments – which should include measurements of exposure, sensitivity, and adaptive capacity – does not present the same stability between studies. This instability is not surprising given the diversity of processes examined under the vulnerability umbrella (Brooks *et al.* 2005). Nonetheless, vulnerability researchers are still faced with the challenge of structuring vulnerability assessments (particularly local-scale studies) such that post-hoc comparisons with the results of other assessments are possible without sacrificing local-scale relevance and validity.

Where a basis for comparisons already exists and does not exist

Some vulnerability assessments already allow for some comparability and therefore will not necessarily benefit from using the Vulnerability Scoping Diagram (VSD) proposed in this chapter. Such studies may not have been designed with the goal of post-hoc comparisons with other studies, but by virtue of the type of data used, the studies are, in theory, comparable. Independent vulnerability assessments that construct indicators using data collected by national and international statistical bureaus such as the United States Bureau of the Census or the United Nations Development Programme are probably already comparable. For example, Moss *et al.* (2001) and Brooks *et al.* (2005) use such datasets to construct indicators for a *global*-scale vulnerability assessment where the country is the unit of analysis. This commonality in data and approaches suggests that readers could readily compare vulnerabilities of different countries not only within these two studies but also between the studies. Researchers could inspect the two studies, for example, for indicators of gross domestic product (GDP) per capita or literacy rates – or some user-defined combination thereof – to calculate and compare vulnerabilities across countries.

Importantly, this advantage of constructing indicators from standardized data should be scale-independent. For example, the *national*-scale vulnerability assessment of 204 US watersheds using data available from the United States Geological Survey by Hurd *et al.* (1999) permits direct inter-watershed comparisons. Similalry, the analysis of 466 Indian "districts" by O'Brien *et al.* (2004) uses government statistics, which permits direct inter-district comparisons. At the *local*-scale, vulnerability assessments that use standardized secondary data should also allow for some comparability – such as the studies by Clark *et al.* (1998), Cutter *et al.* (2000), and Wu *et al.* (2002), which each examine "Census Block" data in their respective US counties. In all of these cases, regardless of the spatial scale at which the study is cast, the indicators constructed could, in theory, be used for direct comparisons of vulnerabilities of the units of analysis, either within or between assessments, because the data and associated indicators allow for comparisons.

Although using standardized data does offer comparability advantages, there is also at least one important associated disadvantage: the data used must correspond to the data provided by the statistical bureau – *not necessarily* to the data demanded by the researchers' theoretical or conceptual frameworks. In this way, vulnerability assessments must satisfy and simply use the best data available. In the context of a given vulnerability assessment, this satisfying may assume one or both of two forms: using unsynchronized data (e.g., using a "road network" data layer for 2000 in an analysis where the other variables reflect 1990 simply because the 1990 road data layer could not be found), or using proxy data (e.g., using "national GDP per capita" to reflect "individual well-being" even though theory suggests that well-being is defined by much more than that simple macro-economic indicator). Although the former case is clearly theoretically problematic, it may not produce significant error provided the features of the road network did not change much between 1990 and 2000. By contrast, the latter case presents potentially severe problems in both theoretical and empirical terms.

The advantages and disadvantages of using standardized data, as discussed above, are the inverse of the case where researchers design and implement their own data collection instruments. Such "primary" data, most commonly reflecting local-scale dynamics, often involve information gained from interviews, surveys, and archival analysis. When researchers design and conduct the collection of their own data, they can ensure that the data are both synchronized and non-proxy, but in so doing they also produce a potential comparability problem. Recall the example of comparing the assessments by Mustafa (1998) and Hill and Polsky (2005). Because the data collection instruments are designed with the specifics of the local-scale dynamics in mind, the underlying indicators differ significantly between studies. As such, it is unclear if it is possible to compare, for example, how adaptive capacity contributes to vulnerability between the two studies. Yet, such a post-hoc vulnerability comparison might be relatively straightforward if these authors had adopted the indicator structuring technique proposed here: the Vulnerability Scoping Diagram (VSD).

Facilitating the comparison of vulnerability assessments

An effective vehicle for facilitating vulnerability assessment comparisons, we argue, is the Vulnerability Scoping Diagram (VSD) and the associated set of indicators collected by each assessment. This diagram is grounded in the "Place Diagram" developed by the Project for Public Spaces (PPS: www.pps.org/topics/gps/gr_place_feat). PPS is a non-profit organization dedicated to creating and sustaining public places that support flourishing communities (PPS 2005). PPS found that successful public places have four key dimensions in common: access and linkages; uses and activities; comfort and image; and sociability. At any place, it is possible to evaluate, qualitatively and/or quantitatively, important features of each of these key attributes. In turn, one or more concrete measurements reflect each feature. For example, associated with the key attribute of "sociability" is the feature of "friendly," which in turn may be reflected by measurements of the extent of "volunteerism." A diagram facilitates the identification and organization of these findings by displaying the four key dimensions (as opposed to three, in the case of vulnerability) on a surface resembling an archery target, with a "bull's-eye" and three concentric rings circling the bull's-eye. The ring nearest the bull's-eye represents the dimensions, the middle ring contains place-based features of each dimension, and the outer ring contains measurements associated with each feature of each dimension.

We find the Place Diagram an effective mechanism for guiding the collection and organization of data, concepts, and indicators needed to assess the success or

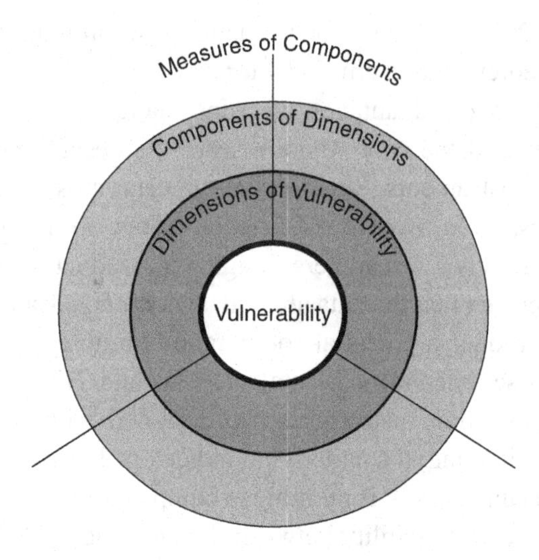

Figure 5.2. General form of the Vulnerability Scoping Diagram (hazard and exposure unit unspecified).

failure of a place along multiple, potentially overlapping principal dimensions. We adopt the main features of this diagram and adapt them to the case of scoping vulnerability. The template for the VSD appears in Figure 5.2. The center of the diagram represents the vulnerability of a given human–environment system. Similar to the PPS diagram, the level of abstraction decreases with distance from the center. The first ring parses vulnerability into its three fundamental *dimensions*, or primary axes along which vulnerability is defined: exposure, sensitivity, and adaptive capacity. The intermediate ring represents the *components*, or the abstract features on which to evaluate each of the three vulnerability dimensions for a given human–environment system. The outer ring includes the *measurements*, or the observable characteristics of the components of the dimensions.

The vulnerability indicators that may be used for inter-assessment comparisons, then, are the measurements and components for each of the three dimensions. These indicators and components reflect what the vulnerability assessment found to be the most salient features of exposure, sensitivity, and adaptive capacity, for each exposure unit and hazard. This diagram and indicator set serve two functions: (1) to build a basis for making comparisons of vulnerability from assessments performed at different places and times, and (2) to provide a starting point for understanding the details of vulnerability in a single exposure unit that may be examined in greater detail using additional research. Indeed, the S in VSD – Scoping – indicates that this diagram is intended to be populated at the early stages of a vulnerability assessment. In the early stages, the VSD would represent the results of an initial scoping, consistent with Step 5 ("Find indicators of vulnerability") of the Eight Steps (Schröter *et al.* 2005). The "final" VSD for an assessment will likely differ from the original scoping diagram on the basis of insights gained from conducting the vulnerability assessment. (See below for an elaboration of the interrelationships between the scoping exercise and the Eight Steps).

To scope vulnerability using the VSD, researchers need to specify five elements of their proposed research: (1) the hazard and associated outcome(s) of interest, (2) the exposure unit, and the (3) dimensions, (4) components, and (5) measures of the vulnerability process in question. The *hazard* refers to the event(s) that threaten people or things people value and that, therefore, may affect the coupled human–environment system. For researchers interested in systems where climate change is a factor, to specify "climate change" as the hazard would be too broad. Instead, this VSD would need to specify an explicit symptom of climate change. For example, a "climate-change-induced increase in the magnitude of hurricane-related flood damages" would be an example of a hazard well suited to a VSD. In this way, the *outcomes* that people wish to eliminate or mitigate (in this case, flood damages) are clearly specified, thereby giving the notion of vulnerability a concrete basis for evaluation. Note that the hazard and outcomes can be multidimensional to reflect simultaneously operating stresses (O'Brien and Leichenko 2000; Turner *et al.* 2003).

The *exposure unit* is the coupled human–environment system that may be vulnerable to the hazard in question. Exposure units should include identifiable assemblages of people and things people value, plus the dominant environmental features of the place; these systems should be relatively clearly bounded in space and time (Kates 1985; Turner *et al.* 2003; Easterling and Polsky 2004). For example, the exposure unit for the work by Turner *et al.* (2004) is the land system – i.e., the collective people, crops, soil, forest, and regional atmospheric conditions – in the Southern Yucatán for the period ~1950 to present. Note that an exposure unit does not need to include all people, institutions, or ecosystems that have any relation whatsoever to the human–environment system in question. In the Yucatán example, Turner *et al.* (2004) restricted the exposure unit to a geographically and temporally delimited land system. Doing so did not prevent them from addressing hazards, exposures, sensitivities, and adaptive capacities that involved agents and processes external to that geographic area. Examples of factors that originate beyond the borders of the Southern Yucatán include tropical storms, social and environmental effects of the North American Free Trade Agreement, and trends in the international chili market. In sum, for the purpose of maintaining analytical focus it is important to bound the specific exposure unit in space and time, even if some of the factors analyzed originate beyond the spatial or temporal borders of the exposure unit.

The *dimensions* of vulnerability in the VSD (Figure 5.2) include the three dimensions discussed above (p. 84): exposure, sensitivity, and adaptive capacity

(Adger and Kelly 1999; Downing *et al.* 2001; Turner *et al.* 2003). From this perspective, to be vulnerable to the effects of global change, human–environment systems not only must be exposed and sensitive to the effects, but also must exhibit limited ability to adapt. Conversely, systems are less vulnerable if they possess strong adaptive capacity (Smit *et al.* 1999; Finan *et al.* 2002). The large number of research papers in recent years (Janssen *et al.* 2006) using this vulnerability perspective reflects an apparently widespread interest by researchers to avoid the criticism that global change studies too often employ simplifying assumptions that systematically under- or overestimate the range of likely impacts from global change. This criticism often involves charging that the impacts – broadly understood as the expression of sensitivity to one or more stresses – are estimated on the basis of overstylized representations of exposure and adaptive capacity. For example, the simplifying assumptions that all farmers idly watch climate change reduce their crop yields without actively responding to mitigate the losses (the implicit assumption of zero adaptive capacity), or alternatively, that farmers presciently and optimally respond to present and future changes (the implicit assumption of perfect adaptive capacity), are both problematic (Schneider *et al.* 2000; Polsky 2004). In the former case, estimated impacts will be too large (because in fact farmers will respond to mitigate some of their losses) and in the latter case the estimated impacts will be too small (because despite farmers' desires to mitigate their losses fully, they will only have limited information about present and future climate and market conditions). The hope driving much of the recent vulnerability literature is that researchers can avoid such biases by adopting a conceptual framework that complements detailed assessments of sensitivity with textured and tailored understandings of exposure and adaptive capacity.

The *components* of the vulnerability dimensions in the VSD (Figure 5.2) are the abstract characteristics of the three vulnerability dimensions for a given human–environment system. Exposure components characterize the stressors and the entities under stress; sensitivity components characterize the first-order effects of the stresses; and adaptive capacity components characterize responses to the effects of the stresses. In theory, one could include a massive number of components for each dimension in any VSD. The task of populating the components ring of the VSD, however, is greatly simplified by first restricting the analytical focus to a particular exposure unit and set of hazards and outcomes.

Finally, the *measures* in the VSD (Figure 5.2) are the recorded observations of the specified components. These measures, parsed into the three "dimensions" categories (see below for a discussion of the problem of overlaps between pairs of dimensions) collectively constitute the vulnerability dataset. These measures can be quantitative (e.g., precipitation variability, distance to market) or qualitative (e.g., political party affiliation, environmental preservation ethic).

An illustration of the Vulnerability Scoping Diagram

To illustrate the *components* and *measurements* rings of the VSD, we draw from our ongoing research on the Human–Environment Regional Observatory (HERO: http://hero.geog.psu.edu/index.jsp) project. This multi-institutional research effort involves four study sites in the USA exhibiting varying human and natural landscapes (Figure 1.1 in Chapter 1). In all four sites, the HERO researchers have studied the vulnerability of local water supply systems to the effects of drought (Sorrensen *et al.* 2005).[2] Thus there is a single, common hazard for the four exposure units; the outcomes to be avoided vary across the sites based on local conditions and interests (e.g., loss of agricultural productivity, loss of access to drinking water, increased cost of providing drinking water). For efficiency in presentation, Figure 5.3 combines features from the four exposure units into a

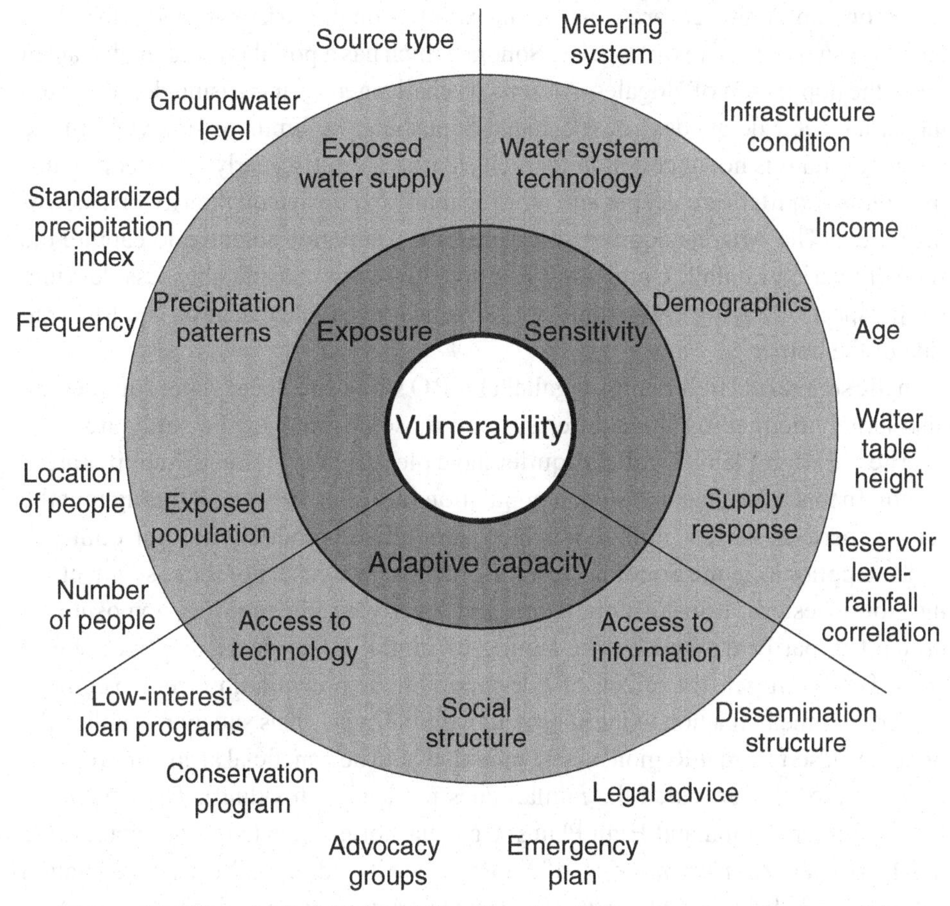

Figure 5.3. Hypothetical Vulnerability Scoping Diagram, where the hazard is drought and the exposure unit is a generic community water system (CWS).

single VSD. (Normally, investigators would populate the diagram with features of only one exposure unit.) In particular, Figure 5.3 presents components and measurements of: exposure from two of the four HERO exposure units, the Southwest and Mexico border region and southwest Kansas; sensitivity from the central Pennsylvania exposure unit; and adaptive capacity from the central and eastern Massachusetts exposure unit.

Exposure in the Sonoran Desert Border Region and High Plains–Ogallala HEROs

The Sonoran Desert Border Region HERO study site includes the semi-arid twin cities of Nogales, Sonora, Mexico and Nogales, Arizona, United States. All residents use aquifers within the Santa Cruz River watershed for drinking water, but these aquifers are much shallower on the Sonoran side of the border and thus more reliant on rainfall for recharge than are the aquifers on the Arizonan side. Moreover, the water supply system of Nogales, Sonora, which has a population estimated at ten times the population of Nogales, Arizona, is challenged by increasing domestic and industrial water demands. Local demand is pushing the limits of safe yield of the resource, and it is not uncommon for wells to run dry during early summer months. In contrast, aquifers are deeper and more spatially expansive on the Arizona side of the border. The Arizona aquifers therefore have a superior potential to capture the spatially varied rainfall. Combining this potential with considerably less demand, the prospect of having wells run dry is significantly smaller than on the Mexican side of the border.

In the semi-arid High Plains–Ogallala HERO study site, local water supplies are also largely defined by the aquifer system: those portions of the study area that overlie the High Plains Ogallala aquifer have plentiful water, but off-aquifer areas do not. In this case, because human populations are relatively small throughout the region, periods of low rainfall threaten agricultural productivity, not domestic water supplies as in the Sonoran Desert Border Region case. For those segments of the southwestern Kansas population not overlying the aquifer, exposure to drought is partly defined by the timing of rainfall. The onset of a wet period could fail to mitigate the effects of a dry period if the precipitation were to come at the wrong time in relation to the annual agricultural cycle. This situation is unlike the Sonoran Desert Border Region, where rainfall would be beneficial at any time of year during a drought (provided the rainfall does not lead to flooding). These Sonoran Desert Border Region and High Plains–Ogallala stories have led us to populate the VSD *exposure components* ring of Figure 5.3 with "drought" (i.e., precipitation patterns) and the principal parts of the water supply system dependent on precipitation, that is, the "exposed population" (both human and animal populations) and the "exposed water supply." In this case, *exposure measurements* could involve such quantities as the "intensity" of the drought as reflected by the Standardized Precipitation Index (SPI), and maps of the "location of people" relative to water sources.

Sensitivity in central Pennsylvania

In the humid central Pennsylvania HERO study site, some small, low-income towns have such enormous aquifer-based water supplies that the residents and water managers have historically had little need to monitor the water use of households and businesses. Even during the most severe periods of low precipitation in the past century, the downturns in local water supplies were not sufficient to trigger monitoring or conservation measures. Recently, however, because of the aging of pipes, some of which are from the nineteenth century and made of wood, massive leaks have developed and caused water deliveries to dwindle to as little as 30% of the total amount of water pumped. Moreover, the local income base cannot support repairing or replacing the aging infrastructure. These towns are therefore now extremely sensitive to sudden drops in water supply from downturns in precipitation, even if the sudden drop is small, i.e., not much below the long-term average of approximately 40 inches of precipitation per year. To grapple with this problem, these towns have had to install water meters for monitoring personal consumption and to institute permanent water-use restrictions even though the total annual precipitation is plentiful. Thus an important *sensitivity component* (Figure 5.3) for this study area includes a characterization of the "water system technology." Associated *sensitivity measurements* could include, for example, the "infrastructure age" and "physical size of the water distribution network."

Adaptive capacity in central Massachusetts

The humid central Massachusetts HERO study site focuses on small suburban towns in the Boston metropolitan area that are not on the metropolitan water supply system. One feature of this exposure unit is that in addition to the usual set of officials involved in water management decision-making, small groups of concerned citizens have organized into highly active and effective non-profit, volunteer advocacy groups (Hill and Polsky 2005). The purpose of these advocacy groups is to improve the ability of local water systems – and the communities the systems serve – to respond effectively to periodic variations in water demand and/or supply. Thus, an important *adaptive capacity component* (Figure 5.3) to include for this study area is a characterization of the resources, such as the influence of the advocacy groups, that water consumers and water system managers can access in times of perceived need (e.g., "access to information" and "access to technology"), and of the "management structures" that facilitate or impede such access. Important *adaptive capacity measurements* would therefore revolve around the observable features of these components, such as the "dissemination structure" that officials use to announce and enforce restrictions on water consumption or that advocacy groups use to attempt to change the local culture of water use.

In general, the VSD guides the collection of indicators in the form of the set of measurements of each component of exposure, sensitivity, and adaptive capacity. We assert that a basis for making inferences and generalizations about vulnerability would emerge if multiple, independent vulnerability assessments each produce a VSD en route to the collection of data to produce indicators. For example, an adaptive capacity measurement for the Central Massachusetts HERO exposure unit is the "revenue stream" of the community water supply system (Figure 5.3). In the study area, it is common for water systems to be funded by aggregate consumption: the higher the consumption, the higher the revenues. In these cases, the water system has a financial *dis*-incentive to promote water conservation even during times when water demand is high relative to available supply, i.e., times when conservation would alleviate the "drought" conditions. If inspection of a large number of independent VSDs revealed that "revenue stream" appeared frequently as an adaptive capacity measurement in those datasets, then a tentative generalization about the importance of this adaptive capacity feature could be offered. This generalization could be cast as a hypothesis to be confirmed or discredited by further investigation. Thus, using the VSD encourages the systematic comparison of independent vulnerability assessments so that inferences and generalizations may emerge.

The role of the VSD in vulnerability assessments

The principal function of the VSD is to provide a basis for comparing results across multiple, independent vulnerability assessments and exposure units. The VSD also plays a direct role in the overall assessment of vulnerability for a single exposure unit by facilitating Step 5 of the Eight Steps, "Find indicators for the components of vulnerability." Thus, as the name of the VSD suggests, a populated VSD may represent an early product – a scoping – of the vulnerability assessment that is subsequently revised to reflect insights gained from engaging the rest of the Eight Steps. We have found the VSD to play four additional important (if indirect) roles in our engaging with the Eight Steps in the HERO vulnerability research. Below, we briefly describe the set of specific activities used or proposed in the HERO project that have either informed, or been informed by, the HERO VSDs, and how these specific activities relate to the general Eight Steps methodological protocol of

Schröter *et al.* (2005). To this end, we include the number(s) of the step(s) from the Eight Steps to which each set of activities corresponds.

Characterizing the local human–environment context (Steps 1 – 2 from the Eight Steps)

An important activity prior to constructing a VSD is learning about the local context of the exposure unit. To this end, prior to the construction of the VSD, HERO researchers engaged in archival analysis and interviews with stakeholders and subject matter experts to provide the local context and knowledge necessary for embarking on the assessment. This background research involved characterizing the local human–environment interactions, climatic variations, and land-use/land-cover changes that were most important in each study site in the past several decades. In this way, for example, we learned of an important feature of one of our study sites, namely that the shape of the aquifer in the Sonoran Desert Border Region exposure unit helps explain geographic variations in groundwater supply, which in turn produces important variations in exposure to drought. More detail on the local contexts of the four HERO sites are provided in Chapters 6 and 8.

Conducting a grounded vulnerability assessment and adopting a conceptual framework of vulnerability (Steps 2 – 4 from the Eight Steps)

There are several conceptual frameworks available to vulnerability researchers; some of the more prominent ones include those from the traditions of risk and hazards (e.g., Kates *et al.* 1985; Cutter 1996), political ecology (e.g., Blaikie and Brookfield 1987; Böhle *et al.* 1994), and ecology (e.g., Holling 1973; Gunderson 2001), as well as recent attempts to synthesize across these traditions (e.g., Turner *et al.* 2003; Adger 2006). These frameworks differ in their primary focus of analysis. In some cases, the focus is on the individual who is at risk for losses associated with a natural or anthropogenic stress and on how that person may or may not cope with the stress. In other cases, the focus is on the structural reasons, at times imposed on the individual, for a lack of coping options. In yet other cases, the primary emphasis is on the likely responses expected from the non-human world.

The HERO project has adopted an iterative approach to conceptualizing vulnerability to give us the freedom to modify the conceptualization to accord with our findings as they emerge. We term the activity to achieve this goal "Grounded Vulnerability Assessment," which informs and is informed by scoping exercises guided by the VSD and its initial synthesis about vulnerability. Grounded assessment draws its name from grounded theory, a well-established qualitative method for developing theories and conceptual frameworks in a way that is both inductive and deductive (Strauss and Corbin 1998) based on long-term fieldwork. Grounded theory research begins with an explorative and iterative approach to data collection and analysis aimed at producing theory, particularly in realms where no clear theoretical framework exists, or where there are multiple competing theoretical frameworks in the literature. If gaps appear in the theoretical models developed from initial interviews, documents, and observations, then the investigators continue "sampling" to fill the gaps and explore inconsistencies in those models. Grounded assessment advances and modifies grounded theory, a qualitative method, by incorporating a mixed-methods approach, grounding itself both in empirical qualitative data about those who experience and manage vulnerability to environmental change, and in quantitative data describing the exposure to, and direct effects of, environmental change.

Adopting this approach has given us the flexibility and methodological rigor needed to adjust the theoretical framework guiding our research to fit the data collected in the field. The HERO project began with a risk/hazards-based conceptualization focused largely on the range of choice of individual decision-making (e.g., farmers' access to aquifer irrigation water in southwestern Kansas) and the consequences associated with individuals' decisions. However, through the process of constructing our VSD, we have been led to include a structurally based understanding of the factors leading to vulnerability in our exposure units, where local and extra-local governmental regulations and other constraints (e.g., the presence or absence of federal agricultural programs to compensate farmers for retiring land from production) assume an important role in constructing vulnerability. Thus, the vulnerability dataset can be informed by, and in turn can also inform, the basic conceptual framework used in a vulnerability assessment.

The HERO team has also worked to create a space for constructivist approaches to vulnerability that we feel has been lacking in the global change literature to date. In this way we hope to report not only on the causes of vulnerability, but also on how peoples' personal interpretations of, and reactions to, their vulnerabilities have evolved as a result of their experiences with global change. Since its introduction (Glaser and Strauss 1967), grounded theory has evolved and been used by researchers with diverse perspectives, from its objectivist and positivist roots (e.g., Glaser 1992; see also Guba and Lincoln 1994), to postpositivist methodologies (Strauss and Corbin 1998), to constructivist approaches (e.g., Charmaz 2000). In our own work, we have found that elements of all of these perspectives are useful in explaining vulnerability. For instance, while local and extra-local government actions play a strong role in determining vulnerabilities and sustainable practices, decision-makers react to those actions differently based on their perceptions, attitudes, and beliefs, all of which are socially constructed over time. A prime example is central Pennsylvania, where local decision-makers have reacted to the

Federal Safe Drinking Water Act and its amendments with responses ranging from immediate compliance to outright defiance, producing differential vulnerabilities to droughts and floods within a narrowly bounded geographic area. This variation in local response to governmental regulations can best be explained through detailed descriptions of how attitudes towards governmental regulation and environmental hazards have been constructed over time through complex interactions between cultures, individual and local histories, previous experience with regulatory agencies, perception of environmental hazards, and the perceived efficacy and feasibility of coping and adaptive strategies. Thus, grounded assessment is consistent with Steps 2 – 4 of the Eight Steps approach, but expands the scope of that protocol to a new domain. This contribution is greatly enhanced through scoping exercises guided by the VSD.

Engaging stakeholders to conduct a post-hoc results evaluation (Step 8 from the Eight Steps)

Validation of results is a critical post-hoc activity in classical scientific modeling and other quantitative research based on hypothesis formulation and testing (Oreskes *et al.* 1994). The vulnerability dataset contributes to validation activities insofar as the findings are derived from the dataset. Researchers focusing on qualitative data, however, at times do not engage in post-hoc validation of research results, largely because "validation" implies a single objective truth against which research results can be compared, whereas researchers in the qualitative realm often focus on producing knowledge about human nature and behavior that is provisionally true based upon the multiple and diverse perspectives of the researchers and informants (or "stakeholders") participating in the study (e.g., Denzin and Lincoln 2002). Nevertheless, there are calls within the qualitative research community (e.g., Baxter and Eyles 1997) for more attention to post-hoc assessments of results. Given this tension, we term this post-hoc assessment exercise *evaluation* rather than validation.

Such an activity is not explicitly referenced in the Eight Steps, but the way in which the HERO team has operationalized this evaluation activity overlaps significantly with Step 8, "Communicate vulnerability creatively." HERO investigators have articulated three criteria for evaluating the results of a vulnerability assessment: saturation, credibility, and transferability. *Saturation* seeks to determine if investigators have sufficient qualitative data and can curtail data collection. It measures the level of redundancy in compiled data and associated research findings; it occurs when investigators no longer find new information with each additional interview, focus group, or other qualitative data collection technique. The second criterion, *credibility*, aims to establish the plausibility and salience of the findings with stakeholders. Evaluating credibility involves sharing results with both participants originally involved in the research and new stakeholders. The third criterion addresses the *transferability* of results to other contexts. This criterion is evaluated by comparing results in one place or time to another place or time through historical documentation or conversations with stakeholders, thus answering the question, "Have vulnerabilities similar to those observed in the vulnerability assessment been observed in other cases?" With respect to the first two criteria, in Chapter 8 we elaborate on methods for measuring the criteria, and in Chapter 10 we report results from an application of these measurements to a partial vulnerability assessment.

Discussion: implementation challenges for the Vulnerability Scoping Diagram

It is much easier to prescribe abstractly how to conduct a vulnerability scoping exercise than it is to conduct the exercise because, in many cases, the dimensions – and by extension the components and associated measures – of vulnerability are not perfectly separable. Indeed, exposure, sensitivity, and adaptive capacity are often intimately related. Thus, the process of populating a VSD is akin to the focusing of a camera lens – it may suggest where the borders of one dimension end and others begin, even if the borderline areas cannot be resolved perfectly. For example, the significant decrease in water deliveries resulting from the antiquated and aging water infrastructure of central Pennsylvanian towns noted above could well be discussed in the context either of sensitivity to periods of rainfall deficits, or of associated adaptive capacities: the decreasing deliveries affect the former at the same time as they reflect the latter. The choice to put this component under sensitivity or adaptive capacity turns, therefore, largely on whether the researchers want to emphasize the present (current sensitivities) or the past (the presence or absence of historical adaptive strategies). We see no basis for claiming a priori that one of these choices is superior to the other. Similarly, the source type – groundwater or surface water – for a local water system could indicate the exposure of the community water system's supply to rainfall variations, with groundwater feeling the effects indirectly and surface water feeling them directly. Source type could also serve as a useful measure of sensitivity because groundwater levels respond less in periods of below-average rainfall than do surface water levels. In addition, if one asks *why* a given human–environment system depends on a given source type at a given point in time, the discussion of where to place "water source type" may well turn to adaptive capacity.

In general, our experience suggests that when there is ambiguity in where to place an element of the human–environment system on the VSD, context dictates the ultimate choice. No one choice is necessarily and absolutely better than others;

these decisions will be influenced by spatially and temporally variable local conditions. Thus, populating a VSD is more of a negotiated and exploratory process than a straightforward, objective exercise. Nonetheless, this analytical activity is capable of advancing our understanding of complex, context-dependent local vulnerabilities. To continue one of the examples above, a VSD could be particularly enlightening in those central Pennsylvania towns where groundwater serves portions of the population and surface water serves other portions. The process of reaching some level of consensus on where to place the measure of "water source type" on the VSD – within the dimension of exposure, sensitivity, or adaptive capacity – may suggest to these Pennsylvania researchers and stakeholders which anticipatory mitigative measures, if any, are needed and feasible in not only technical, but also political terms.

Conclusions and future directions

Despite a growing need for information on the vulnerability of coupled human–environment systems, there is little consensus on best practices in the literature. In particular, there is little guidance on how to structure vulnerability assessments so that their findings are comparable and generalizations can be made, or how to implement an initial scoping to suport a broader vulnerabilty assessment. The Vulnerability Scoping Diagram (VSD) discussed in this chapter attempts to address both of these gaps in a way that follows an emerging conceptualization of vulnerability in the literature. The VSD requires that researchers specify and organize, in a particular graphical way, five features of their research project: (1) the hazard and associated outcomes to be mitigated, (2) the exposure unit, and the (3) dimensions, (4) components, and (5) measures of the vulnerability process in question. The outcome of populating a VSD is a set of components and measures (indicators) that satisfy the demands of Step 5 of the Eight Steps (Schröter *et al.*, 2005). These components and measures not only reflect an interim (in the case of a true "scoping" diagram) or a definitive (in the case of a "final" diagram) statement of the structure of exposure, sensitivity, and adaptive capacity for a given coupled human–environment system, but also form the basis for potential comparisons with other vulnerabilty assessments, even if the various assessments use dissimilar measures.

Populating a VSD is challenging and requires subjective judgments on the part of the researchers. Nonetheless, we argue that if a significant number of future vulnerability assessments were to adopt this call to structure their indicator set using this graphical vehicle, then the overarching goal of this chapter would have been met: to facilitate the construction of comparable global change vulnerability assessments. In the coming years, a library of VSDs could be produced, such that a meta-analysis would be relatively straightfoward to conduct. When the vulnerability literature has produced some definitive meta-analyses, people who lack the time or other resources to conduct their own vulnerability assessments would be able to make better-informed decisions about which anticipatory actions are likely to mitigate their vulnerabilities.

Thus, this chapter provides the overarching conceptual foundation and motivation for organizing information from a vulnerability assessment conducted in one place such that the results of that assessment are comparable with vulnerability assessments conducted in other places. However, this VSD approach only represents some of the methodological needs of a vulnerability assessment. In Chapters 8, 9, and 10, we present methods for and results from additional methodological needs of vulnerability assessments.

Notes

1. This chapter is largely drawn from Polsky *et al.* (2007). Much of this discussion on vulnerability also intersects ideas associated with *resilience* (see, e.g., Holling 1973, 1986; Berkes *et al.* 2002; Walker *et al.* 2002), *adaptation* (e.g., Smithers and Smit 1997; Klein and MacIver 1999; Smit *et al.* 1999; Adger 2000), or *coupled human–environment* interactions (e.g., Ojima *et al.* 2005; Young *et al.* 2006). See Adger (2006) for an elaboration. For purposes of presentation clarity, this paper will use only the *vulnerability* concept throughout.
2. For more information on vulnerability and adaptive capacity of community water systems to weather and climate, see O'Connor *et al.* 1999, 2005; Yarnal *et al.* 2005, 2006; Dow *et al.* 2007.

References

Adger, W. N., 2000. Institutional adaptation to environmental risk under transition in Vietnam. *Annals of the Association of American Geographers* **90**(4): 738–758.

Adger, W. N., 2006. Vulnerability. *Global Environmental Change* **16**: 268–281.

Adger, W. N., and P. M. Kelly, 1999. Social vulnerability to climate change and the architecture of entitlements. *Mitigation and Adaptation Strategies for Global Change* **4**(3/4): 253–266.

Baxter, J., and J. Eyles, 1997. Evaluating qualitative research in social geography: establishing 'rigour' in interview anaysis. *Transactions of the Institute of British Geographers* **22**: 505–525.

Berkes, F., J. Colding, and C. Folke (eds.), 2002. *Navigating Social–Ecological Systems: Building Resilience for Complexity and Change.* Cambridge: Cambridge University Press.

Blaikie, P., and H. Brookfield, 1987. *Land Degradation and Society.* New York: Methuen.

Böhle, H. G., T. E. Downing, and M. Watts, 1994. Climate change and social vulnerability: toward a sociology and geography of food insecurity. *Global Environmental Change* **4**(1): 37–48.

Brooks, N., W. N. Adger, and P. M. Kelly, 2005. The determinants of vulnerability and adaptive capacity at the national level and the implications for adaptation. *Global Environmental Change* **15**(2): 151–163.

Carter, T. R., M. L. Parry, H. Harasawa, and S. Nishioka, 1994. *IPCC Technical Guidelines for Assessing Climate Change Impacts and Adaptations*, 59. London: Department of Geography, University College London; and Tsukuba, Japan: Center for Global Environmental Research, National Institute for Environmental Studies.

Charmaz, K., 2000. Grounded theory: objectivist and constructivist methods. In *Handbook of Qualitative Research*, eds. N. K. Denzin and Y. S. Lincoln, pp. 509–536. Thousand Oaks, CA: Sage Publications.

Clark, G. E., S. Moser, S. Ratick, K. Dow, W. B. Meyer, S. Emani, W. Jin, J.X. Kasperson, R. E. Kasperson, and H. E. Schwarz, 1998. Assessing the vulnerability of coastal communities to extreme storm: the case of Revere, MA, USA. *Mitigation and Adaptation Strategies for Global Change* **3**: 59–82.

Clark, W. C., and N. M. Dickson, 2003. Sustainability science: the emerging research program. *Proceedings of the National Academy of Sciences of the USA* **100**(14): 8059–8061.

Cutter, S., 1996. Vulnerability to environmental hazards. *Progress in Human Geography* **20**(4): 529–539.

Cutter, S., 2001. A research agenda for vulnerability science and environmental hazards. *International Human Dimensions Program Update* **01**(2): 8–9.

Cutter, S., 2003. The vulnerability of science and the science of vulnerability. *Annals of the Association of American Geographers* **93**(1): 1–12.

Cutter, S. L., J. T. Mitchell, and M. S. Scott, 2000. Revealing the vulnerability of people and places: a case study of Georgetown County, South Carolina. *Annals of the Association of American Geographers* **90**(4): 713–737.

Denzin, N. K., and Y. S. Lincoln (eds.), 2002. *The Qualitative Inquiry Reader*. Thousand Oaks, CA: Sage Publications.

Dow, K., R. E. O'Connor, B. Yarnal, G. J. Carbone, and C. L. Jocoy, 2007. Why worry? Community water system managers' perceptions of climate vulnerability. *Global Environmental Change* **17**: 228–237.

Downing, T. E., 1991. Vulnerability to hunger in Africa: a climate change perspective. *Global Environmental Change* **1**: 365–380.

Downing, T. E., 2000. Human dimensions research: toward a vulnerability science? *International Human Dimensions Program Update* **00**(3): 16–17.

Downing, T. E., R. Butterfield, S. Cohen, S. Huq, R. Moss, A. Rahman, Y. Sokona, and L. Stephen, 2001. *Vulnerability Indices: Climate Change Impacts and Adaptation*, Policy Series No. 3. Nairobi: United Nations Environment Programme. Accessed at www.unep.org/library/Default.asp?catid=6.

Easterling, W. E. and C. Polsky, 2004. Crossing the complex divide: linking scales for understanding coupled human–environment systems. In *Scale and Geographic Inquiry*, eds. R. McMaster and E. Sheppard, pp. 55–64. Oxford: Blackwell.

Finan, T., C. West, D. Austin, and T. McGuire, 2002. Processes of adaptation to climate variability: a case study from the US Southwest. *Climate Research* **21**(3): 299–310.

Geist, H. J., 2005. *The Causes and Progression of Desertification*. London: Ashgate Publishing.

Geist, H. J., and E. F. Lambin, 2002. Proximate causes and underlying driving forces of tropical deforestation. *BioScience* **52**: 143–150.

Glaser, B. G., 1992. *Basics of Grounded Theory Analysis: Emergence vs. Forcing*. Mill Valley, CA: Sociology Press.

Glaser, B. G., and A. Strauss, 1967. *The Discovery of Grounded Theory: Strategies for Qualitative Research*. Chicago, IL: Aldine.

Guba, E. G., and Lincoln, Y. S., 1994. Competing paradigms in qualitative research. In *Handbook of Qualitative Research*, eds. N. K. Denzin and Y. S. Lincoln, pp. 163–188. Thousand Oaks, CA: Sage Publications.

Gunderson, L. H., 2001. Managing surprising ecosystems in southern Florida. *Ecological Economics* **37**: 371–378.4

Hill, T. and C. Polsky, 2005. Suburbanization and adaptation to the effects of suburban drought in rainy central Massachusetts. *Geographical Bulletin* **47**(2): 85–100.
Holling, C. S., 1973. Resilience and stability of ecological systems. *Annual Review of Ecology and Systematics* **4**: 1–23.
Holling, C. S., 1986. The resilience of terrestrial ecosystems: local surprise and global change. In *Sustainable Development of the Biosphere*, eds. W. C. Clark and R. E. Munn, pp. 292–317. Cambridge: Cambridge University Press.
Hurd, B., N. A. Leary, R. Jones, and J. Smith, 1999. Relative regional vulnerability of water resources to climate change. *Journal of the American Water Resources Association* **35**(6): 1399–1409.
Janssen, M. A., M. L. Schoon, W. Ke, and K. Borner, 2006. Scholarly networks on resilience, vulnerability and adaptation within the human dimensions of global environmental change. *Global Environmental Change* **16**: 240–252.
Kasperson, R., 2001. Vulnerability and global environmental change. *International Human Dimensions Program Update* **01**(2): 2–3.
Kates, R. W., 1985. The interaction of climate and society. In *Climate Impact Assessment: Studies of the Interaction of Climate and Society*, eds. R. W. Kates, J. H. Ausubel, and M. Berberian, pp. 3–36. Chichester: John Wiley.
Kates, R. W., J. H. Ausubel, and M. Berberian (eds.), 1985. *Climate Impact Assessment: Studies of the Interaction of Climate and Society*. Chichester: John Wiley.
Kates, R. W., W. C. Clark, R. Corell, J. M. Hall, C. C. Jaeger, I. Lowe, J. J. McCarthy, H. J. Schellnhuber, B. Bolin, N. M. Dickson, S. Faucheux, G. C. Gallopin, A. Gruebler, B. Huntley, J. Jäger, N. S. Jodha, R. E. Kasperson, A. Mabogunje, P. Matson, H. Mooney, B. Moore III, T. O'Riordan, and U. Svedin, 2001. Sustainability science. *Science* **292**: 641–642.
Kelly, P. M., and W. N. Adger, 2000. Theory and practice in assessing vulnerability to climate change and facilitating adaptation. *Climatic Change* **47**: 325–352.
Klein, R. J. T., and MacIver, D. C., 1999. Adaptation to climate change and variability: methodological issues. *Mitigation and Adaptation Strategies for Global Change* **4**(3/4): 189–198.
Klein, R. J. T., R. J. Nicholls, and N. Mimura, 1999. Coastal adaptation to climate change: can the IPCC Technical Guidelines be applied? *Mitigation and Adaptation Strategies for Global Change* **4**(3/4): 239–252.
Liverman, D. M., 1990. Drought impacts in Mexico: climate, agriculture, technology, and land tenure in Sonora and Puebla. *Annals of the Association of American Geographers* **80**(1): 49–72.
Moss, R. H., A. L. Brenkert, and E. L. Malone, 2001. *Vulnerability to Climate Change: A Quantitative Approach*. Richland, WA: Pacific Northwest National Laboratory.
Mustafa, D., 1998. Structural causes of vulnerability to flood hazard in Pakistan. *Economic Geography* **74**: 289–305.
National Research Council (NRC), 1999. *Our Common Journey: A Transition Toward Sustainability. Board on Sustainable Development*. Washington, D.C.: National Academy Press.
O'Brien, K. L., and R. Leichenko, 2000. Double exposure: assessing the impacts of climate change within the context of economic globalization. *Global Environmental Change* **10**: 221–232.
O'Brien, K., R. Leichenko, U. Kelkar, H. Venema, G. Aandahl, H. Tompkins, A. Javed, S. Bhadwal, S. Barg, L. Nygaard, and J. West, 2004. Mapping vulnerability to multiple stressors: climate change and globalization in India. *Global Environmental Change* **14**: 303–313.

O'Connor, R. E., B. Yarnal, R. Neff, R. Bord, N. Wiefek, C. Reenock, R. Shudak, C. L. Jocoy, P. Pascale, and C. G. Knight, 1999. Weather and climate extremes, climate change, and planning: views of community water system managers in Pennsylvania's Susquehanna River Basin. *Journal of the American Water Resources Association* **35**(6): 1411–1419.

O'Connor, R., B. Yarnal, K. Dow, C. S. Jocoy, and G. Carbone, 2005. Feeling at-risk matters: water managers and the decision to use forecasts. *Risk Analysis* **25**: 1265 1275.

Ojima, D., E. Moran, W. McConnell, M. Stafford Smith, G. Laumann, J. Morais, and B. Young (eds.), 2005. *Land-Use and Land-Cover Change: Science/Research Plan*, IGBP Report No. 53/IHDP Report No. 19. Stockholm: IGBP Secretariat.

Oreskes, N., K. Schrader-Frechette, and K. Belitz, 1994. Verification, validation, and confirmation of numerical models in the Earth sciences. *Science* **263**(5147): 641–646.

Parry, M. L., 2001. Viewpoint: Climate change – where should our research priorities be? *Global Environmental Change* **11**: 257–260.

Parry, M. L., O. F. Canziani, J. P. Palutikof, P. J. van der Linden, and C. E. Hansen (eds.), 2007. *Climate Change 2007: Impacts, Adaptation and Vulnerability*. Cambridge: Cambridge University Press.

Polsky, C., 2004. Putting space and time in Ricardian climate change impact studies: the case of agriculture in the U.S. Great Plains. *Annals of the Association of American Geographers* **94**(3): 549–564.

Polsky, C., and D. Cash, 2005. Reducing vulnerability to the effects of global change: drought management in a multi-scale, multi-stressor world. In *Drought and Water Crises: Science, Technology, and Management Issues*, ed. D. Wilhite, pp. 215–245. Amsterdam: Marcel Dekker.

Polsky, C., R. Neff, and B. Yarnal, 2007. Building comparable global change vulnerability assessments: the Vulnerability Scoping Diagram. *Global Environmental Change* **17**: 472–485.

Project for Public Spaces (PPS), 2005. Accessed at www.pps.org/.

Ragin, C. C., 1987. *The Comparative Method: Moving beyond Qualitative and Quantitative Strategies*. Berkeley, CA: University of California Press.

Rudel, T. K., 2005. *Tropical Forests: Regional Paths of Destruction and Regeneration in the Late 20th Century*. New York: Columbia University Press.

Rudel, T. K., 2008. Capturing regional effects through meta-analyses of case studies: an example from the global change literature. *Global Environmental Change* **18**: 18–25.

Schneider, S. H., W. E. Easterling, and L. O. Mearns, 2000. Adaptation: sensitivity to natural variability, agent assumptions and dynamic climate changes. *Climatic Change* **45**: 203–221.

Schröter, D., C. Polsky, and A. Patt, 2005. Assessing vulnerabilities to the effects of global change: an eight step approach. *Mitigation and Adaptation Strategies for Global Change* **10**(4): 573–595.

Smit, B., I. Burton, R. J. T. Klein, and R. Street, 1999. The science of adaptation: a framework for assessment. *Mitigation and Adaptation Strategies for Global Change* **4**(3/4): 199–213.

Smithers, J., and B. Smit, 1997. Human adaptation to climatic variability and change. *Global Environmental Change* **7**(2): 129–146.

Sorrensen, C., C. Polsky, and R. Neff, 2005. The Human–Environment Regional Observatory (HERO) Project: undergraduate research findings from four study sites. *Geographical Bulletin* **47**(2): 65–72.

Strauss, A. L., and J. Corbin, 1998. *Basics of Qualitative Research: Techniques and Procedures for Developing Grounded Theory*. Thousand Oaks, CA: Sage Publications.

Turner, B. L. II, R. E. Kasperson, P. Matson, J. J. McCarthy, R. W. Corell, L. Christensen, N. Eckley, J. X. Kasperson, A. Luers, M. L. Martello, C. Polsky, A. Pulsipher, and A. Schiller, 2003. A framework for vulnerability analysis in sustainability science. *Proceedings of the National Academy of Sciences of the USA* **100**: 8074–8079.

Turner, B. L. II, J. Geoghegan, and D. R. Foster (eds.), 2004. *Integrated Land-Change Science and Tropical Deforestation in the Southern Yucatán: Final Frontiers*. Oxford: Oxford University Press.

Walker, B., S. Carpenter, J. Anderies, N. Abel, G. Cummings, M. Janssen, L. Lebel, J. Norberg, G. D. Peterson, and R. Pritchard, 2002. Resilience management in social–ecological systems: a working hypothesis for a participatory approach. *Conservation Ecology* **6**(1):14 (online). Accessed at www.consecol.org/vol6/iss1/art14/main.html.

Wolf, F. M., 1986. *Meta-Analysis: Quantitative Methods for Research Synthesis*. London: Sage Publications.

Wu, S.-Y., B. Yarnal, and A. Fisher, 2002. Vulnerability of coastal communities to sea-level rise: a case study of Cape May County, New Jersey. *Climate Research* **22**: 255–270.

Yarnal, B., R. O'Connor, K. Dow, G. Carbone, and C. Jocoy, 2005. Why don't Community Water System managers use weather and climate forecasts? *Preprints of the 15th Conference on Applied Climatology*, Savannah, GA, June 2005. Accessed at www.ametsoc.org/.

Yarnal, B., A. L. Heasley, R. E. O'Connor, K. Dow, and C. L. Jocoy, 2006. The potential use of climate forecasts by Community Water System managers. *Land Use and Water Resources Research*. Accessed at www.luwrr.com.

Young, O., E. F. Lambin, F. Alcock, H. Haberl, S. I. Karlssone, W. J. McConnell, T. Myint, C. Pahl-Wostl, C. Polsky, P. S. Ramakrishnan, M. Scouvart, H. Schroeder, and P. H. Verburg, 2006. A portfolio approach to analyzing complex human–environment interactions. *Ecology and Society* **11**(2): 31.

6

Comparative assessment of human–environment landscape change

JOHN HARRINGTON, JR., BRENT YARNAL, DIANA LIVERMAN, AND B. L. TURNER II

Introduction

Humans acting to change Earth away from hypothetical pristine conditions is one of three key themes on human–environment relationships identified in Clarence Glacken's (1967) classic work, *Traces on the Rhodian Shore*. A century earlier, George Perkins Marsh (1864) helped create awareness and elucidate concerns regarding the nature and magnitude of human-induced changes to the planet. More recent compilations (e.g., Thomas 1956; Turner *et al.* 1990a; Foley *et al.* 2005) have continued to expand our knowledge of the complex and multiple pathways in which human actions alter the Earth system.

A key issue in human dimensions of global change research (NRC 1999) and in sustainability science (Kates *et al.* 2001) is a need to understand how the specifics of human structure and agency[1] interact (Sorrensen *et al.* 2005) with the natural environment in disparate places. In theory, local transformations could then be accumulated to produce the cumulative impact on the planet (Turner *et al.* 1990b; NRC 1992). What similarities and differences exist in the human activities, what are the socioeconomic drivers of those activities, and what are the impacts of those activities in forested, grassland, and desert environments? And, how can scholars compare and contrast these human actions in areas where very different natural resources and settlement histories exist?

The HERO transect of North American research sites, from humid central Massachusetts and central Pennsylvania, to semi-arid southwestern Kansas, to the arid border region between Arizona and Sonora, provides the opportunity for a comparative examination of human–environment interactions over time – especially those forces that have altered land cover and land use. Inspired by a paper in the Sauerian tradition on 15 events that shaped the California landscape (Dilsaver *et al.* 2000), informed by national and regional environmental histories (e.g., Worster 1979; Cronon 1983, 1993; Williams 2003), and framed by agendas for understanding the local causes of land-use/land-cover change (Lambin *et al.* 2001; Turner *et al.* 2007),

Sustainable Communities on a Sustainable Planet: The Human–Environment Regional Observatory Project, eds. Brent Yarnal, Colin Polsky, and James O'Brien. Published by Cambridge University Press.

we analyzed the HERO knowledge base to identify the key human–environment interactions that have influenced today's landscapes in central Massachusetts, central Pennsylvania, southwestern Kansas, and the Arizona–Sonora border region. In establishing this research effort, we wondered what similarities would lead to cross-site generalizations about land-use/land-cover change and what differences would point to site-specific trajectories.

The driving forces for environmental and land-use/land-cover change in the United States are usually considered to include changes in physical environmental conditions (such as climate variations), population, technology, economic activity, institutions (including laws and government policies), and behavior (often associated with cultural values or consumption habits) (NRC 1992). While the effects of some of these drivers can be tracked in census and other data records, others are recorded in histories of environmental policy or economic development, or in long-term environmental datasets that include both direct observations of climate and vegetation and proxies of climate and vegetation such as pollen and tree rings.

In each of the HERO regions, there is considerable evidence of the processes of long-term human adaptation within environmental constraints. Examples include the processes traditionally documented by cultural ecologists including the growth of irrigation systems in drier areas and the slow flows of population and ideas between regions. There is also evidence of more abrupt moments of transformation, often associated with rapid technological shifts, extreme natural events, or major political and economic changes. The opening of the Erie Canal in 1825, the 1938 hurricane in New England, and the 1985 Farm Bill are examples of events that resulted in significant, rapid local land-cover transformations in one or more of the HERO study sites.

This chapter provides a comparative method for analysis of human–environment interactions that focuses on the impacts of larger processes and events on the material landscape and that provides a framework for assessing pivotal changes in human–environment interactions. For each location in the HERO case study sites, we reviewed environmental and land-use histories and consulted local literature and experts about the changing landscape of the region. Our criteria for identifying the key events and processes included the need for evidence of large-scale and enduring material impact on the landscape.

We set out to answer three critical questions about human–environment interactions:

- What were the most important events or processes that changed the landscape, especially land use, in each of the four regions?
- What are the major natural, demographic, political, economic, and cultural drivers of these local events and processes within and outside the region?
- Are there parallels or significant differences in the nature of these events and their drivers across the sites?

A somewhat parallel effort, looking at the human drivers of environmental and landscape change, has been appended recently to the Long-Term Ecological Research (LTER) effort of the United States National Science Foundation (Gragson and Grove 2006; Redman *et al.* 2004). Prompted by a growing knowledge base regarding the global consequences of human land use (Foley *et al.* 2005), feedbacks between human-induced land-cover change and the climate system (Feddema *et al.* 2005), and the relative importance of land use in altering ecosystem services (Costanza *et al.* 1997; DeFries and Bounoua 2004), recent projects within the LTER research framework have been established to examine the types of land-use/land-cover changes that can be assessed with satellite data and to look at the role of agrarian change at selected sites. While this "social science add-on" effort has helped document some important connections between human activities and the legacies that are evident in the current ecological landscape (Foster *et al.* 2003), the types of questions posed by a research team are different when social scientists are present at the formative stages of the effort (Redman *et al.* 2004; Zhou *et al.* 2007). The HERO effort documented in this chapter is based on work that asked human–environment questions from the beginning.

Theoretical framework

Based on findings presented in Intergovernmental Panel on Climate Change assessments, the Millennium Ecosystem Assessment, the now two decades old "Earth Transformed" endeavor (Turner *et al.* 1990a), and other sources, the evidence that human actions have become a fundamental agent of change in Earth system functioning continues to accumulate. From rising emissions and concentrations of greenhouse gases, to changes in planetary albedo, to losses of biodiversity, to declining Arctic sea ice extent, and to significant adjustments in the hydrologic cycle, the results of human actions have altered energy and biogeochemical cycling and prompted the identification of the modern epoch in Earth history as the Anthropocene (Crutzen and Stormer 2000; Ruddiman 2003). Evidence of the human dimensions of global environmental change have become ubiquitous (McKibben 1989; NRC 1992, 1999, among others) so that it is now more necessary than ever before to understand the interactions, feedbacks, and vulnerabilities associated with social–ecological systems (SES) or coupled human–environment systems (CHES) (Turner *et al.* 2003; Redman *et al.* 2004).

This need for improved understanding of the changing functioning of our planetary life-support system comes at a time when we will move through what E. O. Wilson (2002) has identified as "the bottleneck" – that time when a large and growing human population combines with significant stresses on the ability of natural resource systems to provide needed ecosystem goods and services. The ongoing scenarios of coupled human and natural environment change will not have the same script in every place; Liverman (1999), Wilbanks and Kates (1999), Lambin *et al.* (2001), and Adger *et al.* (2005) among others have presented strong arguments for a need to understand how change plays out at the local to regional scale.

Scholars working within geography's human–environment tradition (Turner 2002) have made major strides in identifying the locations, types of changes, and magnitudes of human actions that have altered the patterns across the land (Johnson 1976), including some of the cultural and economic drivers associated with those changes (e.g., Association of American Geographers Global Change in Local Places Research Team 2003). Overlaps among the human–environment tradition within geography, environmental history (Cronon 1993), and the new and rapidly growing field of land-change science (e.g., Rindfuss *et al.* 2004; Foley *et al.* 2005) provide a theoretical foundation for distributed and collaborative efforts to assess similarities and differences in landscape change and associated drivers of change. While there now exists a considerable volume of case studies documenting the complex and intricate nature of local change (e.g., Dilsaver *et al.* 2000), few investigations have attempted to establish a comparative method for local assessment and then synthesize the results across multiple environments (Polsky *et al.* 2007; Geist and Lambin 2002; see Lambin and Geist, 2003; Geist 2005; Misselhorn 2005; Rudel 2005, 2008 for notable examples of post-hoc analyses).

Methods

We based the study on the four parallel regional research projects in central Massachusetts, central Pennsylvania, southwestern Kansas, and the Arizona–Sonora border region that together comprise a series of human–environment research observatories (HEROs). At these HERO sites, we analyzed past and present interactions among human activity, climate variations, and land use. We first documented local human–environment interactions at each HERO and then performed a cross-site comparison of those interactions.

Documentation of local human–environment interactions

The first step was to develop lists of significant events for each of the four HERO study areas. This task made it necessary to identify the types of events with which we would be concerned and the relevant timescale for our investigation. We decided to select events that produced a visible imprint on the landscape and to limit our selections to those that came with or followed initial Euro-American contact with the place. In general, we followed the lead of Dilsaver *et al.* (2000) and worked chronologically forward from initial Euro-American contact to assemble lists of approximately ten events that shaped the look of our local areas. In part because we were aware that natural events, such as a major drought, could leave a visible mark on the landscape for decades, we chose to include events such as the 1938 hurricane in central Massachusetts (Foster and Aber 2004).

As our local area experts assembled their lists of significant events and chronicled the period associated with each event, it became clear that some processes took shape over several years to decades. Thus, we concluded that the term "interactions" (rather than "events") is a more appropriate choice to characterize the processes that resulted in a pronounced change to "the look of the land." For example, although the December 1985 US Farm Bill established the Conservation Reserve Program (CRP), which produced significant changes to the southwestern Kansas HERO study site, the impact on the landscape evolved over many years as more and more farmers and ranchers took advantage of the new federal land conservation policy (Leathers and Harrington 2000; Gersmehl and Brown 2004).

At each of the four HERO study areas, an individual investigator worked independently to develop an initial set of important interactions (which was not limited to a specific number). This lengthy list was then shared with local HERO team members to distill the list and strengthen the ideas associated with the relative importance of each of the events/interactions identified. For example, the post-Dust Bowl creation of the Cimarron National Grassland (Duram 1995) was an early entry for southwestern Kansas. That event was later combined with other actions of the same period and generalized into a category labeled "modern soil conservation." In several cases, local area experts who were not HERO research team members, such as other geographers, environmental historians, and rural sociologists, were asked for their insights regarding the relative importance of human influences on the visible landscape. Through this process, we distilled the lists to about ten important natural events and human–environment interactions.

Rather than simply present four lists of ten seemingly disconnected events and interactions, we decided to generate narratives that chronicled the evolution of human–environment interactions for each of the HEROs. Following Cronon (1993), this activity sought to tell the local story of human–environment interactions and to identify serial connections between interactions (e.g., suburbanization following road-network development) or simultaneous occurrence of two or more interactions (e.g., the growth of feedlots and the spread of center pivot irrigation throughout southwestern Kansas).

Cross-site comparison

Following the independent documentation of the most important events and human–environment interactions at each HERO site, the authors shared lists and ideas. A goal in this cross-site comparative activity was a higher-order synthesis. In working toward a fusion of ideas, we addressed three important questions:

- What similarities among the four local areas could we identify?
- What drivers of change were present at all four sites?
- Could the interactions be grouped into the classic drivers associated with human–environment interactions (i.e., population change, technological innovation, economic change, institutions and policies, and behavior and culture)?

An important step in this synthesis involved a face-to-face session wherein the site representatives discussed a table of drivers, processes, and related events. A synthesis section later in this chapter presents this table and our general findings on cross-site comparability.

Local human–environment narratives

Central Massachusetts HERO

As it was at European contact, the central Massachusetts landscape is predominantly forested today. Nevertheless, the environmental characteristics of the landscape have been significantly transformed by over 300 years of interactions between land-use activities and natural processes (Hall *et al.* 2002). The legacies of this history continue to shape contemporary patterns of ecosystems and of human interactions with the landscape.

Evidence suggests that, at the time of European contact, the Native Americans heavily managed ecosystems through selective burning to produce widely spaced, "park-like" forests (Cronon 1983). With American Indian populations decimated by smallpox and other diseases, European settlers rapidly established agricultural communities on the coastal plains and in the Connecticut River valley. The region presented a difficult agricultural environment: north–south ridges and valleys overlay nutrient-poor bedrock, with coarse, stony soils covering all but the organic-rich drainage areas. The modern climate supports an average of 137 frost-free days per year, but was cooler and had a shorter growing season at European contact, which occurred during the Little Ice Age.

With great labor, colonial farmers introduced the English agricultural system, marked by the close integration of tilled crops and livestock and by new species such as English hay, cattle, sheep, horses, and orchard plants (Donohue *et al.* 2000). Perhaps most important, a land-ownership pattern of privately held lots emanating from town centers was established. The legacy of this ownership pattern continues today, as management decisions concerning the fate of forests are made by hundreds of thousands of individuals, making forest lands susceptible to land conversion and to parcelization into smaller ownership units (Foster 1999).

Following the American War of Independence, agriculture expanded rapidly. Population increases, availability of better farm tools, improvements in crop varieties and livestock breeds, and growth of commercial opportunities created a "market revolution" in agriculture (Donohue *et al.* 2000). Land-use conversion in this period, driven largely by the need for pasture and to produce hay for livestock, filled the valleys before moving to the uplands. The agricultural expansion peaked by mid nineteenth century, covering roughly 75% of the Massachusetts landscape and driving a major environmental transformation, marked by a steep decline in wildlife populations (Foster *et al.* 2002) and by a simplification and homogenization of soil conditions, vegetation patterns, and micro-environmental conditions (Compton *et al.* 2003).

In the midst of the agricultural expansion, a new economic structure began to emerge. When the tariff war with Europe suspended the Atlantic trade for manufactured goods in 1807, mill villages proliferated (Howe 1960). New towns emerged to access water power at locations along major rivers and streams, and many existing towns, which had been established on well-drained hilltops during the early agricultural period, developed new industrial villages in the valleys (O'Keefe and Foster 1998). The lasting landscape impacts of the mill towns followed from their infrastructural developments, particularly roads, which would shape transportation and land use in the following centuries as urbanization moved into the interior uplands (Kulik *et al.* 1982).

The construction of dams and railroads initiated the abandonment of central Massachusetts's agricultural economy, as they foretold the pressures of an integrated, industrializing national economy. The opening of the Erie Canal in 1825, connecting Lake Erie and the Hudson River, reduced the transport cost of food supplies from the Midwest to one-thirtieth of its previous cost (Van Royen 1928). Marginal pastures were abandoned and farming was intensified in the production of specialty crops as rural communities began to meet much of their material needs through imports.

The abandonment of the agricultural economy and associated emigration of the rural population resulted in the re-establishment of forest ecosystems. Early successional species such as white pine, cherry, birch, and red maple dominated the abandoned farmland, and some of these species have persisted. Particular compositional and structural characteristics of contemporary forests are related to the specific land-use types of this period – woodlot, pasture, or tillage – and the date of field abandonment or last harvest (Foster 1999).

Thoroughly connected to the national economy by rail and coastal steamer by mid nineteenth century, central Massachusetts's industrial age was imminent. The adoption of steam power in the early 1860s fostered the rise of multiple factory complexes, supported by immigrant labor, which filled in the industrial sections of large cities such as Worcester (Balk 1944). With the increase in production and population, the industrial cities of southern New England became the center of manufacturing in the country, giving rise to complex human–environment legacies. For example, although forests continued to re-establish across the landscape, these industrial centers left legacy brownfields, with more than 1100 contaminated former industrial sites currently in central Massachusetts alone. This contamination, and associated liability issues, has inhibited redevelopment of the state's older cities and contributed to suburban expansion.

The return of forests, particularly white pine, precipitated a boom in the timber industry, which peaked in late nineteenth century. Much of the timber came from white pine stands that had established on old agricultural lands 50–70 years earlier (Foster 1992). The heavy harvest of softwood timber left large amounts of flammable material and resulted in increased fire frequency and intensity. The net consequence of these activities was thick, brushy woodland dominated by stump sprouts and the pre-agriculture species composition of mixed hardwoods (Hall *et al.* 2002).

By late nineteenth century, concerns about excessive logging gave rise to the state's early environmental movement. A host of private conservation organizations emerged to address specific interests, such as birding or favored forest spots. Communities were authorized to own and manage Town Forests in 1913, and in 1914 a State Forest Commission was created, charged with acquiring land suitable for timber cultivation and forest reclamation (Rivers 1998). By 1920, forest cover had returned to nearly 50% of the landscape.

The dramatic rise in urban–industrial populations in the early twentieth century, especially in Boston, necessitated the major impoundment and diversion of water from central Massachusetts. The Wachusett and Quabbin Reservoirs were opened in 1928 and 1946, respectively, totaling a combined 44 miles^2 (114 km^2) of water with an associated 100 000 acres (40 000 ha) of protected watershed forest (Greene 1981). The protected land surrounding the Quabbin Reservoir has become an "accidental wilderness," as flora and fauna have increased in the absence of human disturbance (Conuel 1981).

Forests continued to expand until about 1950, but since then the ongoing expansion of forest in the highlands has roughly balanced the loss of forest to development in the coastal lowlands (Hall *et al.* 2002). This suburban development pattern was amplified when the Massachusetts economy experienced resurgence in the mid-1970s. Fueled by a venture capitalist community working in collaboration with the universities and research centers in the Boston area, the economic base shifted to services and high-tech manufacturing. Landscape impacts from this new economy have been great. Peri-urban residential developments, characterized by 2-acre (1 ha) residential lots located near major highways and junctions, are increasingly common in the spaces between Boston and Worcester, the state's two largest population centers. Over the last few decades, this pattern of development has driven a decrease in forest cover and increase in fragmentation, with associated edge effects in forest ecosystems.

On the one hand, landform and climate still provide the strongest explanation for broad vegetation zones at the state scale, although the history of agriculture and farming have homogenized forest composition and structure within these zones. On the other hand, land-use history has created a patchier, more heterogeneous pattern at landscape and parcel scales. Natural disturbances, such as the devastating hurricane of 1938, which felled up to 75% of trees in some central Massachusetts towns, continue to affect local ecological dynamics, although most often at a chronic, local level. Invasive species, such at the chestnut blight and Dutch elm disease of the early twentieth century, continue to threaten forests, as today the hemlock wooly adelgid advances north and threatens the region's hemlocks.

In the past 20 years, distress over forest conversion to other uses has led to planning tools, such as conservation easements, to counteract widespread development and the loss of open space. State agencies and conservation groups are buying private land or the development rights to the property to protect forests, open space, and wetlands and to promote sustainable forestry and farming. The perpetuity of these contracts ensures that the landscape will remain partly forested and imbued with the legacies of social processes on long timescales.

Central Pennsylvania HERO

Native Americans farmed central Pennsylvania, but their numbers were low and impacts on the landscape were relatively trivial. Early European farmers started filtering into the area at the end of the eighteenth and beginning of the nineteenth century, resulting in localized deforestation in the valleys of the Ridge and Valley physiographic province. With continued in-migration throughout the nineteenth century and with the coming of the iron industry, the valleys were completely deforested and devoted to agriculture, except in locations where soils were unsuited for cultivation. Two forms of agriculture developed: traditional agriculture practiced by Anabaptist (Amish and Mennonite) farmers, and modern industrial agriculture practiced by all other farmers (Miller 1995a). Farmland loss is an environmental concern today.

High-grade iron ores appear at the surface in many parts of central Pennsylvania. The iron industry flourished during the middle third of the nineteenth century, supplying pig iron to the urban areas of the mid-Atlantic region (Eggert 1994). The technology of the time called for iron furnaces powered by charcoal. Although agriculture partially deforested the valleys before the advent of the iron industry, the ravenous furnaces completed valley deforestation and resulted in total deforestation of the ridges (Bining 1973). After that, farmers kept valley farmlands clear of trees, but the forests on ridges grew back.

The geology of the Ridge and Valley province and of significant portions of the Allegheny Plateau makes the rivers essentially unnavigable. The coming of the Pennsylvania Canal during the mid nineteenth century started to open the territory (Stranahan 1993). Nevertheless, this canal was difficult and expensive to build in the rugged landscape and was restricted to river floodplains, so the impacts were not great (Shank 1965) More important were the roads that connected the Pennsylvania Canal to settlements in the area. Even more important was the coming of the railroads, which soon followed construction of the canal and eliminated its value (Taber 1972). Railroads had localized environmental impact, but made possible the large-scale conveyance of people and goods. Railroads became critical for the growth of the mining industry, bringing miners and provisions into and taking coal out of central Pennsylvania (Shank 1990).

Nearly all of central Pennsylvania was deforested by the mid nineteenth century through the combined effects of agriculture and the iron furnace industry (Stranahan 1993), but the forests grew back on the ridges of central Pennsylvania by the late nineteenth and early twentieth centuries. Advances in mechanization at that time made it possible for rapid clear-cutting of these areas for a variety of purposes, including pulp and paper operations to feed demand for paper in eastern cities. In addition, stands of white pine in hard-to-reach pockets of the Allegheny Plateau that were avoided by iron-furnace foresters became more accessible because of mechanization and more desirable because of the high value of the timber (Miller and Schein 1995). The result was the second round of complete deforestation in the area.

The forests and ornamental trees of central Pennsylvania have suffered considerably from invasive species and pathogens. Notable among these have been chestnut blight, Dutch elm disease, oak leaf roller, and gypsy moths (Abrams and Nowacki 1992). Chestnut trees were the largest, most majestic trees in the central Pennsylvanian forest; the blight eliminated them by the mid twentieth century (Peattie 1964). Similarly, Dutch elm disease wiped out the most popular ornamental tree in the region in two waves, with one wave in the 1920s and 1930s (Faull 1938) and another starting in the 1960s (Gibbs 1978) and continuing today. Oak leaf roller struck hard and fast, starting in 1967 and lasting about one decade, and was the most destructive tree-killing insect in the twentieth century (Department of Conservation and Natural Resources 2004). At the peak of this outbreak in 1971, over 1 million acres were defoliated and over $100 000 000 of oak timber was lost (Abrams and Nowacki 1992). Gypsy moths destroyed wide swaths of central Pennsylvania forest in the late twentieth century (Abrams and Nowacki 1992). New species and pathogens continue to invade the area – native and ornamental hemlocks and rhododendrons, for instance, are currently under siege. Combined with forest management practices – chiefly fire suppression – the presence of invasive species and pathogens means that the composition, look, and functioning of the central Pennsylvania forest are radically different than they would be without them (Abrams and Nowacki 1992).

Large-scale mining of bituminous coal became important in the western portion of central Pennsylvania around the beginning of the twentieth century (DiCiccio 1996). In-migration to the area via the railroads soared and small mining towns sprung up everywhere. Soils were too poor for agriculture, so trains brought food and other goods into mining towns, but left with coal (Stranahan 1993). Not only did mining operations totally devastate large tracts of land, but acid mine drainage significantly affected nearly every mile of stream (Casner 1994). Poorly planned transportation networks and settlements added to the environmental destruction. Mining continues today, but output and employment are much smaller than in previous decades because of the rapid decline of steelmaking in Pennsylvania and the influence of the Clean Air Act on the desirability of the area's high-sulfur coal (DiCiccio 1996). In addition to coal mining, smaller-scale mining for materials needed to make refractory bricks – which were used to line the steel blast furnaces of Pittsburgh, Bethlehem, and Steelton, Pennsylvania – resulted in settlements around the refractories, which also died off with the decline of steelmaking and coal mining. In all cases, the outcome is abandoned or remnant towns, poverty, and squalor (Miller 1995b). Today, mine reclamation is restoring some of the landscape, but acid mine drainage is difficult to stem and continues to ravage local streams (Casner 1994). Other major mining operations, such as limestone mining and calcining, continue to be important to the area.

Because of (not in spite of) its relative isolation, major government institutions sprang up in central Pennsylvania starting in the mid nineteenth century and continuing today. Most notable among these are the Pennsylvania State University (Penn State) and a string of high-security state prisons. These institutions transformed large expanses of land through their buildings, agricultural lands, and forestlands. They continue to draw tens of thousands of people directly to live, work, and visit there, and thousands more to service them, all of whom put significant pressures on the environment through demands for space, resources, and infrastructure and through solid, liquid, and gaseous wastes.

Road building in central Pennsylvania was slow until recently. The rugged terrain made road building difficult and expensive; the low population density precluded the need for major roads (Miller 1995c). Improving construction technologies, growing population pressures, and increasing interstate truck traffic have led to a recent explosion in major highway construction. Perhaps as important is the region's centrality, lying at the crossroads of major interstates from the Northeast, mid-Atlantic, Southeast, and Great Lakes regions. Road construction is consuming the limited prime farmland and important wildlife habitats and is making possible the rapid development of the region.

Urbanization also has been slow in central Pennsylvania. Many small towns and villages are suffering population loss as primary industries, such as coal mines, are shutting down (Simpkins 1995). More than offsetting these losses are growth of urban centers like State College, which until recently had only one significant industry – education at Penn State. Now, high-tech and other light industries and service industries are coming into the area because of its relative quiet and low cost of living, as well as the many advantages and amenities afforded by a major university. With continued growth of the university, the expanded road network, and improved air transportation, urbanization and suburbanization are rapid and could be explosive. Farmland loss to suburban, commercial, and transportation development is significant in valleys, but farmland preservation is growing (Kelsey and Kreahling 1994).

Environmental protection efforts are starting to have profound impacts on people's relationship to nature. For example, reclamation of coal mines and streams affected by acid mine drainage is starting the slow process of recovery for hundreds of acres and several miles of stream each year (Casner 1994). The Clean Air Act is reducing the acidity of rainfall, which in central Pennsylvania is some of the most acid in the world (but local limestone buffers much of the acid rain, so the impact of this pollution is not as severe as in many areas with less acidic rainfall, but no limestone). Illegal dumping of solid and liquid wastes is declining, thereby improving the quality of local drinking supplies. Tightened regulations on community water systems and on industry also are improving the quality of water. The list is long and results are encouraging. Nevertheless, development and associated rapid population growth are increasing pressures on the environment, so it is difficult to claim that the overall environment of central Pennsylvania is improving.

High Plains–Ogallala HERO

In the High Plains–Ogallala HERO region, climate variability has played an important role in shaping an environment that seems to be in a constant state of flux (e.g., Kuchler 1972; Muhs and Holliday 1995). Human interactions with the semi-arid southwestern Kansas landscape have greatly influenced these landscape changes throughout prehistory and recorded history. While the impacts of Native Americans and Euro-American explorers may have been modest when compared to more recent happenings, these early interactions helped start a mindset of local resource extraction and utilization. Native Americans were important in establishing trade routes that are still used today by railroads and modern highways.

They also influenced the co-evolving native ecosystems (Brown and Gersmehl 1985) by regulating fire frequency and hunting bison. Spanish explorers brought with them the horse (which led to the Native American "horse culture" of the Great Plains) and new diseases (which nearly erased the existing human population). Bison numbers are thought to have increased greatly with the decline in hunting pressure on these native browsers.

Travelers and traders using the Santa Fe Trail learned that southwestern Kansas is an isolated area despite its good connections to other places. With the founding of the transcontinental railroads, the large bison herds were all but eliminated and a Euro-American cattle and ranching culture emerged. When Euro-American settlers moved into railroad towns and surrounding areas in the second half of the nineteenth century, cattle trails and the combined implementation of the Public Land Survey System (PLSS) and the Homestead Act established additional lines and patterns on the land. Dodge City, the "Queen of the Cowtowns," became an important regional center with its outside connection being the Atchison, Topeka, and Santa Fe Railroad. Animal power and the plow were used to etch the PLSS grid into the shortgrass prairie sod.

Climatic variations continue to be a significant environmental constraint in southwestern Kansas (Palmer 1965; Skaggs 1978; Worster 1979) with suggestions of drought cycles (Borchert 1971) and a teleconnection to tropical Atlantic sea surface temperature fluctuations (McCabe *et al.* 2004). Kincer (1923) pointed out that agriculture was a precarious proposition in this semi-arid environment without a supplemental water source and Sherow (1990) documented the establishment of ditch irrigation agriculture in the region using waters diverted from the major through-flowing Cimarron and Arkansas Rivers.

The early twentieth century was a time of agricultural expansion because of the high wheat prices driven by World War I. At the same time, the United States Geological Survey was documenting the magnitude and extent of important subterranean natural resources in the region. Southwestern Kansas suffered a major setback when a major drought combined with poor land management practices to lead to the Dust Bowl during the "dirty thirties." Land abandonment and out-migration were all too common in these desperate times. Federal intervention, in the form of New Deal conservation and economic stabilization programs, helped the people and the ecosystems recover and led to the establishment of the Cimarron National Grassland in Morton County (Duram 1995).

By mid twentieth century, several factors aligned to facilitate a new and radical transformation of the southwestern Kansas landscape. Technological developments made it possible to pump vast quantities of fossil Ogallala aquifer groundwater for flood irrigation of level lands or for center-pivot irrigation systems that could navigate across uneven terrain. Cheap power for lifting the water from deep underground came from the local Hugoton natural gas field – the largest natural gas field in North America. The expansion of irrigated agriculture occurred at a time when confined animal feeding operations were transforming the livestock industry. The arrival of several large meat-packing plants and the corresponding growth of the refrigerated trucking industry helped establish a vertically integrated, local agribusiness economy that was driven by outside demand for beef. The 1941 to 1964 Bracero (guest worker) Program, which helped establish physical and cultural pathways for Mexican workers to come to the region, and the 1975 Indochinese Migration and Refugee Assistance Act, which led to the immigration of Laotian and Vietnamese workers, helped provide a labor market for the meat-packing industries. As the green circles of center-pivot irrigation began to dot the landscape in ever-increasing numbers, improvements in transportation and ongoing societal trends saw a shift from a dispersed population to an increasing concentration of local residents in the major "Ogallala Oases" (White 1994) of Garden City, Liberal, and Dodge City. Technological improvements in irrigation technology, from drop tubes to subsurface drip, have improved the efficiency with which water is delivered to the crops, helped "preserve" the resource, and helped keep energy costs down.

While the region today certainly does not fit the literal view of a "Buffalo Commons"[2] (Popper and Popper 1987), the Poppers' metaphorical idea of landscape change is playing out (Popper and Popper 1999). Many local producers have chosen to take advantage of government subsidies from the Conservation Reserve Program to "retire" highly erodible farmland from production, while at the same time bringing other lands into production (Leathers and Harrington 2000). In areas where the available groundwater resource has been depleted, some agricultural producers have begun the conversion back to dryland farming (Kettle *et al.* 2007) – the agriculture practiced before center-pivot irrigation swept through the region. In other areas, economic diversification is adding confined feeding facilities for hogs and feedlot dairies to the prevailing beef feedlot standard. As new "farms" of wind turbines tap a renewable resource, the ongoing story of a region in a state of nearly continuous flux continues. While most local experts doubt that the area will become a "Buffalo Commons," it seems safe to conclude that major human transformations of the High Plains–Ogallala HERO will continue for some time to come.

Sonoran Desert Border Region HERO

From the seventeenth century onward, European contact has resulted in an increasingly transformed Sonoran Desert environment. The arrival of Spanish peoples, ecologies, ideas, and technologies in about 1600 resulted in the settlement of missions and presidios. The new desert residents imposed their religious and military culture, Moorish architecture, land and water private property right systems, and land-use preferences on this arid land. The transformed landscape had urban areas associated with religious and military cores and irrigated agriculture along the rivers, extensive cattle ranching on large landholdings, and sites of mineral resource extraction with their associated spoils.

The Spanish imprint remains on the Sonoran Desert Border Region HERO today, but the policies of the subsequent federal governments – the United States and Mexico – have also had profound effects on the land. The Gadsden Purchase, an 1853 agreement between the United States and Mexico for the United States to acquire lands south of the Gila River and west of the Rio Grande, put in place the current international boundary between the two countries. Motivation for the deal was to acquire the land for a southern transcontinental rail line. A local result of this boundary demarcation was the founding of the twin cities of Nogales, Mexico and Nogales, Arizona (see Chapters 9, 10, and especially 14). Putting the border in this location provided the origin for today's spatial division of natural resources and institutions across the international boundary; different types and intensities of land use and cover are now clearly evident at the border. Additional boundary lines were created on the United States side of the Sonoran Desert Border Region HERO in the nineteenth century when lands were set aside for Native Americans. These reservations marginalized a distinctive culture group and completed a process started by the Spanish, which set in place a land-use trajectory away from gathering and hunting to grazing and irrigated agriculture in areas with sufficient water rights.

Critical to the land transformation of the border region in the late nineteenth and twentieth centuries were several government initiatives, including the Homestead Act of 1862, the Desert Land Act of 1877, and the construction of railroads, all of them promoting the settlement and development of arid lands. Additional federal efforts went into developing water resources at the beginning of the twentieth century, such as Bureau of Reclamation-assisted construction of dams and irrigation works. In Mexico, parallel federal investment in resource development, railroad building, and irrigation district expansion occurred during the Presidency of José de la Cruz Porfirio Díaz Mori. Further growth occurred during the period of import substitution following the Mexican Revolution (which ended about 1920), later in association with agricultural intensification during the Green Revolution (which took place from the 1940s through 1960s), and most recently in relation to parastatal manufacturing enterprises ranging from the Border Industrialization Program to the North American Free Trade Agreement (which extended from the 1960s to today).

War also had its effect on the landscape. The Mexican Revolution of 1910–20 sent thousands of Mexicans to the north fleeing the political instability. This period dramatically transformed the structure of rights to land and resources in Mexico. Formerly landless farmers received tenure in the form of the *ejido* system of communally held land that could not be sold or rented. In Sonora, the new agricultural lands were often former desert or grasslands.

Use of these more marginal lands comes at a price, however. The arid climate of southern Arizona and Sonora is subject to significant inter-annual variation in precipitation. Major multi-year droughts occurred in the 1890s, the 1950s, and the early 2000s. The 1890s drought coincided with a period of significant expansion of the cattle industry, resulting in overgrazing and the eventual collapse of the industry. The combination of drought and overgrazing also resulted in desertification; a natural threshold was crossed, which fostered a change in vegetation from grassland to shrublands in many areas by promoting the invasion and establishment of mesquite and acacia. Although herd sizes were much lower on the United States side of the border by the 1950s, in Mexico the coincidence of the 1950s drought with the Bracero (guest worker) Program in the United States resulted in large-scale out-migration from the inland Sierra Madre Occidental Mountains of Sonora and associated decline of inland agriculture and ranching.

Roughly coincident with these land transformations and migrations, duty- and tariff-free manufacturing plants (known as *maquiladoras*) began to form along the border with the Border Industrialization Program in 1965. Maquiladoras are associated with industrial land uses, migration to urban–industrial centers such as Nogales, urbanization (often in the form of informal settlements), and increases in per capita consumption of resources. The gradual decline of protectionist economic policies in North America was eventually formalized in the North American Free Trade Agreement (NAFTA). Some analysts believe that NAFTA prompted a set of other neoliberal policies in Mexico, including privatization of land and water, reductions in government support for agriculture, and decentralization of water and environmental management. However, the reforms to the ejido sector may result in the sale of ejidos and consolidation of landholdings, the privatization of water may make it impossible for some residents to afford water, and the reduction of government support for agriculture may threaten the viability of some agricultural enterprises. In rural areas, maize is becoming a less competitive crop and, where water is available, production is shifting to irrigated forage (e.g., alfalfa) and to vegetables. The land and water of this HERO region continue to reel in response to federal government policies.

There is some hope, however, that the environment can recover. The rapid use of natural resources in the western United States during the late nineteenth century and the associated scars on the landscape led to conservation and forest management initiatives. The ideas expounded by John Muir and Gifford Pinchot helped establish several new federal programs. Formation of the National Park Service and of the United States Forest Service placed a large proportion of southwestern United States land under federal management, which continues to the present. Following the publication of Rachel Carson's *Silent Spring* in September 1962, the environmental movement expanded, resulting in a change in public attitudes, institutions, and legislation that affected landscapes of the southwestern United States. For example, the Endangered Species Act has constrained land development in some areas, and the Clean Air Act has contributed to the closure or cleanup of the copper smelters in Arizona. Not to be outdone by their American counterparts, the growth of environmental consciousness in Mexico since the 1970s has helped protect ecosystems in the border area.

Synthesis

The protocol we established for assessing historical landscape change and producing local environmental change narratives reveals consistent sets of human activities that shaped the land. Landscape change resulted from (Table 6.1):

- Traditional agriculture and resource extraction
- Transportation development
- Industrialization and urbanization
- Global, national, and state-level economic and legislative processes.

Examples of change are numerous. Forest clearing for agriculture changed central Massachusetts and central Pennsylvania, whereas livestock grazing altered the surface cover of southwestern Kansas and southern Arizona–northern Sonora. Mineral mining changed large areas of Pennsylvania and Arizona, water mining modified the look of the High Plains, and reservoir building and subsequent water extraction transformed Massachusetts. Transportation networks, such as railroads, allowed the migration of people into the four study areas and the removal of resources or manufactured goods from them. Industrialization and the associated urban growth of the population and built environment disturbed the land in myriad ways in all four places, from the classic industrialized landscape of central Massachusetts, to the landscape of "industrialized education" in central Pennsylvania, to the landscape of industrialized agriculture in southwestern Kansas, to the industrialized free-trade zone of the maquiladoras along the southern Arizona–northern Sonora border. The coming – and sometimes the going – of agrarian, industrial, and post-industrial economic systems modified and continue to modify the landscape in each locale. Distant law-making changed the landscapes of the HERO sites, e.g., recent environmental legislation or programs have protected green space in Massachusetts, helped push coal mining out of Pennsylvania to the western United States, conserved erodible land in Kansas, and halted land conversion in Sonora.

Surprisingly, the role of natural environmental stress through major events was not the most significant source of landscape change across the four regions. This

Table 6.1 *A grouping of landscape processes and transformative events among the HERO study areas by major and specific drivers of change, with more important processes and events in bold font*

Drivers		Processes and events			
Major driver	Specific driver	Central Massachusetts HERO	Central Pennsylvania HERO	High Plains–Ogallala HERO	Sonora Desert Border Region HERO
Biophysical	Climate variability	Hurricane of 1938	Floods	**Kansas droughts of 1930s (Dust Bowl) and 1950s (adaptations reduced landscape impact)**	**Droughts of 1890s (collapse of ranching, change in vegetation from grassland to shrub), 1950s (collapse of ranching, further vegetation change), and 1990s**
	Invasive species and disease	Exotic vegetation and pests	**Pennsylvania forest pathogens include chestnut blight, Dutch elm disease and gypsy moths**		Valley fever, dengue
	Fire and fire suppression		**Fire suppression causes oak to be replaced by maple**		Frequent fires associated with drought and expansion of human activity, intensity increases with fire suppression policies in twentieth century
Demography	Migration and immigration	**European settlement and clearing the forests from 1650**	**European settlement and clearing the forests from 1800**	European settlement in 1800s in railroad towns – boosterism	**European colonization – introduction of cattle, Catholicism, capitalism and private property (land grants)**

	Migration and immigration			Increase in migrant labor to work in intensive livestock	Growth of population post 1950; Mexican migration associated especially with the Bracero Program 1948–65
	Population density and distribution within region	**Suburbanization from 1970**	Urban growth of State College attracted by quality of life "Happy Valley"	Ogallala oases redistribute population from rural to urban areas	Suburban sprawl?
Politics (sometimes driven by agency)	International boundaries				**Gadsden Purchase in 1863 moves southern Arizona from Mexico to USA**
	Boundaries				**Native American reservations from 1859 and water rights from 1990 promote land use change from hunter–gathering to grazing to irrigated agriculture**
	Revolution				Mexican Revolution 1910–20 creates *ejido* landholding system and nationalization of resources such as water and minerals

Table 6.1 (*cont.*)

Drivers		Processes and events			
Major driver	Specific driver	Central Massachusetts HERO	Central Pennsylvania HERO	High Plains–Ogallala HERO	Sonora Desert Border Region HERO
	Federal incentives for land conversion and settlement	**Decline of farming and regrowth of forests as a result of agricultural settlement of the west**		**Homestead Act promotes conversion of grassland to crops and livestock**	**Homestead Act 1862 and Desert Lands Act promote settlement of the west**
	Federal incentives for land conversion and settlement			**Conservation Reserve Program responds to problems in agricultural economy and environment**	
	Beginnings of conservation	Transcendentalist and nature protection			**Muir, Pinchot and the establishment of National Forest Service and National Park Service in early twentieth century result in some protection of ecosystems in southwestern USA**
	1930s New Deal Conservation Programs			**Creation of National Grasslands in response to the Dust Bowl and poverty**	Taylor Grazing Act promotes managed rangeland and Civilian Conservation Corps builds infrastructure

	Environmental movement	**protects some green space**	**Water Act constrains mining, promotes reclamation, and pushes shift to low-sulfur coal in west**		**Act halts land conversion in Sonoran Desert, protects riparian areas, and other legislation controls air pollution**
Economy	Agricultural restructuring	**Agricultural boom 1800–60 to meet urban demands**	**Settlement of European farmers from 1780s onwards including Amish and Mennonite**	**Rise of corporate agriculture after 1950 and decline of family farms**	Decline of cotton production from 1960
	Mineral development		**Iron industry in mid nineteenth century causes deforestation and subsequent opportunities for agriculture**	**Extraction of natural gas from the Hugoton Field**	**Mining boom at end of nineteenth century with landscape impacts on forests, water, waste**
	Mineral development		**Coal industry in twentieth century associated with pollution and deforestation**		
	Rise of manufacturing	**Rise of industrial economy from about 1840. Textile mills until 1920 then heavy manufacturing. Results in decline of agriculture and growth of secondary forest**		**Local establishment of major meat-packing plants**	

Table 6.1 (*cont.*)

Drivers		Processes and events			
Major driver	Specific driver	Central Massachusetts HERO	Central Pennsylvania HERO	High Plains–Ogallala HERO	Sonora Desert Border Region HERO
	Rise of the service economy	**Decline of manufacturing post 1950 and rise of peri-urban settlement and recreation**	**Educational and prison facilities grow as agriculture and mining decline after 1950**		Increase in tourism from 1950
	Border industrialization and NAFTA				**1962 Border Industrialization Program and 1994 NAFTA stimulates growth of maquilas (manufacturing) and urban areas in border region**
Technology	Water development	**Quabbin Reservoir 1946 floods forests**		**River basin irrigation in early twentieth century**	**Late nineteenth century: J. W. Powell, the Bureau of Reclamation, and the development of water infrastrature in the west. Parallel growth in irrigation in Mexico in late nineteenth century with Porfiriato**

Water development			**Center-pivot irrigation post 1950**	Expansion of Sonoran irrigation districts with Green Revolution of 1950–90
Intensive livestock production			**Development of confined animal feeding systems post 1950**	Growth of alfalfa production?
Transportation	**Creation of highway network post 1945 promoted suburbanization**	**Canals and railroads in nineteenth century supported resource extraction**	**Railroads in nineteenth century promote settlement and agricultural trade**	1890s expansion of railroads supports mining and ranching. Highway from Tucson to Hermosillo through Nogales makes region a major trucking and trade route
Transportation		**New highway construction since 1990 alters land use and increases accessibility and settlement**	Highways in twentieth century promoted expansion of agriculture	
Refrigeration and air conditioning			Refrigerated trucks promote vertically integrated agricultural economy	Widespread air conditioning after 1950 encourages urbanization and retirement migration from snowbelt

finding does not imply that natural drivers have no impact on these places. To the contrary, the natural environment sets the baseline conditions for all four regions, ranging from the humid east, to the semi-arid Great Plains, to the arid United States–Mexico border region, and all four biophysical and socioeconomic systems are attuned to those conditions and their normal range of variations. Moreover, significant landscape-altering events have occurred in all four regions, e.g., the 1938 hurricane in central Massachusetts, repeated pests and pathogen outbreaks in the forests of central Pennsylvania, the Dust Bowl in southwestern Kansas, and the 1890s droughts in the Arizona–Sonora border region. Our finding also does not imply that major events will not change the regional landscapes of the future. For instance, a mega-drought engulfing the Sonoran Desert Border Region HERO could profoundly alter the way that humans use that already arid landscape. Our finding simply states that in the time since Euro-Americans inhabited the four study sites, although natural events have occasionally precipitated landscape changes, they have not been the principal driver of those changes. The principal drivers have been human.

Despite the common themes that did emerge from the four narratives, each region has distinctive biophysical and socioeconomic attributes that made that area unique. It is important to note that different socioeconomic processes sometimes led to similar landscape changes. For instance, reforestation of central Massachusetts resulted from nineteenth- and early-twentieth-century agricultural abandonment, whereas the first reforestation of the ridges of central Pennsylvania was a consequence of the collapse of the iron furnace industry in the mid nineteenth century and the second reforestation occurred because of the collapse of the logging industry in the early to mid twentieth century.

Higher-order commonalities exist among the four study sites and relate directly to the drivers of environmental change identified earlier. Population change, in terms of the absolute number of people carrying out their activities, has profoundly altered all four places. Migration into the study sites continues to increase populations and influence human–environment interactions. Technological change enabled central Massachusetts to industrialize, central Pennsylvania to extract its resources, southwestern Kansas to mine its groundwater, and the Arizona–Sonora Border Region to transfer huge quantities of Mexican produce to the United States. Economic forces drove manufacturing in central Massachusetts, coal mining in central Pennsylvania, vertical integration of the agricultural system in southwestern Kansas, and the formation of the maquiladoras along the Arizona–Sonora border. Institutional drivers facilitated the State Forests in central Massachusetts, the founding and continued growth of Penn State in central Pennsylvania, the rise of the Conservation Reserve Program in southwestern Kansas, and the rapid urban growth along the Arizona–Sonora border. Human desire for and perceptions of a better life brought – and continues to bring – people to the four study regions, and modern North American culture drives individuals and communities to use energy, materials, and land intensively. As noted in the theory section of this chapter and as evident in the examples given above, these drivers of environmental change are far from mutually exclusive, working together in most cases to drive the human activities that changed the landscapes of the study sites.

In summary, this study demonstrated that at the four HEROs the most important human–environment interactions or events promoting landscape change went through a predictable sequence (DeFries *et al.* 2006), starting with pre-industrial resource extraction and agriculture. That period, which took place at different times depending on when the areas were settled, was always associated with enabling institutions and technologies that then promoted industrialization, post-industrialization, and urbanization, all in the name of economic growth. Transportation technologies were especially significant in moving people into, and removing materials from these places. The pursuit of personal and social betterment was fundamental to the landscape transformations of these North American regions. Thus, the classic drivers of human–environment interactions – population change, technological innovation, economic change, institutions and policies, and behavior and culture – were present at all times in all HEROs and, in the end, were at the heart of the observed landscape changes.

Notes

1. The influence of structure versus agency on human behavior is a key debate in the social sciences. "Agency" addresses human ability to act independently and choose freely, whereas "structure" refers to socioeconomic factors, such as class, religion, gender, ethnicity, institutions, etc., that restrict the opportunities people have to act or choose. Although structure and agency operate simultaneously, the debate concerns their relative roles in any given context.
2. The "Buffalo Commons" is a proposal aimed at creating a massive nature preserve. The plan reintroduces the American bison and native shortgrass prairie to 140 000 square miles (360 000 km^2) of the most arid parts of the Great Plains. The proposal affects ten western states, including Kansas.

References

Abrams, M. D., and G. J. Nowacki, 1992. Historical variation in fire, oak recruitment, and post-logging accelerated succession in central Pennsylvania. *Bulletin of the Torrey Botanical Club* **119**: 19–28.

Adger, N., K. Brown, and H. Hulme, 2005. Redefining global environmental change. *Global Environmental Change* **15**: 1–4.

Association of American Geographers Global Change in Local Places Research Team (eds.), 2003. *Global Change in Local Places: Estimating, Understanding, and Reducing Greenhouse Gases*. Cambridge: Cambridge University Press.

Balk, W., 1944. *The Expansion of Worcester and Its Effect on the Surrounding Towns*. Worcester, MA: Clark University.

Bining, A. C., 1973. *Pennsylvania Iron Manufacture in the Eighteenth Century*, 2nd edn. Harrisburg, PA: Pennsylvania Historical and Museum Commission.

Borchert, J., 1971. The Dust Bowl of the 1970s. *Annals of the Association of American Geographers* **61**: 1–22.

Brown, D., and P. Gersmehl, 1985. Migration models for grasses in the American midcontinent. *Annals of the Association of American Geographers* **75**: 383–394.

Casner, N. A., 1994. Acid water: a history of coal mine pollution in western Pennsylvania, 1880–1950. Ph.D. dissertation. Pittsburgh, PA: Department of History, Carnegie–Mellon University.

Compton, J. E., M. R. Church, S. T. Larned, and W. E. Hogsett, 2003. Nitrogen export from forested watersheds in the Oregon Coast Range: the role of N_2-fixing red alder. *Ecosystems* **6**: 773–785.

Conuel, T., 1981. *Quabbin: The Accidental Wilderness*. Lincoln, MA: Massachusetts Audubon Society.

Costanza, R., R. d'Arge, R. de Groot, S. Farber, M. Grasso, B. Hannon, K. Limburg, S. Naeem, R. V. O'Neill, J. Paruelo, R. G. Raskin, P. Sutton, and M. van den Belt, 1997. The value of the world's ecosystem services and natural capital. *Nature* **387**: 253–260.

Cronon, W., 1983. *Changes in the Land: Indians, Colonists, and the Ecology of New England*. New York: Hill and Wang.

Cronon, W., 1993. The uses of environmental history. *Environmental History Review* **17**: 1–22.

Crutzen, P. J., and E. F. Stormer, 2000. The "Anthropogene." *IGBP NewsLetter* **41**: 17–18.

DeFries, R., and L. Bounoua, 2004. Consequences of land use change for ecosystem services: a future unlike the past. *GeoJournal* **61**: 345–351.

DeFries, R., G. P. Asner, and J. Foley, 2006. A glimpse out the window: what landscapes reveal about livelihoods, land use, and environmental consequences. *Environment* **48** (8): 22–36.

Department of Conservation and Natural Resources, Pennsylvania, 2004. Important insect and disease pests of Pennsylvania forests: Leafrollers 2004. Accessed at www.dcnr.state.pa.us/forestry/pests/leaf.htm.

DiCiccio, C., 1996. *Coal and Coke in Pennsylvania*. Harrisburg, PA: Pennsylvania Historical and Museum Commission.

Dilsaver, L., W. Wyckoff, and W. Preston, 2000. Fifteen events that have shaped California's human landscape. *The California Geographer* **40**: 1–78.

Donohue, K., D. R. Foster, and G. Motzkin, 2000. Effects of the past and the present on species distributions: land-use history and demography of wintergreen. *Journal of Ecology* **88**: 303–316.

Duram, L. A., 1995. Public land management: The National Grasslands. Part 1: Historical development; Part 2: Case study of Pawnee National Grassland; Part 3: Future management scenarios. *Rangelands* **17**: 36–42.

Eggert, G. G., 1994. *The Iron Industry in Pennsylvania*. Camp Hill, PA: Plank's Suburban Press.

Faull, J. H., 1938. The Dutch elm disease situation in the United States at the close of 1938. *Arnold Arboretum Harvard University Bulletin of Popular Information, Series 4* **6**: 75–78.

Feddema, J. J., K. W. Oleson, G. B. Bonan, L. O. Mearns, L. E. Buja, G. A. Meehl, and W. M. Washington, 2005. The importance of land-cover change in simulating future climates. *Science* **310**: 1674–1678.

Foley, J. A., R. DeFries, G. P. Asner, C. Barford, G. Bonan, S. R. Carpenter, F. S. Chapin, M. T. Coe, G. C. Daily, H. K. Gibbs, J. H. Helkowski, T. Holloway, E. A. Howard,

C. J. Kucharik, C. Monfreda, J. A. Patz, I. C. Prentice, N. Ramankutty, and P. K. Snyder, 2005. Global consequences of land use. *Science* **309**: 570–574.

Foster, D., 1992. Land-use history (1730–1990) and vegetation dynamics in central New England, USA. *Journal of Ecology* **80**: 753–772.

Foster, D., 1999. *Thoreau's Country: Journey through a Transformed Landscape.* Cambridge, MA: Harvard University Press.

Foster, D., and J. Aber (eds.), 2004. *Forests in Time: The Environmental Consequences of 1000 Years of Change in New England.* New Haven, CT: Yale University Press.

Foster, D., G. Motzkin, D. Bernardos, and J. Cardoza, 2002. Wildlife dynamics in the changing New England landscape. *Journal of Biogeography* **29**: 1337–1357.

Foster, D., F. Swanson, J. Aber, I. Burke, N. Brokaw, D. Tillman, and A. Knapp, 2003. The importance of legacies to ecology and conservation. *BioScience* **53**: 77–88.

Geist, H. J., 2005. *The Causes and Progression of Desertification.* London: Ashgate.

Geist, H. J., and E. F. Lambin, 2002. Proximate causes and underlying driving forces of tropical deforestation. *BioScience* **52**: 143–150.

Gersmehl, P. J., and D. A. Brown, 2004. The Conservation Reserve Program: a solution to the problem of agricultural overproduction? In *Worldminds: Geographical Perspectives on 100 Problems*, eds. D. G. Jannelle, B. Wharf, and K. Hansen, pp. 381–386. Dordrecht: Kluwer.

Gibbs, J. N., 1978. Intercontinental epidemiology of Dutch elm disease. *Annual Review of Phytopathology* **16**: 287–307.

Glacken, C. 1967. *Traces on the Rhodian Shore: Nature and Culture in Western Thought from Ancient Times to the End of the Eighteenth Century.* Berkeley, CA: University of California Press.

Gragson, T. L., and M. Grove, 2006. Social science in the context of the Long-Term Ecological Research program. *Society and Natural Resources* **19**: 93–100.

Greene, J. R., 1981. *The Creation of Quabbin Reservoir: The Death of the Swift River Valley.* Athol, MA: Performance Press.

Hall, B., G. Motzkin, and D. Foster, 2002. Three hundred years of forest and land-use change in Massachusetts, USA. *Journal of Biogeography* **29**: 1319–1335.

Howe, H. F., 1960. *Massachusetts: There She Is Behold Her.* New York: Harper and Brothers.

Johnson, H. B., 1976. *Order upon the Land: The U. S. Rectangular Survey and the Upper Mississippi Country.* New York: Oxford University Press.

Kates, R. W., W. C. Clark, R. Corell, J. M. Hall, C. C. Jaeger, I. Lowe, J. J. McCarthy, H. J. Schellnhuber, B. Bolin, N. M. Dickson, S. Faucheux, G. C. Gallopin, A. Gruebler, B. Huntley, J. Jäger, N. S. Jodha, R. E. Kasperson, A. Mabogunje, P. Matson, H. Mooney, B. Moore III, T. O'Riordan, and U. Svedin, 2001. Sustainability science. *Science* **292**: 641–642.

Kelsey, T. W., and K. Kreahling, 1994. *Farmland Preservation in Pennsylvania: The Impact of 'Clean and Green' on Local Governments and Taxpayers*, Extension Circular No. 411. University Park, PA: Penn State College of Agricultural Sciences.

Kettle, N., L. Harrington, and J. Harrington, Jr., 2007. Groundwater depletion and agricultural land use change in Wichita County, Kansas. *Professional Geographer* **59**: 221–235.

Kincer, J., 1923. The climate of the Great Plains as a factor in their utilization. *Annals of the Association of American Geographers* **13**: 67–80.

Kuchler, A., 1972. The oscillation of the mixed prairie in Kansas. *Erdkunde* **26**: 120–129.

Kulik, G., R. Parks, and T. Z. Penn (eds.), 1982. *The New England Mill Village, 1790–1860.* Cambridge, MA: MIT Press.

Lambin, E. F., and H. J. Geist, 2003. Regional differences in tropical deforestation. *Environment* **45**(6): 22–36.

Lambin, E., B. Turner II, H. Geist, S. Agbola, A. Angelsen, J. Bruce, O. Coomes, R. Dirzo, G. Fischer, C. Folke, P. George, K. Homewood, J. Imbernon, R. Leemans, X. Li, E. Moran, M. Mortimore, P. Ramakrishnan, J. Richards, H. Skånes, W. Steffen, G. Stone, U. Svedin, T. Veldkamp, C. Vogel, and J. Xu, 2001. The causes of land-use and -cover change: moving beyond the myths. *Global Environmental Change* **11**: 261–269.

Leathers, N., and L. M. B. Harrington, 2000. Effectiveness of Conservation Reserve Programs and land slippage in southwestern Kansas. *Professional Geographer* **52**: 83–93.

Liverman, D. M., 1999. Geography and the global environment. *Annals of the Association of American Geographers* **89**: 107–120.

Marsh, G. P., 1864 [1965]. *Man and Nature: or, Physical Geography as Modified by Human Action*. Cambridge, MA: Belknap Press of Harvard University.

McCabe, G., M. Palecki, and J. Betancourt, 2004. Pacific and Atlantic Ocean influences on multidecadal drought frequency in the United States. *Proceedings of the National Academy of Sciences of the USA* **101** 4136–4141.

McKibben, B., 1989. *The End of Nature*. New York: Random House.

Miller, E. W., 1995a. Agriculture. In *A Geography of Pennsylvania*, ed. E. W. Miller, pp. 183–202. University Park, PA: The Pennsylvania State University Press.

Miller, E. W., 1995b. Mineral resources. In *A Geography of Pennsylvania*, ed. E. W. Miller, pp. 203–233. University Park, PA: The Pennsylvania State University Press.

Miller, E. W., 1995c. Transportation. In *A Geography of Pennsylvania*, ed. E. W. Miller, pp. 234–251. University Park, PA: The Pennsylvania State University Press.

Miller, E. W., and R. D. Schein, 1995. Forest resources. In *A Geography of Pennsylvania*, ed. E. W. Miller, pp. 74–86. University Park, PA: The Pennsylvania State University Press.

Misselhorn, A., 2005. What drives food security in southern Africa? A meta-analysis of household economy studies. *Global Environmental Change* **15**: 33–43.

Muhs, D. R., and V. T. Holliday, 1995. Evidence of active dune sand on the Great Plains in the 19th century from accounts of early explorers. *Quaternary Research* **43**: 198–208.

National Research Council (NRC), 1992. *Global Environmental Change: Understanding the Human Dimensions*, eds. P. Stern, O. Young, and D. Druckman. Washington, D. C.: National Academies Press.

National Research Council (NRC), 1999. *Global Environmental Change: Research Pathways for the Next Decade*. Washington, D. C.: National Academies Press.

O'Keefe, J. F., and D. R. Foster, 1998. Ecological history of Massachusetts forests. In *Stepping Back to Look Forward*, ed. C. H. W. Foster, pp. 19–66. Cambridge, MA: Harvard University Press.

Palmer, W. C., 1965. *Meteorological Drought*, Research Paper No. 45. Washington, D. C.: United States Weather Bureau.

Peattie, D. C., 1964. *A Natural History of Trees of Eastern and Central North America*. Boston, MA: Houghton Mifflin.

Polsky, C., R. Neff, and B. Yarnal, 2007. Building comparable global change vulnerability assessments: the Vulnerability Scoping Diagram. *Global Environmental Change* **17**: 472–485.

Popper, D., and F. Popper, 1987. The Great Plains: from dust to dust. *Planning* **53**(12): 12–18.

Popper, D., and F. Popper, 1999. The Buffalo Commons: metaphor as method. *Geographical Review* **89**: 491–510.

Redman, C. L., J. M. Grove, and L. H. Kuby, 2004. Integrating social science into the Long-Term Ecological Research (LTER) network: social dimensions of ecological change and ecological dimensions of social change. *Ecosystems* **7**: 161–171.

Rindfuss, R. R., S. J. Walsh, B. L. Turner II, J. Fox, and V. Mishra, 2004. Developing a science of land change: challenges and methodological issues. *Proceedings of the National Academy of Sciences of the USA* **101**: 1 3976–13 981.

Rivers, W. H., 1998. Massachusetts state forestry programs. In *Stepping Back to Look Forward*, ed. C. H. W. Foster, pp. 147–219. Cambridge, MA: Harvard University Press.

Ruddiman, W., 2003. The anthropogenic greenhouse era began thousands of years ago. *Climatic Change* **61**: 261–293.

Rudel, T. K., 2005. *Tropical Forests: Regional Paths of Destruction and Regeneration in the Late Twentieth Century*. New York: Columbia University Press.

Rudel, T. K., 2008. Capturing regional effects through meta-analyses of case studies: an example from the global change literature. *Global Environmental Change* **18**: 18–25.

Shank, W. H., 1965. *Amazing Pennsylvania Canals*. York, PA: Historical Society of York County.

Shank, W. H., 1990. *Pennsylvania Transportation History: A Supplement*. York, PA: American Canal and Transportation Center.

Sherow, J. E., 1990. *Watering the Valley: Development along the High Plains Arkansas River, 1870–1950*. Lawrence, KS: University of Kansas Press.

Simpkins, P. D. 1995. Growth and characteristics of Pennsylvania's population. In *A Geography of Pennsylvania*, ed. E. W. Miller, pp. 87–112. University Park, PA: The Pennsylvania State University Press.

Skaggs, R., 1978. Climatic change and persistence in western Kansas. *Annals of the Association of American Geographers* **68**: 73–80.

Sorrensen, C., C. Polsky, and R. Neff, 2005. The HERO REU experience: undergraduate research on vulnerability to climate change in local places. *Geographical Bulletin* **47**: 65–72.

Stranahan, S. Q., 1993. *Susquehanna: River of Dreams*. Baltimore, MD: Johns Hopkins University Press.

Taber, T. T. III, 1972. *Logging Railroad Era of Lumbering in Pennsylvania,* Vol. 4, *Sunset along Susquehanna Waters*. Williamsport, PA: Lycoming Printing Co.

Thomas, W. (ed.), 1956. *Man's Role in Changing the Face of the Earth*. Chicago, IL: University of Chicago Press.

Turner, B. L. II, 2002. Contested identities: human–environment geography and disciplinary implications in a restructuring academy. *Annals of the Association of American Geographers* **92**: 52–74.

Turner, B. L. II, W. C. Clark, R. W. Kates, J. F. Richards, J. T. Mathews, and W. B. Meyer (eds.), 1990a. *The Earth as Transformed by Human Action: Global and Regional Changes in the Biosphere over the Past 300 Years*. Cambridge: Cambridge University Press.

Turner, B. L. II, R. E. Kasperson, W. B. Meyer, K. M. Dow, D. Golding, J. X. Kasperson, R. C. Mitchell, and S. J. Ratick, 1990b. Two types of environmental change: definitional and spatial-scale issues in their human dimensions. *Global Environmental Change* **1**: 14–22.

Turner, B. L. II, R. E. Kasperson, P. Matson, J. J. McCarthy, R. W. Corell, L. Christensen, N. Eckley, J. X. Kasperson, A. Luers, M. L. Martello, C. Polsky, A. Pulsipher, and

A. Schiller, 2003. A framework for vulnerability analysis in sustainability science. *Proceedings of the National Academy of Sciences of the USA* **100**: 8074–8079.

Turner, B. L. II, E. Lambin, and A. Reenberg, 2007. The emergence of land change science for global environmental change and sustainability. *Proceedings of the National Academy of Sciences of the USA* **104**: 20 666–20 671.

Van Royen, W., 1928. *Geographic Studies of the Population and Settlements in Worcester County, Massachusetts*. Worcester, MA: Clark University.

White, S., 1994. Ogallala oases: water use, population redistribution, and policy implications in the High Plains of western Kansas, 1980–1990. *Annals of the Association of American Geographers* **84**: 29–45.

Wilbanks, T., and R. Kates, 1999. Global change in local places: how scale matters. *Climatic Change* **43**: 601–628.

Williams, M., 2003. *Deforesting the Earth: From Prehistory to Global Crisis*. Chicago, IL: University of Chicago Press.

Wilson, E. O., 2002. *The Future of Life*. New York: Knopf.

Worster, D., 1979. *Dust Bowl: The Southern Plains in the 1930s*. Oxford: Oxford University Press.

Zhou, W., A. Troy, and J. M. Grove, 2007. Modeling residential lawn fertilization practices: integrating high resolution remote sensing with socioeconomic data. *Environmental Management* **41**: 742–752.

7

Landsat mapping of local landscape change: the satellite-era context

RACHEL M. K. HEADLEY, ROBERT GILMORE PONTIUS, JR., JOHN HARRINGTON, JR., AND CYNTHIA SORRENSEN

Introduction

To set the stage for a vulnerability analysis, investigators must describe and understand the geographic context, including physical characteristics of the landscape and the political and socioeconomic milieu of the population (Jianchu *et al.* 2005). Vulnerability studies focus on a particular place, at a specific time through its three dimensions, exposure, sensitivity, and adaptive capacity; therefore, understanding place is essential to analyzing vulnerability.

Land-use studies are essential to understanding place because they generalize human activities on the physical landscape. Essentially, land use indicates past human decisions and actions, environmental constraints, and, in some cases, gives insight into subsequent change. Like vulnerability, land use is particular to a place at a certain time, and the analysis of that land use can be used as a baseline for future change and its implications. Vulnerability and land use are linked by the concept of place and are fundamental to contemporary research on human–environment interactions.

Although the literature on land use, land-use change, and climate change is extensive, the land-use component of vulnerability is usually conceptualized as a feedback mechanism to climate change: forest cutting releases carbon dioxide, which increases atmospheric carbon dioxide concentrations, which increases radiative forcing, which changes climate, and which ultimately changes land cover and subsequent land use (e.g. DeFries and Bounoua 2004; Jianchu *et al.* 2005; Salinger *et al.* 2005; Watson 2005). Moreover, land use is rarely specifically identified as a component of vulnerability. For example, the location of forested hillsides or the patterning of impermeable surfaces and urban sprawl can characterize exposure to flooding. Thus, a literature gap exists at the intersection of vulnerability, land use, and land-use change where land use and land-use change directly alter the vulnerability of the population that resides in or depends on that landscape.

Sustainable Communities on a Sustainable Planet: The Human–Environment Regional Observatory Project, eds. Brent Yarnal, Colin Polsky, and James O'Brien. Published by Cambridge University Press.

In one of the rare studies that have investigated the direct consequence of land use on vulnerability, Mustafa (1998) looked at how political and social structures can drive infrastructure, which can increase the vulnerability of certain portions of a population. In another such study, Bankoff (2003) explored how urban expansion can lead to increased vulnerability to floods. These accounts tend to address issues such as social justice and access to technological advancements. Strict land-use change analysis may incorporate some of these ideas, measure the changes quantitatively, and interpret the consequences or significance of those changes. Research questions might be: What are the consequences when exurban growth encroaches into floodplains? What is the significance of grassland areas disappearing and suburban development expanding?

Research documenting other intersections of land-use change, vulnerability, and climate change have ranged from the economic analysis of irrigated agriculture in the western Great Plains (Polsky and Easterling 2001) to the impacts on forest-system biodiversity in the fragmented hardwood forests of the eastern United States (MacIver and Wheaton 2005). These investigations are valuable in expanding our understanding of these intersections, but they do not address a fundamental question of concern to us here: in which dimension of vulnerability – exposure, sensitivity, or adaptive capacity – does land use fit? The bulk of this chapter explores this question. We find that land use appears to influence every dimension of vulnerability in some places, whereas it affects only one dimension in other areas. A key finding is that we did not explore any landscape where land use did not influence exposure, sensitivity, or adaptive capacity in some way.

Land-use change

While land use is often used as context for vulnerability analysis, land-use *change* may be a better metric for developing our understanding of exposure, sensitivity, and adaptive capacity (Table 7.1). For instance, land-use change results in alterations to ecosystem services, such as net primary production. Net primary production often increases when grassland is converted to agricultural uses due to the increased use of water and nutrients (DeFries and Bounoua 2004). In this case, the land-use change is an adaptation that decreases sensitivity to drought by increasing water and nutrients supplied to the ecosystem. In contrast to such straightforward cause and effect relationships, incorporating the vulnerability of humans to flood or drought events with studies of land-use change gives a more robust understanding of the complex consequences of human influence on the landscape. For example, the combination of increased floods and land-use change can exacerbate problems of erosion and runoff (Nearing *et al.* 2005), which can lead to a decrease in the quality of water sources identified for human use.

Table 7.1 *Consequences of land-use change to the "structural, compositional, and functional components of the ecosystem"*

Alteration of disturbance regimes
Alteration of water regimes
Alteration of nutrient cycles
Loss of topsoil
Disrupted dispersal of plant and animal species
Effects of biochemicals (fertilizers and pesticides)
Loss of pollinators, seed dispersers, and fungal symbionts
Pollution of ground and surface water
Change in microclimate (affects seedling establishment, growth rates, disturbance regimes)

Source: Neke and du Plessis (2004).

Because of complex feedbacks across scales, it can be difficult to determine if land-use change is a result or a driver of local environmental change. Land-use change in arid regions can alter surface albedo, water exchange, and nutrient cycles, which can then feed back into the regional and global climate systems. In these areas, trends caused by local land-use changes, such as massive irrigation or overgrazing, are difficult to differentiate from consequences of global climate trends (Lioubimtseva *et al.* 2005).

Land-use change is tied to human decisions and intentions, which are the result of a host of factors. These political and socioeconomic forces change over time in relation to the landscape and its use, which can alter vulnerability (Bankoff 2003). Jianchu *et al.* (2005) suggest that land use can even be used as an indicator of political and economic drivers of vulnerability.

Land use and the dimensions of vulnerability

It is important to put land-use change in the context of the three dimensions of vulnerability: exposure, sensitivity, and adaptive capacity. As noted above, land-use change can be situated in all of these dimensions, depending on the type of land-use change and the consequences of that change to the human and natural environments. For example, increases or decreases in the frequencies or intensities of floods and droughts can increase or decrease exposure to these events and can drive vulnerability. A socioeconomic change that results in an alteration of land use, however, can increase vulnerability independent of a change in the physical exposure (Kakembo and Rowntree 2003). Other forces, such as drivers that encourage rural to urban migration, can result in the shortage of safe housing, which can lead to establishment of communities in areas of higher exposure (Bankoff 2003).

To begin analysis where land use and land-use change intersects vulnerability, we began with the Vulnerability Scoping Diagram (VSD; described in Chapter 5). We analyzed how knowledge of land use and land-use change could contribute to or enhance the VSD's three dimensions: exposure, sensitivity, and adaptive capacity.

Land-use change can be considered a component of exposure because it leads to the population's exposure to the hazard and exposure of things on the landscape that the population values. Vulnerability resulting from exposure of the population concerns several elements related to land use, including population location, population density, permanence of settlement, and intensity of land use. The location of the population can shift, but often in ways that can be forecasted. Areas that are adjacent to large populations and that people find amenable are more likely to be settled than those areas that are more remote or considered hostile. When land-use analysis is applied to an area, these patterns are easily recognized and mapped for incorporation into exposure analysis. The density of settlement also has consequences for exposure. A densely settled suburban area, with wide extents of impermeable surfaces and high biochemical inputs, may have greater exposure to floods than lightly settled housing amongst woodlots.

The permanence of the population further contributes to the level of exposure. If people inhabit an area for one or two seasons per year because of its amenities, such as skiing or beachfront activities, then exposure would change each season. Finally, the intensity of how a population uses the land is important. If vegetables are grown on agricultural land, for instance, the impact to exposure can vary, depending on whether the vegetables are grown in greenhouses, rainfed fields, or irrigated fields. Greenhouses require less area and release fewer chemicals into the environment because of the controlled setting. They are a form of impermeable surface, however, which has a direct consequence on runoff and the reduction of habitat for wildlife. Soini (2005) argues that the challenge is to maintain and increase agricultural productivity in an ecologically sustainable, and therefore less vulnerable, manner. These are just a few examples of how land use and exposure intersect; the details of this intersection depend on the landscape, the people living there, and the hazard. Climate change will add an additional layer of exposure (Sivakumar *et al.* 2005).

Interactions between exposure and land use also apply to sensitivity and land use, although the relationship is less distinct. The way in which land use contributes to sensitivity is largely dependent upon the hazard under study; as the HERO proof-of-concept research focuses on floods, droughts, and agriculture, we will discuss land use and sensitivity in that context. People that use land for agriculture can be sensitive to floods or droughts, depending on the timing, magnitude, and the damage of the events. In general, variations in climate regimes have a direct influence on quantity and quality of agricultural production, often in an adverse way (Salinger *et al.* 2005; Walker 2005). Consequently, the agricultural sector may be an early indicator of climate change because of its this sensitivity to climate (Walker 2005), particularly where crops are near their environmental limits (Burton and Lim 2005). Sensitivity has a temporal component tied to land use: a farmer might have low sensitivity to repeated short droughts, but might be highly sensitive to one prolonged drought. Floods and droughts directly affect plant growth, which then affects farmer income, which, in turn, affects numerous economic sectors dependent on farmer success. In addition, considerations beyond the farm, and outside the realm of climate, influence farmer success. For instance, local farmer decisions are sensitive to their political and socioeconomic environments (Polsky and Easterling 2001). Climate change may not adversely affect agricultural production at the global scale, but severe inequalities among farmers, both positive and negative, will occur at finer scales (Salinger *et al.* 2005; Thomson *et al.* 2005a); this differential sensitivity will lead to uneven vulnerability in different geographic locations (Polsky and Easterling 2001).

To survive and prosper, farmers have always had to adjust to short-term exposures and adapt to longer-term changes in exposure. Because climate change and its impacts on droughts and floods will add an additional layer of exposure, the agricultural community may need to alter the normal range of adaptive capabilities that it presently uses (Sivakumar *et al.* 2005). Adjustments commonly employed by agriculture are expansion, intensification, and diversification, and include methods of conservation tillage, proper management of irrigation, and changes in land allocation (Salinger *et al.* 2005; Soini 2005). Long-term climate change can result in long-term adaptation measures such as alterations in cropping patterns, crop varieties, or crop-cycle adjustments (Salinger *et al.* 2005; Thomson *et al.* 2005b; Walker 2005).

Adaptive capacity differs from adaptation in that it stresses the ability of individuals, groups, or populations to apply these and other adaptation measures. Differing adaptive capacities result in a wide variation of vulnerabilities in the agricultural sector (Burton and Lim 2005). The geographic distribution of productive cropland must be considered when analyzing agricultural adaptive capacity (Thomson *et al.* 2005a). The diversity of soil types places limits on adaptation strategies that include land-use change or land-area adjustments; i.e., although the climate might be appropriate for a certain agricultural type, the soil may not be. Moreover, while increased warmth and moisture in some areas could lead to increased yields, soil erosion and nutrient leaching might accelerate, counteracting these benefits. Productivity in other areas could decrease due to increased heat stress and drought tendencies (Salinger *et al.* 2005). In addition to changes in the temperature and precipitation regimes, overall increases in weeds, diseases, and pests, in addition to expansions of their traditional ranges, could reduce the capacity of agriculturalists to adapt to the new climatic environment (Motha and Baier 2005; Salinger *et al.* 2005; Walker 2005).

Adaptation by the agricultural sector is complicated by national and international policy and by the globalization of agribusiness (Burton and Lim 2005; Salinger *et al.* 2005). Farm subsidies often do not reward innovation and adaptation (Burton and Lim 2005). If subsidies are not responsive to changes in climate conditions suitable for growing certain crop types, the adaptive capacity of individual farmers, specific subsectors, and entire regions may be stifled.

Land value in the agricultural sector is a primary driver for agricultural abandonment, expansion, or conversion to higher land values (e.g., urban). Land value therefore has significant implications for the capacity of farmers to adapt to climate variation and change by expansion or relocation. The relocation of certain crop types must take into account the disruptive impact on existing infrastructure and rural communities, both at the place of origin and the destination (Motha and Baier 2005). This idea of relocation is likely to be limited by existing land-ownership and land-use patterns.

Transfer of scientific knowledge of mitigation measures, crop–climate matching, crop indices, crop modeling, and risk assessment can increase the adaptive capacity of land-owners (Walker 2005). Introduction of technological advancement can also increase adaptive capacity (Liverman 1990; Salinger *et al.* 2005). For some land-owners, technological advancement may decrease adaptive capacity for those that live on the margins of that technology (Liverman 1990). In addition, dependence on technology may result in slower acceptance of long-term trends because the effects are largely hidden by still-high yields (Warrick 1980). Agricultural sectors or regions that rely on irrigation may be doubly disadvantaged because there will be more people relying on the same finite resource (Thomson *et al.* 2005b); conversely, some geographic areas may experience a decrease in demand for irrigation because of increased rainfall, thereby improving adaptive capacity (Edmonds and Rosenberg 2005).

Adaptive capacity can change with scale. The focus of agricultural stresses differs at a regional scale because other demands for resources must be considered (Thomson *et al.* 2005b). At such broader scales, agriculture must compete for resources with other non-agricultural users, such as residential and industrial needs (Motha and Baier 2005).

The connection between land use and vulnerability is complex. When land use changes, the dimension that is most directly related to land use can also change. For instance, if forest is cut and a parcel of land is converted to agriculture, the exposure and sensitivity to drought may increase, yet adaptive capacity may also increase. Thus, the overall vulnerability might remain about the same, although the drivers behind the vulnerability have changed.

Cross-site comparison

The above discussion of the intersection of land-use change, climate change, and vulnerability suggests that each geographic location is distinct in the ways these factors interrelate, and that comparison across sites is difficult. This section will attempt to answer questions such as: Can land use and land-use change be a metric of vulnerability that can be compared across sites? Does vulnerability have different implications at each site, thereby limiting comparability?

For this section of the chapter, land-use maps derived from remotely sensed data of each HERO study site were used to determine land-use change. Descriptions of how its land-use change might influence vulnerability is discussed in the context of exposure, sensivity, and adaptive capacity. These components of vulnerability are then compared across sites.

There are many potential difficulties to such a cross-site comparison, including the maps' categories, spatial extents, spatial resolutions, temporal extents, and temporal resolutions. Methodological concerns could include techniques to measure change over time in each land-use category, transitions among categories, pattern metrics, and relationships between land cover and other variables such as topography.

One of the biggest challenges in performing cross-site analysis lies in determining the level of detail to include in the analysis. It is helpful to describe this problem using two axes: cross-site comparability and site-specific detail (Figure 7.1). The

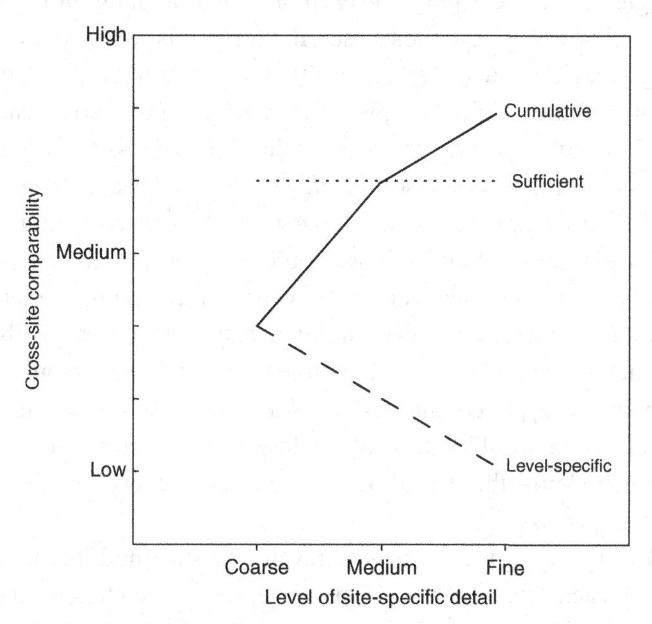

Figure 7.1. As details of site analysis increase, cross-site comparison is increasingly difficult.

dashed line denotes a decreasing relationship between the two variables. When the level of site-specific detail is coarse, a single variable allows straightforward comparison among sites. For example, if each site were to descibe its landscape in terms of forest or non-forest, then the cross-site comparison would be simple and the analysis would possess only limited site-specific detail. When the level of site-specific detail is fine, then it is more difficult to perform cross-site comparison. If each site were to use a different set of detailed categories, for instance, then the analysis would possess fine site-specific detail, but cross-site comparisons would be complex and difficult.

The point of a cross-site analysis is to detect important differences, or similarities, among sites, if they exist. It is reasonable to begin with a coarse level of detail. If the resulting comparison reveals important differences among the sites, then the coarse level of detail may be sufficient. However, if coarse-level analysis does not provide enough information to discriminate among the sites, then it is necessary to delve into finer detail such that the deeper analysis complements and supplements the coarser analysis. If we construct a method such that the analysis accumulates information as it passes from coarser levels to finer levels, then the cumulative cross-site comparability grows as the research progresses (as the solid increasing curve of Figure 7.1 shows). Our strategy, therefore, is to begin with a coarse analysis, then to supplement it with more detailed analyses (Pontius *et al.* 2004).

To demonstrate this principle, if we were to compute only the net change over time of a single "forest" category for each site, then a straightforward cross-site comparison would be simple. In most cases, however, this coarse analysis would not be sufficiently nuanced to uncover important cross-site similarities and differences in terms of vulnerability, which is typically a complex, multivariate entity. Moving to a moderate level of detail, we can compute the gain and loss of each of the United States Geological Survey (Anderson *et al.* 1976) Level I categories, so that the medium-level analysis adds detail to the coarse-level analysis (Pontius and Malizia 2004). At an even finer level of detail, we could analyze pattern metrics such as the average land-use patch size. We would then need to interpret the pattern metrics in the context of the coarse- and medium-level results in order for the fine-level analysis to generate an increase in cumulative cross-site comparability. If we were to analyze the fine-scale pattern metrics in isolation from the coarser results, then the pattern metrics alone would offer only a low level of cross-site comparability because we need to know the change in areas of each category so we could interpret the changes in patch size.

In the HERO research, we ultimately developed a method that uses a medium level of detail at each of the four proof-of-concept sites such that we used the same categories and spatial resolution, and temporal resolutions, but did not use the same spatial extent or years. We found that our approach offered a sufficiently rich analysis to capture the broad differences among the sites given their varying vulnerabilities, while providing the reader with a digestible and informative report in terms of length and complexity. We summarize that work in the following section, describing the four HERO sites in regard to land-use change and the implications of that change to exposure, sensitivity, and adaptive capacity.

Land-use change of the HERO sites

The complexities of land-use change at different geographic locations make comparisons between places difficult (Civco *et al.* 2002; Vasconcelos *et al.* 2002). Humans manipulate their landscape in many ways, and each can be expressed differently in terms of land use. Change can be conceptualized as one-directional and permanent – farmland is converted into suburbs, forests into farmland, or desert scrub into urban sprawl but in many cases, however, a land use can be transitory, the change cyclic in nature. Few can argue that the direction of change in agricultural areas in northern New Jersey is toward increasing urban areas, which is a unidirectional change. Conversely, considering the rates of timber harvest in western Oregon, a parcel that is bare in 1980 is likely to be forested by 2000. Consequently, the forested areas in 1980 may be bare in 2000.

Some HERO sites have higher rates of change than others, and some of that change was a permanent, one-directional change (e.g., forest to urban), whereas other changes were cyclic, or transitory (e.g., an aspect of a forest–clearcut–forest regrowth cycle). We hypothesize that these differences can be used to compare sites, giving insight into possible future land-use change, the constraints that may result, and the implications of such constraints for exposure, sensitivity, and adaptive capacity.

In this section, we explore an analytical technique that simplifies the dynamics of a changing landscape into *gain, loss, total change, net change,* and *swap* from a single statistical methodology based on Pontius *et al.* (2004). Gain, loss, and total change are the absolute values of the areas that have changed. Net change can either be a gain or a loss such that one land-use change can result in a net gain, while another results in a net loss. For example, when agricultural land is changed to urban, agriculture has a loss while urban has a gain. Swap occurs when a land use has loss in one area (forest harvest), but a gain in another area (forest regrowth). For instance, if 500 ha of forest are converted to cropland, but 500 ha of cropland becomes forest in a different location in the same time period, then the change is 100% swap (there is no net gain or loss in either forest or cropland). When combined with an understanding of land-use change specific to geographical context, this information allows for an initial comparison of exposure, sensitivity, and adaptive capacity across sites.

Each project team independently completed a post-classification, land-use change analysis derived from remotely sensed data for each study site. Although land use varied at each site, all classifications used satellite imagery or aerial photography from two dates. Land uses were reduced to the Anderson Level I classification (Anderson *et al.* 1972) and were resampled to 30-m resolution. For this analysis, only a portion of each of the full HERO sites was mapped and is described in the following paragraphs.

Central Massachusetts HERO

The portion of the central Massachusetts HERO (CM-HERO) site used for land-use change analysis includes the city of Worcester and surrounding towns. The region has experienced several important shifts in its economy and land use. The heavily deforested agricultural landscape during the early nineteenth century became a center of industrial manufacturing by the early twentieth century. As agricultural areas were abandoned, the forested landscape returned (Foster 1993).

The Worcester area contains hundreds of contaminated commercial and industrial sites, known as brownfields. The presence of brownfields in the study site, with their high toxicity levels and blighted landscapes, inhibits redevelopment of the metropolitan area and pushes new industry and residential development further from the urban center. However, the reorientation of the Worcester economy to biotechnology and health services in recent decades has prompted a sprawling pattern of land use that directly results in forest loss. These challenges to land-use management have led to the nearly irreversible conversion of forest and agricultural lands to urban areas.

The land-cover data used for the CM-HERO originated from 1:25 000-scale aerial photographs from 1971 and 1999. The Resource Mapping Project (RMP) at the University of Massachusetts, Amherst, created the maps using visual photo interpretation. The 1999 image was interpreted from aerial color infrared photography, and was originally interpreted into 37 land-use classifications.

The CM-HERO landscape experienced a total change of 10.70% between 1971 and 1999. The landscape experienced a net change of 8.28%, the largest net change observed at any of the study sites. This change was confined mostly to the urban and forest categories. Urban experienced a net gain of 8.23% of the total landscape, whereas forest had a net loss 6.12%. The net change experienced at CM-HERO indicates a unidirectional transition to an urban landscape.

High Plains–Ogallala HERO

The High Plains–Ogallala HERO (HPO-HERO) in southwestern Kansas is characterized by an agriculturally driven and groundwater-dependent economy (Kromm and White 1992). Center-pivot irrigation agriculture has placed a heavy demand on groundwater reserves, causing substantial depletions (Schloss *et al.* 2000; Goodin *et al.* 2002). The region has been further transformed by the addition of several large meat-packing plants, confined animal feeding operations, and associated crop restructuring to produce more feed grains.

Gray County, Kansas, the area chosen within the HPO-HERO region, provides an example of the effects of the Conservation Reserve Program, which was enacted in 1985. This program had an overarching goal of reducing erosion on environmentally sensitive lands. Retired lands usually are planted with native grass.

Landsat Thematic Mapper (TM) imagery was used to map Gray County land cover in 1985 and 2001. Twelve-band composite images for each year were created by stacking spring and summer TM images (bands 1–5, 7). A supervised approach to signature development was used and a maximum-likelihood algorithm was applied to produce modified Anderson Level II classifications (Anderson *et al.* 1972) for 1985 and 2001. The resultant nine-category classifications were reclassified using a contextual filter and collapsed into five of the seven Anderson Level I categories (urban, cropland, range, forest, and water).

The HPO-HERO landscape experienced 13.41% total change between 1985 and 2001. Swap accounted for 9.55% of the landscape change during the time period. Almost all change on the landscape occurred between the agriculture and range categories. Agriculture experienced a net loss of 3.85% while rangeland experienced a net gain of 3.68%.

Susquehanna River Basin HERO

The Susquehanna River Basin HERO (SRB-HERO) encompasses the Susquehanna River Basin in the eastern United States. This largely rural watershed has had two episodes of near-complete deforestation in the last two centuries, as well as intensive underground and surface coal extraction. The Susquehanna River Basin contains the most densely settled rural areas and some of the most productive agricultural lands in the United States.

Within the SRB-HERO, we assessed land-use change for Centre County, Pennsylvania. The major land-use change in Centre County is the reclamation of more than 4400 ha of mined lands from 1973 to 2000. In addition, 3700 ha of timberland was harvested or replanted in Centre County between 1972 and 2000. Urban expansion is a small component of the overall land-use/land-cover change in Centre County, increasing only 0.56% (1627 ha) from 1973 to 2000.

The SRB-HERO land-use maps were derived from a TM image from 1993. Aerial photography and ground-truthing aided the manual interpretation of the imagery. The 1993 map was used to identify changes in Landsat Multi-Spectral

Scanner data for 1972 and Landsat Enhanced Thematic Mapper Plus data for 2000. Only the maps from 1973 and 2000 have been used in the analysis presented here.

The total change on the Centre County landscape was 3.70%, with net change accounting for 2.14% of the landscape, and swap accounting for the remaining change of 1.56%. The most important categories at this site are rangeland and barren. The net gain of rangeland (reclaimed mined lands) was 1.59% of the total landscape while the net gain for barren (mined lands) was 1.98%. Forest experienced a 1.75% swap, almost all of which is transitory as timber is harvested in some locations while being allowed to grow back in others.

Sonoran Desert Border Region HERO

The Sonoran Desert Border Region HERO encompasses the Santa Cruz watershed, which traverses the border of Arizona, USA, and Sonora, Mexico. In rural areas, federal and state policies relating to land tenure structures affect land cover, with rangeland degradation more pronounced on the Mexican side. In Ambos Nogales, the major urban area in this study site, international policies designed to encourage industrialization are driving urban expansion on both sides of the border.

Between 1965 and 2000, urban expansion on the Mexican side increased dramatically in comparison to the US side, with makeshift and industrial park development responsible for urban infilling and expansion. On the US side, urban expansion progressed relatively slowly with minimal buildup of industrial park and transport facilities.

At this HERO site, two Landsat TM scenes were merged and calibrated for each date, 1985 and 1999. Land-use maps were derived from these satellite mosaics, and aerial photography was used for increased classification accuracy of the 1999 image.

The total change on the border region landscape was low, with 0.68% of the landscape experiencing change. The net change on the landscape was 0.60%, while only 0.08% of the landscape experienced swap. Loss of rangeland resulted from the growth of Ambos Nogales and the expansion of a small mine.

A cross-site comparison of the HERO sites

The land-use results from all sites convey the degree of dynamics that occurred in each landscape, where the HPO-HERO site had the most total change and swap, and the CM-HERO had the most net change. In contrast, the Sonoran Desert Border Region site had the smallest amount of total change, net change, and swap. The relative amounts of net change and swap (Figure 7.2), can be used as a proxy indicator of what proportion of change at each site is likely to be more

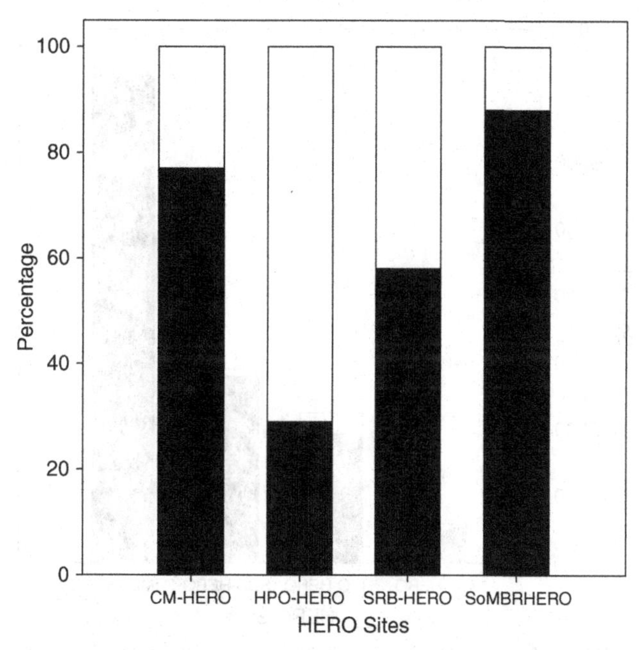

Figure 7.2. The relative amounts of net change and swap may indicate the influence of land-use change on vulnerability, with solid shading denoting net change and the unshaded area denoting swap.

unidirectional and potentially more permanent (net change), or more transitory (swap). It is important to note that while each site has different land-use change dynamics, landscape persistence is the largest component at each site (Figure 7.3).

Land-use changes often are tied to policy and economy, which fact indicates that different date selections may result in very different land uses (Pearson *et al.* 1999; Vasconcelos *et al.* 2002). The time-span between the two land-use/land-cover maps then becomes important for capturing the full suite of possible change. The time span at the CM-HERO site was 29 years, and as a result, some short-term cyclic change may have been missed. Conversely, long-term changes may have been missed in the 15-year span at the Sonoran Desert Border Region HERO site. The time-span was 28 years for SRB-HERO, and 17 years for HPO-HERO. Unlike spatial extent, the temporal scale of analysis can be driven by data availability instead of ideal time-spans. This suggests that the effectiveness of our methods still relies on our understanding of geographical context and the selection of time-frames that are representative of change patterns in our study sites.

Land-use maps are also constrained by the classification system used. For this project, all independently derived maps were collapsed into Anderson Level I land-use/land-cover categories (water, urban/built-up, wetland, barren, agriculture,

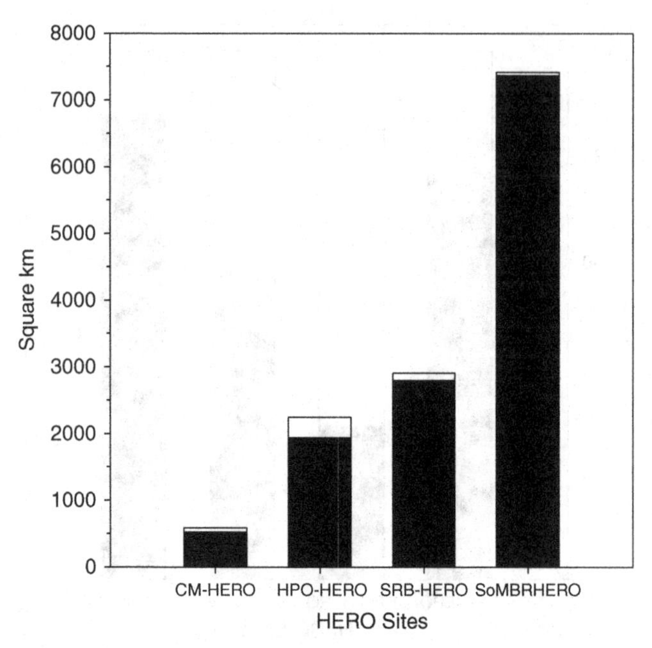

Figure 7.3. The total area of persistence and change in all sites. A majority of the each landscape at all sites remained stable, with solid shading denoting persistence and the unshaded area denoting change.

forest, and rangeland). Because the Anderson Level I classification allows for generalizations, locally significant change may be lost. For instance, the Level I categories could not capture the specific agricultural change on the HPO-HERO landscape. Crop-rotation practices, estimated to occur on 81% of agricultural lands, and the 100% increase in confined animal feeding operations are important, yet undetectable, at the coarsest Anderson level. Classification systems are chosen based on the ability to answer research questions. In this case, where we are measuring broad land-use changes in directionality, exchangeability, and permanence, Anderson Level I captured sufficient detail.

A vast majority of the landscapes at each site did not change during the study periods, regardless of time-span or dates. This stability is tied to policy and ownership patterns combined with the physical geographical constraints of each landscape. Policies that restrict land-use change (e.g., city zoning, farmland protection, wildlife reserve) can be the predominant mechanisms that shape land-use change. At the local level, the driving forces behind changes are more often individual ownership choices, which are difficult to generalize at a regional scale. Our methodological approach provides concepts that, when used *in addition to* knowledge of specific geographical context, provide a mode of useful comparison.

The implication of the types of comparison introduced in this chapter is that perhaps we can begin to understand the intersection of land-use change and vulnerability. With this in mind, the largely one-directional change in CM-HERO implies that vulnerability is increased because large areas of forested land are being converted to urban areas. Urban expansion creates multiple vulnerabilities because of the inequitable distribution of wealth. Wealthier portions of the population may be less sensitive and have higher adaptive capacity, whereas lower-income areas may be very sensitive to climate change and have little adaptive capacity. Therefore, under a climate-change scenario, CM-HERO has more people that will be exposed, and there will be fewer natural areas to mitigate the alterations in the moisture regime.

The land-use change at HPO-HERO was mostly a transitory type of change, where lands enrolled in federal programs can be taken out just as swiftly. Active cropping of land has higher sensitivity to climate change than rangeland enrolled in a government reservation program. However, because as much land was taken out of production as was put in, the increase in sensitivity is balanced by the reduction. Therefore, the high amount of swap at HPO-HERO may be an indicator that the people in HPO-HERO have high adaptive capacity, although their exposure and sensitivity to climate change has not been greatly affected.

SRB-HERO has nearly equal amounts of net change (surface mine reclamation and urban expansion) and transitory change (timber harvest). Although the mine reclamation is a positive change for environmental reasons, it does not particularly affect vulnerability at the regional scale. The increase in urban area is insufficient to have an impact at the regional level. Therefore, SRB-HERO has slightly increased exposure due to the increased urban area, but its sensitivity and adaptive capacity remain the same. The land-use change in SRB-HERO has only slightly increased its overall vulnerability.

As for the Sonoran Desert Border Region HERO site, although there has been some land-use change, the landscape is in a persistent state. While exposure may be increased at specific locations, such as the increased population density of Ambos Nogales, the region as a whole is no more or less vulnerable to climate change as a result of recent land-use change.

Conclusions

Our results and subsequent analysis show that among the four HERO sites, the aggressive expansion of urban areas into forested lands at the central Massachusetts site is making it more vulnerable to climate change than the other sites. The Susquehanna River Basin HERO has the second-highest increase in vulnerability due to a slightly increased urban population. While timber harvesting is a significant land-use change, the rotation of cutting and forest regrowth does not affect the vulnerability of the region because they essentially cancel each other out. The High Plains–Ogallala HERO, while very sensitive to climate change in general, is actually quite highly adaptive in regard to land-use change. While this implies a decreasing vulnerability, there are also definite thresholds of that adaptive capacity. For instance, if groundwater depletion becomes too severe, the vulnerability of the HPO-HERO could change dramatically. The Sonoran Desert Border Region HERO has had such a minor amount of change that the vulnerability is largely stable. This region is already highly dependent upon groundwater and its scant annual rainfall, so climate change that results in severe drought could impact land use drastically, thereby altering vulnerability.

Throughout this analysis, certain commonalities are found across the regions. One commonality is that the same type of change found in all regions tends to result in similar changes in vulnerability. Unidirectional (net) land-use change, such as forest to urban, tends to increase the vulnerability of the region. Cyclic change, or swap, does not appear to affect vulnerability, and, in fact, may indicate an area with high adaptive capacity. Another common link appears to be that the amount of change in a region must reach a threshold before land-use change becomes a significant driver to vulnerability. This idea is tantalizing in its implications, but more research is required to determine its significance.

The efforts of this chapter have been focused on the intersection of land-use change and vulnerability with an emphasis on cross-site comparison. Each location has complex decision-making and driving forces that create region-specific landscape patterns. Events and policies occur in non-linear patterns that may, at any time, affect the trajectory of land-use change. At each site, the complexity of land use, land-use change, and the implications to vulnerability are staggering. By using the techniques outlined in this chapter, the vulnerability of sites can be generalized for the sake of comparison, while retaining important geographic context. Here, we have only begun to attempt to better define how to compare vulnerability across sites.

References

Anderson, J. R., E. E. Hardy, and J. T. Roach, 1972. *A Land-Use Classification System for Use with Remote-Sensor Data*, United States Geological Survey Circular No. 671. Washington, D.C.: U.S. Government Printing Office.

Anderson, J. R., E. E. Hardy, J. T. Roach, and R. E. Witmer, 1976. *A Land Use and Land Cover Classification System for Use with Remote Sensor Data*, United States Geological Survey Professional Paper No. 964. Washington, D.C.: US Government Printing Office.

Bankoff, G., 2003. Constructing vulnerability: the historical, natural, and social generation of flooding in metropolitan Manila. *Disasters* **27**(3): 224–238.

Burton, I., and B. Lim, 2005. Achieving adequate adaptation in agriculture. *Climatic Change* **70**: 191–200.

Civco, D. L., J. D. Hurd, E. H. Wilson, C. L. Arnold, and M. Prisloe, 2002. Quantifying and describing urbanizing landscapes in the Northeast United States. *Photogrammetry Engineering and Remote Sensing* **68**(10): 1083–1090.

DeFries, R., and L. Bounoua, 2004. Consequences of land use change for ecosystem services: a future unlike the past. *GeoJournal* **61**: 345–351.

Edmonds, J. A., and N. J. Rosenberg, 2005. Climate change impacts for the conterminous USA: an integrated assessment summary. *Climatic Change* **69**: 151–162.

Foster, D. R. 1993. Land-use history and forest transformations in Central New England. In *Humans as Components of Ecosystems*, eds. M. J. McDonnell and S. T. A. Pickett, pp. 91–110. New York: Springer.

Goodin, D. G., J. A. Harrington, Jr., and B. C. Rundquist, 2002. Land cover change and associated trends in surface reflectivity and vegetation index in Southwest Kansas: 1972–1992. *Geocarto International* **17**(1): 43–50.

Jianchu, X., J. Fox, J. B. Volger, Z. Peifang, F. Yongshou, Y. Lixin, Q. Jie, and S. Leisz, 2005. Land-use and land-cover change and farmer vulnerability in Xishuangbanna Prefecture in southwestern China. *Environmental Management* **36**(3): 404–413.

Kakembo, V. and K. M. Rowntree, 2003. The relationship between land use and soil erosion in the communal lands near Peddie Town, Eastern Cape, South Africa. *Land Degradation and Development* **14**: 39–49.

Kromm, D. E. and S. E. White (eds.), 1992. *Groundwater Exploitation in the High Plains.* Lawrence, KS: University of Kansas Press.

Lioubimtseva, E., R. Cole, J. M. Adams, and G. Kapustin, 2005. Impacts of climate and land-cover changes in arid lands of Central Asia. *Journal of Arid Environments* **62**: 285–308.

Liverman, D. M., 1990. Drought impacts in Mexico: climate, agriculture, technology, and land tenure in Sonora and Puebla. *Annals of the Association of American Geographers* **80**(1): 49–72.

MacIver, D. C., and E. Wheaton, 2005. Tomorrow's forests: adapting to a changing climate. *Climatic Change* **70**: 273–282.

Motha, R. P., and W. Baier, 2005. Impacts of present and future climate change and climate variability on agriculture in the temperate regions: North America. *Climatic Change* **70**: 137–164.

Mustafa, D., 1998. Structural causes of vulnerability to flood hazard in Pakistan. *Economic Geography* **74**(3): 289–305.

Nearing, M. A., V. Jetten, C. Baffaut, O. Cerdan, A. Couturier, M. Hernandez, Y. Le Bissonnais, M. H. Nichols, J. P. Nunes, C. S. Renschler, V. Souchère, and K. van Oost, 2005. Modeling response of soil erosion and runoff to changes in precipitation and cover. *Catena* **61**: 131–154.

Neke, K. S., and M. A. Du Plessis, 2004. The threat of transformation: quantifying the vulnerability of grasslands in South Africa. *Conservation Biology* **18**(2): 466–477.

Pearson, S. M., M. G. Turner, and J. B. Drake, 1999. Landscape change and habitat availability in the Southern Appalachian Highlands and the Olympic Peninsula. *Ecological Applications* **9**: 1288–1304.

Polsky, C. and W. E. Easterling III, 2001. Adaptation to climate variability and change in the US Great Plains: a multi-scale analysis of Ricardian climate sensitivities. *Agriculture, Ecosystems and Environment* **85**: 133–144.

Pontius, R. G., Jr., and N. Malizia, 2004. Effect of category aggregation on map comparison. In *Lecture Notes in Computer Science 3234*, eds. M. J. Egenhofer, C. Freksa, and H. J. Miller, pp. 251–268. Berlin: Springer.

Pontius, R. G., Jr., E. Shusas, and M. McEachern, 2004. Detecting important categorical land changes while accounting for persistence. *Agriculture, Ecosystems and Environment* **101**(2/3): 251–268.
Salinger, M. J., M. V. K. Sivakumar, and R. Motha, 2005. Reducing vulnerability of agriculture and forestry to climate variability and change: workshop summary and recommendations. *Climatic Change* **70**: 341–362.
Schloss, J. A., R. A. Buddemeier, and B. B. Wilson, 2000. *An Atlas of the Kansas High Plains Aquifer*, Kansas Geological Survey Educational Series 14. Accessed at www.kgs.ku.edu/HighPlains/atlas/.
Sivakumar, M. V. K., H. P. Das, and O. Brunini, 2005. Impacts of present and future climate variability and change on agriculture and forestry in the arid and semi-arid tropics. *Climatic Change* **70**: 31–72.
Soini, E., 2005. Land use change patterns and livelihood dynamics on the slopes of Mt Kilimanjaro, Tanzania. *Agricultural Systems* **85**: 306–323.
Thomson, A. M., R. A. Brown, N. J. Rosenberg, R. C. Izaurralde, and V. Benson, 2005a. Climate change impacts for the conterminous USA: an integrated assessment. 3: Dryland production of grain and forage crops. *Climatic Change* **69**: 43–65.
Thomson, A. M., N. J. Rosenberg, R. C. Izaurralde, and R. A. Brown, 2005b. Climate change impacts for the conterminous USA: an integrated assessment. 5: Irrigated agriculture and national grain crop production. *Climatic Change* **69**: 89–105.
Vasconcelos, M. J. P., J. C. Musa Biai, A. Araujo, and M. A. Diniz, 2002. Land cover change in two protected areas of Guinea-Bissau (1956–1998), *Applied Geography* **22** 139–156.
Walker, S., 2005. Role of education and training in agricultural meteorology to reduce vulnerability to climate variability. *Climatic Change* **70**: 311–318.
Warrick, R. A., 1980. Drought in the Great Plains: a case study of research on climate and society in the USA. In *Climatic Constraints and Human Activities*, eds. J. Ausubel and A. K. Biswas, pp. 93–123. New York: Pergamon Press.
Watson, R. T., 2005. Turning science into policy: challenges and experiences from the science–policy interface. *Philosophical Transactions of the Royal Society of London B* **360**: 471–477.

Part IV

8

Assessing local vulnerabilities: methodological approaches and regional contexts

COLIN POLSKY, CYNTHIA SORRENSEN, JESSICA WHITEHEAD, AND ROB NEFF

Introduction

The global change literature has experienced a significant, growing interest in vulnerability since the early 1990s. Although a host of authors have provided overarching conceptual discussions of vulnerability,[1] they have largely ignored methodological issues. Consequently, several papers have recently addressed the topic of operationalizing vulnerability.[2] Despite this growing number of vulnerability studies privileging methodology, we are unaware of published papers tackling the challenge of how to conduct a coordinated vulnerability assessment in multiple places. The HERO project, with this chapter, responds to that gap in the literature.

By addressing the topic of replicable protocols, the HERO project seeks to advance the science of vulnerability. Coupled human–environment systems are dynamic, which means that the vulnerability estimated at one point in time or space may not be a faithful predictor of vulnerability at a later point in time or another point in space. Therefore, as argued elsewhere in similar contexts (e.g., Chapter 5 of this volume; Redman *et al.* 2004; Gragson and Grove 2006; Haberl *et al.* 2006; Polsky *et al.* 2007), replicable protocols must be consistently applied to vulnerability assessments over time and space to develop a database sufficiently sensitive to distinguish trends from anomalies. Thus, HERO is aiming to contribute to a larger discussion on scientific infrastructure development (see Chapters 1, 3, and 5).

In the remainder of this chapter, we first elaborate on methodological issues associated with conducting vulnerability assessments in the context of a multi-site network. We then present an overview of the regional contexts of the vulnerability assessment research presented in Chapters 9 and 10.

Rapid Vulnerability Assessment

Two overarching methodological approaches are helpful for vulnerability studies: the "Eight Steps" described by Schröter *et al.* (2005) and the grounded vulnerability

Sustainable Communities on a Sustainable Planet: The Human–Environment Regional Observatory Project, eds. Brent Yarnal, Colin Polsky, and James O'Brien. Published by Cambridge University Press.

framework, both of which are described in Chapter 5. These two complementary approaches guide the researcher to a comprehensive assessment of the effects of and responses to multiple, interacting stresses in multi-scaled, coupled human–environment systems. Yet, such a comprehensive approach requires significant time, personnel, and financial investments, so the overarching methodological approaches described in Chapter 5 may be viewed as ideal types that researchers should aim to achieve. In practice, however, resource constraints will limit how much time may be devoted to a vulnerability assessment (Polsky *et al.* 2007). Thus, a natural question is: how can practitioners conduct a coordinated, multi-site vulnerability assessment when resources are limited?

To answer this question, HERO presents in this chapter a Rapid Vulnerability Assessment approach. Rapid Vulnerability Assessment draws heavily from the concept of Rapid Rural Appraisal (Chambers 1983). Rapid Rural Appraisal and its subsequent elaborations, e.g., Participatory Rural Appraisal (Chambers 1993) and Rapid Ecological Assessment (Sayre *et al.* 2000), utilize insights from the cultural anthropology of the early to mid twentieth century, emphasizing contextual local knowledge and achieving broad understanding rather than the potentially misleading statistical estimations of large-scale surveys (IISD 2008). Rapid Rural Appraisal and offshoots use multidisciplinary teams to visit localities and observe and participate in activities, study secondary sources, interview individuals and groups, compile maps, assemble local histories and case studies, develop timelines, and administer short questionnaires (IISD 2008).

Like Rapid Rural Appraisal, Rapid Vulnerability Assessment seeks to use the researchers' time efficiently. The HERO Rapid Vulnerability Assessment is distinguished by an almost exclusive reliance on secondary data rather than on primary data, by its interest in fostering cross-site data collection and analysis, and by its focus on the three vulnerability dimensions (exposure, sensitivity, and adaptive capacity).

The Rapid Vulnerability Assessment approach

We developed a nine-step approach to rapid cross-site assessment of vulnerability (Table 8.1), and, for purposes of illustration, applied it to the single vulnerability dimension of adaptive capacity in the four study areas. These steps are grouped into three stages. In the *planning stage*, we defined the scope of our assessment and developed our working framework for adaptive capacity. During the *data acquisition* stage, we determined the data available for each site and arranged it into standard data forms for each county of each site. For this exercise and based on the available data, we selected four counties to analyze (one from each HERO study site – Worcester County, Massachusetts; Centre County, Pennsylvania; Ness

Table 8.1 *Nine-step process for Rapid Vulnerability Assessment, grouped by three rapid assessment stages*

Planning
Select sectors to investigate in each site
Conduct literature search
Develop framework for adaptive capacity assesment
Data acquisition
Generate list of possible indicators
Discuss and prioritize indicators
Gather data
Analysis
Evaluate individual site matrices
Cross-evaluate all site matrices
Draw conclusions for each site based on matrix sets

County, Kansas; and Santa Cruz County, Arizona) using data matrices. In the *analysis* stage, we had representatives from each site rank the indicators for their own county and for two other counties, which resulted in a set of three ranked indicator matrices for each county. We used these matrix sets to draw conclusions about adaptive capacity across the four sites.

Planning stage

The determinants and characteristics of adaptive capacity may be place-, sector-, and hazard-specific. We therefore narrowed our inquiry to the agricultural and domestic water supply sectors for select counties in Massachusetts, Pennsylvania, Kansas, and Arizona, where in each case the hazard of interest was hydroclimatic variability (i.e., droughts and floods). To test the ability of this approach to produce useful output with minimal expense, we used only free or low-cost secondary data. We also excluded maps as a data source because not all researchers attempting a Rapid Vulnerability Assessment would have experience with or access to digital (GIS) mapping or the time to learn how to map. We built our framework of adaptive capacity assessment on six determinants that are commonly accepted as important in the literature (Klein and MacIver, 1999; Smit *et al.* 1999, 2000, 2001; Yohe and Tol 2002; Adger *et al.* 2004) and that are conducive to Rapid Vulnerability Assessment:

- Social capital, including property rights, communities, networks, and bonding
- Human capital, including education and skills
- Access to and range of available technology
- Economics
- Natural resources
- Institutional factors.

A seventh common determinant encompassing perceptions of vulnerability and adaptation also appears in the literature. We anticipated that secondary data on perceptions would be difficult (if not impossible) to obtain and consequently left it out of the assessment.

Data acquisition

With the framework in place, we created a bank of possible quantitative and qualitative indicators for each determinant. Through a series of cross-site conversations, we compiled a common list of indicators to use for the cross-site analysis, but also allowed each site to include site-specific indicators. This approach produced a data set that exhibits some cross-site comparability while retaining some place-based specificity. We used videoconferencing (Chapter 3) to evaluate collectively each potential indicator based on its relationship to adaptive capacity, its data availability, and its potential to be ranked on a subjective, qualitative scale. Some suggested indicators, such as land-cover change, were rejected because they have more impact on exposure and sensitivity than on adaptive capacity. Moreover, such indicators typically require the sophisticated, time-consuming analysis of digital maps, which, as described above, is an activity we rejected in an effort to emulate the research environment of practitioners working under severe resource constraints. Other possible indicators, such as the degree of cooperation between sector-related community organizations, could have been useful but were clearly beyond the scope of Rapid Vulnerability Assessment. The remaining indicators were evaluated based on whether we would be able to rank their relative contributions to agricultural and water supply adaptive capacity.

The next step involved gathering data for each site. To accelerate the process, we organized the indicators by priority, placing more time and resources into the data search for higher-priority indicators. Lower-priority indicator data were pursued when higher-priority data were unavailable or as time permitted. We gathered county-level data because the county is a common scale across many datasets, e.g., the 2000 Decennial Census, the 1997 Economic Census (United States Census Bureau 1997), and the 2002 Agricultural Census (United States Department of Agriculture, National Agriculture Statistics Service 2002). Some of these datasets also present data at finer spatial scales. In some cases, one dataset provides information on a variable of interest to the project – but for only a subset of the four study sites. For example, the American Community Survey includes information on language spoken at home, but these data are generally only collected for places with populations larger than some of our study areas (United States Census Bureau 2004). We also searched for more locally available data, but data availability often depends upon state and county resources and priorities. Some states, e.g., Pennsylvania, provide information on public water system capacities (PA DEP

2005), but similar data are unavailable in Arizona. Other states, e.g., Massachusetts, maintain a data warehouse for various social, economic, and environmental variables on the places within their state boundaries (MASSGIS 2008), but those measures are not necessarily repeated consistently over time, or replicated in other states.

Analysis

After we gathered data for the four counties representing the HERO sites, we organized it into a common matrix. Matrices were divided into indicator lists for each determinant, and two judgments were made for each indicator: one for droughts and one for floods. This methodology was based on the livelihood sensitivity matrix of Downing *et al.* (2003). Instead of using a quantitative five-point scale to rank each indicator, we used each indicator's value or description to make qualitative judgments on its contribution to adaptive capacity for the site (positive, negative, or neutral).

There were three major difficulties in ranking the indicators. First, some indicators are "double-edged" – depending on the context, such indicators could represent either a positive or a negative contribution to adaptive capacity (see Polsky *et al.* 2007). For example, a large population growth rate could stress local water resources in a time of drought, representing a negative contribution to adaptive capacity. However, that large population growth may bring an influx of new resources (tax dollars, water system customers, new local skills, etc.), which may add to the adaptive capacity of the area through the development of new infrastructure. Second, and related to the first difficulty, we found that the values of some indicators are difficult to interpret. Returning to the population growth example, is there a threshold beyond which a positive contribution becomes a negative contribution to adaptive capacity? Finally, we were concerned that familiarity with the site through other HERO research would skew each site's analysis of its county matrix.

To address these issues, we added three descriptive columns to the data analysis matrix for each ranking to justify our reasoning. In the first column, the researcher specified whether he or she used some context-specific knowledge to judge that indicator. The second column included a list of any other indicators whose values the researcher referred to in order to make the judgment; this input allowed us to examine some of the double-edged indicators and their connections. The final column asked the researcher to give a brief justification of his or her indicator ranking.

The final step in our Rapid Vulnerability Assessment methodology involved cross-checking each site's matrices to address potential bias from site familiarity. We compiled a set of three matrices for each county. One was the matrix outlined above compiled by a representative of a given county. Then we attempted to have representatives from two other sites also generate a matrix for that county. We compared the results in each matrix set to write a consensus assessment of the adaptive capacity for each site, carefully outlining any uncertainties or inconsistencies.

Concerns raised by Rapid Vulnerability Assessment

Rapid Vulnerability Assessment has several benefits, but its results must be used with caution. Depth of analysis is sacrificed for a quick, relatively cheap and simple survey. Indicators are difficult to interpret at times, and clarifying information is often unavailable through readily obtainable sources. Moreover, adaptive capacity is a complex concept, encompassing several factors that often cannot be assessed rapidly from the comfort of an office. For example, social networks are too complex to understand thoroughly using a Rapid Vulnerability Assessment approach. A farmer or community water system with few economic resources may have tremendous connections to other agents who can provide assistance when it is needed, often at little or no cost; such connections are not apparent in secondary data, yet they may constitute a significant positive contribution to adaptive capacity. Similarly, we could accumulate information on regulations affecting agriculture and water supply, but it would be too difficult – from a "rapid" perspective – to gather and analyze information on the scope of these regulations and the degree to which the regulations are enforced. In addition, selecting a geographically broad unit of analysis like a county can obscure variations in adaptive capacity at the level of municipalities or individuals. In short, Rapid Vulnerability Assessment is not a substitute for either the eight-step approach by Schröter *et al.* (2005) or the grounded vulnerability assessment approach presented in Chapter 5. When such in-depth approaches are not feasible – as is commonly the case – there are still some general conclusions about adaptive capacity that we can draw from Rapid Vulnerability Assessment.

Vulnerability Assessment Evaluation

Once researchers conducting a Rapid Vulnerability Assessment generate findings, they should ask themselves if the results are credible. There is a rich tradition in social science on this topic, discussed under the headings of *validity* and *reliability* (e.g., King *et al.* 1994; Creswell 2002; Singleton and Straits 2005; see also Oreskes *et al.* 1994; Rastetter 1996). Interestingly, post-hoc studies of vulnerability assessment results are relatively uncommon in the literature on human–environment interactions. This gap in the literature is understandable insofar as researchers may be unable to validate their own research in the classical scientific sense of the word *validate*; the lines between the subject and the object are often blurry in social scientific work (McDowell 1992; England 1994; Barnes and Gregory 1997. Yet such a challenge does not mean that researchers should abandon attempts at criticizing their results, even if such an attempt does not qualify as validation in the classic sense. At a minimum, vulnerability researchers should strive to determine whether they asked the right research questions, whether they defined and conceptualized the problems and concepts appropriately, and whether their research design and methods were sufficiently rigorous and sensitive.

Our reading of the vulnerability and allied literatures suggests that such evaluations are executed extremely rarely. We suspect that this final step is often ignored because the project's funding finishes, the researchers submit their last publications, and new projects are already in development or under way.[3] As a result, evaluative comments or criticisms of vulnerability assessments typically come from reviewer comments on papers submitted for scholarly publication rather than from the careful application of a priori defined (and transparently presented) criteria for quality.

Despite the philosophical and practical challenges associated with validating social science research, there is a real need to articulate the legitimacy of the results. To this end, the HERO team has produced a slightly different approach to validation – what we term *Vulnerability Assessment Evaluation* – for critiquing our results. Even if the product of this Vulnerability Assessment Evaluation cannot be defended on objective scientific grounds, conducting such an evaluation will help the project's results gain recognition outside of the researcher's niche – including the stakeholders – by being explicit about research design, motives, methods, and assumptions (Baxter and Eyles 1997). Moreover, the greater the confidence in research results, the greater the likelihood of producing policy recommendations for reducing vulnerabilities that achieve their stated goals (Kelly and Adger 2000; Wilbanks *et al.* 2003; Schröter *et al.* 2005). This action-oriented goal is a fundamental part of the global change vulnerability assessment paradigm.

An approach to Vulnerability Assessment Evaluation

The evaluative strategy we propose draws on the rich discussion of rigor in qualitative research (see Lincoln and Guba 1985; Strauss and Corbin 1998) by evaluating research results against two criteria. The first criterion, *saturation*, considers whether we obtained enough data to answer our research questions. This criterion therefore roughly corresponds to what social scientists term "internal validity," a measure of the extent to which the results are internally consistent. During the research process, saturation concerns the level of redundancy in data collected and in research findings. Saturation is attained when investigators no longer find new information with each additional data-gathering exercise

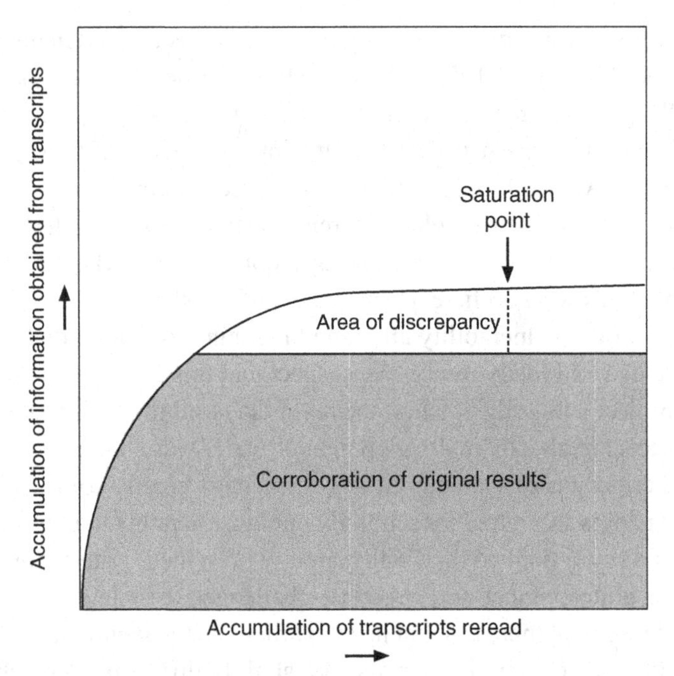

Figure 8.1. A hypothetical saturation curve.

(e.g., interview, focus group) conducted. Increasing redundancy in the emerging dataset indicates that the researcher has likely reached a point of diminishing returns; further data collection is not likely to produce additional insight to the questions being asked.

Applying this criterion requires investigators to engage in a reflexive analysis of the research. For our case study (see Chapter 10), we evaluated our work against this criterion by rereading all interview transcripts. With each transcript, we tallied all insights that corroborated our original study results and noted any additional pertinent findings that had been overlooked in the original analysis. If we continued to find new information and results from the rereading of the transcripts, we would conclude that our original effort was insufficient; if we came to a point at which no new information came from the transcripts, then we could stop the process and feel comfortable about our original conclusions.

To visualize how this process works in general terms, Figure 8.1 plots the cumulative number of findings obtained by rereading the transcripts for a hypothetical "saturation evaluation." The x-axis represents each transcript and the y-axis represents the cumulative number of findings. When the curve levels off, it indicates the saturation point beyond which there are few or no additional findings with each additional interview. If the leveling point is above the area of original findings, then there is an area of discrepancy, which reflects either misinterpretation during the original extraction of data from the transcripts or an area of information missed during data extraction. In either case, the saturation curve helps provide evidence on the validity of that research.

The second criterion, *credibility*, evaluates the agreement with the findings by a sample of the original interview participants, which in turn speaks to the generalizability of findings across sites. This criterion therefore roughly corresponds to what social scientists term "external validity," a measure of the extent to which the results correspond to the concept we purport to be measuring. In our case study, we worked to understand the credibility of our findings through a second interview process in which we returned to a sampling of stakeholders who were a part of the original research, presented our original vulnerability assessment results to them, and solicited their reactions to the findings. At each of the four HERO study sites, we presented stakeholders a list of statements reflecting the findings, so these statements could be site-specific or could pertain to two, three, or four sites. We sought stakeholder agreement or disagreement with simplified summary statements written in plain language. By considering their responses to these statements, we could identify specific findings not supported by this second sample of stakeholders, evaluate the overall credibility of the original work, and appraise which results were more general across the sites and which ones were site-specific.

Regional contexts

The four HEROs vary greatly in their physical and social characteristics, and these differences hypothetically influence the ways that drought-related exposure, sensitivity, and adaptive capacity manifest at each site. If these common methodologies of the Rapid Vulnerability Assessment and Vulnerability Assessment Evaluation are well designed, then the influences of site-specific characteristics will be present and recognizable in the results, but generalizations (if any) across sites will also be able to emerge from the analysis.

The location and selected characteristics of these places are presented in Figure 8.2. In this study, the researchers at the central Massachusetts HERO restrict the work to Worcester County. Similarly, investigators at the central Pennsylvania HERO limit this analysis to Centre County. The High Plains–Ogallala HERO consists of 19 counties in southwestern Kansas, and this work uses the entire study region. Finally, the Sonoran Desert Border Region HERO is unique in that it includes portions of both the United States and Mexico. Where possible, this analysis applies to the entire study area, although in most cases data availability limits the work to the United States side of the border. The following subsections introduce each site in terms of its physiographic environment, population and development, and socioeconomic activities.

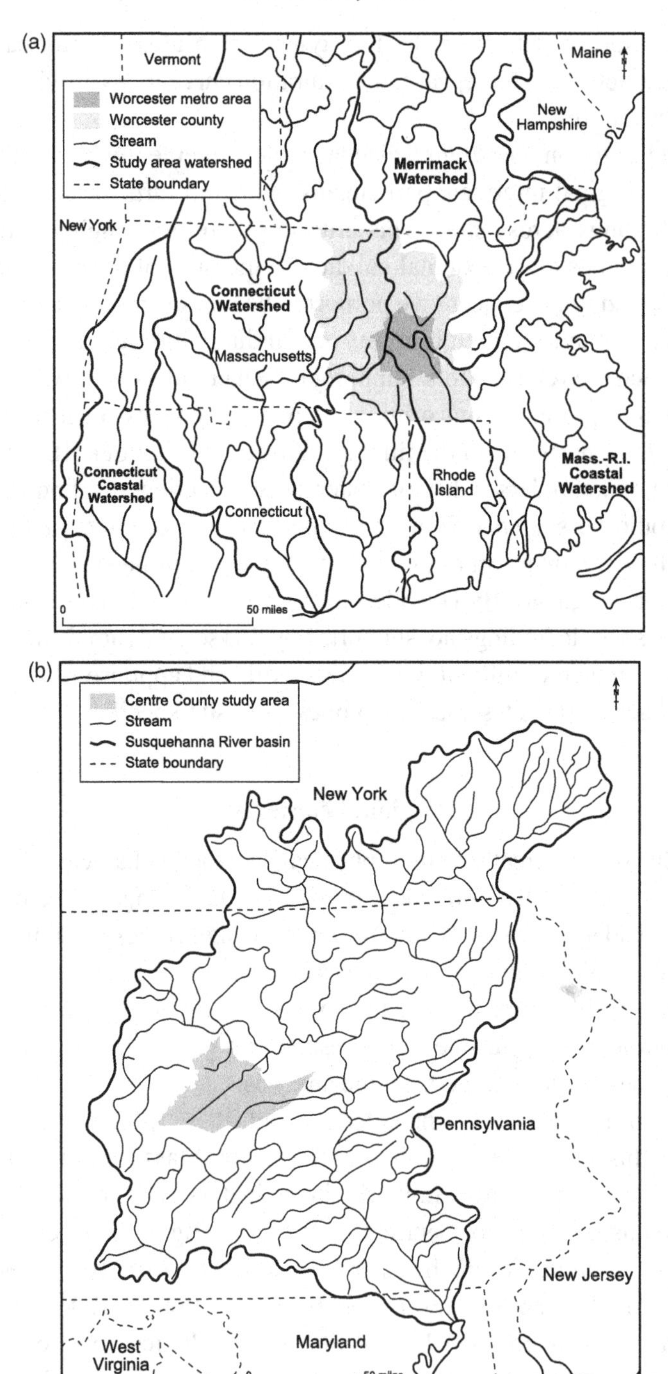

Figure 8.2. Location and features of the four study areas: (a) Central Massachusetts HERO (HERO-CM), (b) Susquehanna River Basin HERO (SRB-HERO), (c) High Plains – Ogallala HERO (HPO-HERO), and (d) Sonoran Desert Border Region HERO.

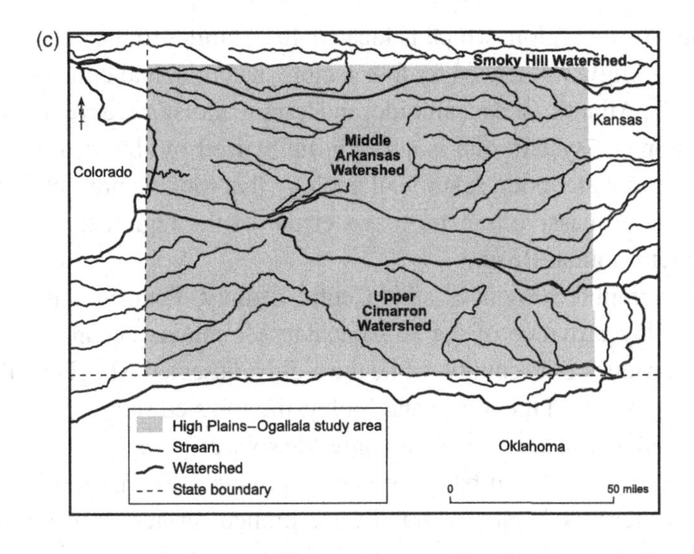

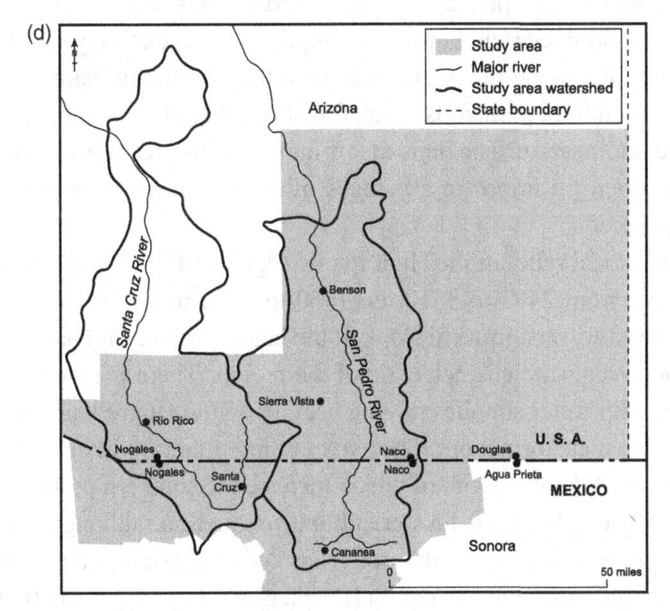

Figure 8.2. (cont.)

Physiographic environments

Interspersed with large patches of woodlands and remnants of an agricultural economy that has been absent for at least one century, small- to medium-sized towns and cities and densely settled peri-urban communities mark the landscape of Worcester County, Massachusetts. The rolling hills of the Worcester Plateau organize the land into four watersheds, which are dotted with a multitude of local ponds and lakes and are superimposed by several large reservoirs that serve the metropolitan Boston region, which is located 40–50 miles to the east. The regional hydrology is strongly influenced by four factors: a continental climate modified by proximity to the Atlantic Ocean, altitude, and terrain; moisture recycling through the land–vegetation–air system that is strongly influenced by the local abundance of wooded land and water bodies; surficial geology that took its present face from the glacial age; and a massive transfer of water out of the region to serve the water demand of metropolitan Boston.

Two physiographic provinces split Centre County, Pennsylvania in half: the Ridge and Valley province of the Appalachian Mountains occupies the southern and eastern portions of the county, while the highly dissected Allegheny Plateau lies in the north and west. In the Ridge and Valley, the forested ridges consist primarily of hard sandstones, which produce thin, infertile soils and rapid runoff to the valleys below. The agricultural and urbanizing valleys principally consist of limestone and have thick, fertile soils, karst topography, and limited surface drainage. Associated with the karst, well-developed cavern and underground stream systems underlie the valley floors and produce rich, but shallow aquifers with relatively rapid response to precipitation inputs. In the rugged landscape of the Allegheny Plateau, steep slopes are associated with thin, poor soils, thick forest cover, and an abundance of streams. Large surface and near-surface deposits of highly sulfurous bituminous coal result in large-scale open pit mines in all stages of development, from active, to abandoned, to reclaimed.

There is little local relief in the High Plains–Ogallala HERO study area, although elevation ranges from 2493 to 3510 feet (760 to 1070 m) above sea level with the imperceptible east-to-west upward slope of the land. Variable climate conditions and severe weather events are characteristic of the region. There is relatively little local surface water, with water supplies relying heavily on the High Plains aquifer system (including the Ogallala formation). The two most important rivers in the region are the Cimarron and the Arkansas, both of which traverse the southern portion of the study area. Soils are generally good for agricultural production, although some areas of sandier soils are generally unsuitable for crops. Wind erosion potential is of greater concern than water erosion in this region (Leathers and Harrington 2000). Oil and gas fields underlie the area, with important but declining production.

The Sonoran Desert Border Region HERO comprises two transborder watersheds traversing the Arizona, United States – Sonora, Mexico boundary. Three counties in southeastern Arizona and four municipios in Sonora, Mexico encompass the majority of the watersheds. The region is part of the Sonoran Desert and contains a diverse mix of environments ranging from fragile upland deserts, to open grasslands and oak woodlands, to spruce–fir forests at elevations reaching 10 000 feet (3000 m) above sea level. The physiography is characterized by broad alluvium-filled valleys bounded by steep mountain ranges.

Population and development

Worcester County is located in the center of Massachusetts, covering the full north–south extent of the state from the southern border of New Hampshire to the northern borders of Connecticut and Rhode Island. The area has hosted European settlement since the mid seventeenth century. The human population is currently approximately 750 000 persons, of which close to one-third live in the metropolitan area of the City of Worcester, located in the center of the county. The current population, while not presently increasing significantly, has more than doubled since the start of the twentieth century (Figure 8.3a). Two main factors drove the population dynamics of the past two centuries. The first was the vitality of the local manufacturing base, which was principally textiles but also included various products important for the Industrial Revolution. The second was transportation access to Boston to the east (via train, with only a small fraction today of the daily trips that were available decades ago) and Providence, Rhode Island to the south (via the Blackstone Canal, which temporarily

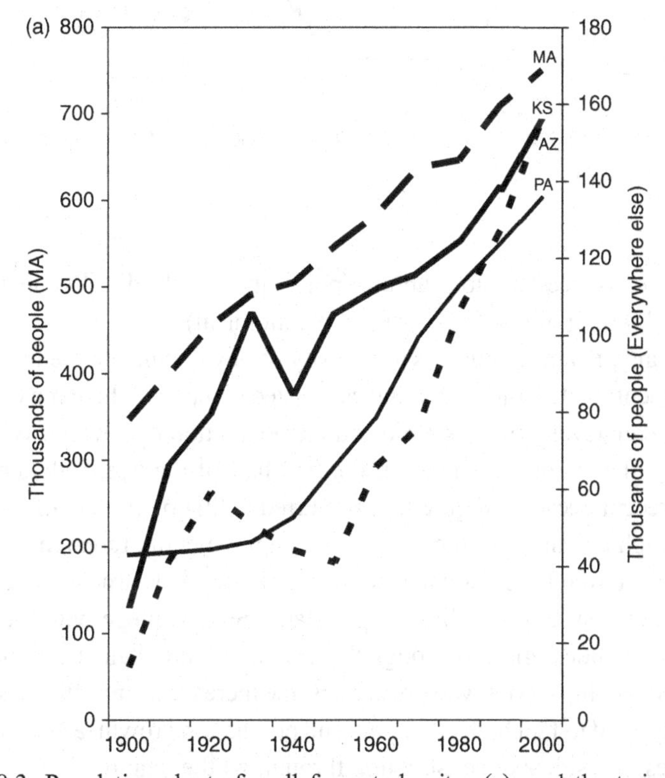

Figure 8.3. Population charts for all four study sites (a), and the twin cities of Nogales (b) in the Arizona–Mexico study area. (Sources: United States Census Bureau 2004; Instituto Nacional Estadística Geografia y Informática (INEGI) 2000.)

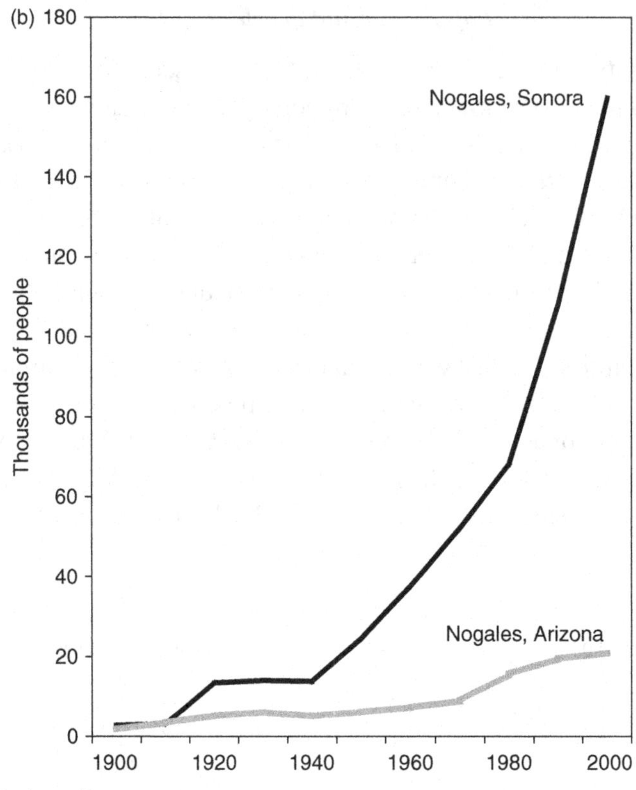

Figure 8.3. (cont.)

turned the city of Worcester into a vibrant "port" city until the development of the train rendered the shipping of goods via canal uneconomical).

Centre County, Pennsylvania has a diversity of development patterns. Around the largest settlement, State College and the associated campus of the Pennsylvania State University, urban development is rapid and the population is growing. While population growth in the county was slow in the first half of the twentieth century, it has increased in recent decades (Figure 8.3a), spurred at first by growth of the university and later through a combination of increased employment in research in and around the university and of other regional economic activities. Construction of spurs on the interstate highway system will link long-isolated State College with the rest of the state and mid-Atlantic region. Although this construction is stalled pending remediation of unintended impacts on water resources, the increased connectivity to the rest of the state is expected to fuel both economic and population growth, as well as to enable increased rates of residential construction throughout the region.

The High Plains–Ogallala region is predominantly rural and agricultural, but also contains three regional population centers: Garden City, Dodge City, and Liberal.

Population densities in the 19 counties are low, but vary from 0.9 km^{-2} to 11.3 km^{-2}. Population increased rapidly from 1900 to 1930, dropped because of out-migration resulting from the Dust Bowl drought of the 1930s, and increased at a lower rate after that event (Figure 8.3a). Recent (1990–2000) population change has been highly variable and has ranged from −14.4% to 22.5%, with seven counties showing declines. During this period, major growth occurred in the three counties containing the three urban centers. Immigrants in the latter part of the twentieth century included both Hispanics and Southeast Asians. Hispanic-identified persons now make up one-third of the study area population; those identified as Asian make up about 2% of the total.

The Sonoran Desert Border Region HERO's main urban area is the border-city complex of Nogales, Sonora and Nogales, Arizona, henceforth referred to Ambos Nogales. In the 1960s, border industrialization programs were initiated in Ambos Nogales and have persisted, fostering very high urbanization rates. Industrialization increased hazardous waste inputs to the watershed, while demand for drinking water soared in tandem with growth. Groundwater resources on the Sonora side, where most population growth has taken place (Figure 8.3b), reside within relatively shallow aquifers that are dependent on surface water recharge and climatic conditions. Water infrastructure is limited, and a significant proportion of households rely on water trucked into their neighborhoods. Aquifers are much deeper on the Arizona side and less responsive to climate variation. During periods of drought, it is common for the city of Nogales, Sonora to send water trucks to the Arizona side to obtain water for its residents.

Socioeconomic activities

Worcester County, Massachusetts represents a transitional socioeconomic landscape in at least two important ways: a transition between three major economic types (the extractive, industrial, and post-industrial) of the past three centuries; and a land-use/land-cover transition manifest as an urban-to-rural gradient under continuing pressure of suburbanization. Currently, agriculture accounts for less than 1% of the total regional employment, whereas the sectors of manufacturing, transportation and warehousing, and educational, health, and social services collectively account for over half of the total employment.

Two types of small-scale agriculture typify the Ridge and Valley physiographic province in the eastern and southern portions of Centre County, Pennsylvania: English-speakers manage modern industrial farms and dairies employing high chemical and energy inputs, while German-speaking Anabaptist (i.e., Amish and Mennonite) populations run pre-industrial farms using few chemical inputs and relying on animal power. Around State College, in the center of the county, the economy is healthy and unemployment is low because of continuous support of and spin-offs from the university. In the Allegheny Plateau to the north and west of State College, large surface and near-surface deposits of highly sulfurous bituminous coal result in large-scale open pit mines in all stages of development, from active, to abandoned, to reclaimed. Acid drainage from the mines severely degrades thousands of miles of streams in the Allegheny Plateau and dozens of miles in Centre County. Communities are scattered and isolated by the topography and the location of mines. Agriculture is practically absent in this part of the county and, because of significant decreases in demand for high-sulfur coal, the economy is in severe decline with high unemployment.

The basis of the economy of southwestern Kansas is agriculture and agricultural support services. The top five counties in agricultural sales for Kansas are in the High Plains–Ogallala HERO study area; they accounted for over $9.2 billion in 1997 (USDA ERS, 2006). With a subhumid climate, most crop production depends on irrigation. Although agriculture represents a regional sensitivity to drought, as demonstrated during the 1930s Dust Bowl event, the current reliance on groundwater extraction mitigates this sensitivity, at least over the short term. Over the longer term, however, reliance on "fossil" water (i.e., deep groundwater with very low recharge rates) is likely to make the region more vulnerable to drought. Not all parts of the study area overlie the High Plains aquifer system; although they may have insufficient groundwater to support farming, even these locations may be dependent on groundwater drinking supplies.

Ranching, irrigated agriculture, and mining are the main economic activities outside the urban areas of the Sonoran Desert Border Region HERO study area. Much of this rural population has no municipal water hook-ups and relies on personal wells for drinking water. On the Sonora side, rangeland degradation and new privatization laws are shifting land-use patterns towards more consolidated ranching operations. Development and rehabilitation of water diversion structures are also on the rise for new fodder and horticultural operations. Approximately 59% of rangeland on the Arizona side is under public ownership (by the Bureau of Land Management, Forest Service, and State of Arizona) and is less degraded. United States ranchers rely on well water and dirt catchments of North American monsoon precipitation to maintain drinking holes for livestock. Although the State of Arizona has managed and restricted the extent of irrigation since 1980, it remains the number one user of water resources.

Recapitulation

This section provided brief biophysical and socioeconomic snapshots of the four HERO study areas. In addition, Chapter 6 presented an overview of the environmental histories of these regions. Taken together, these contexts are crucial for understanding the differential vulnerabilities to the effects of environmental change in these places, which we describe and analyze in Chapters 9 and 10.

Summary

There is a relative dearth of information on methodologies for conducting global change vulnerability assessments, especially for studies involving multi-site networks (Chapter 5; Polsky *et al.* 2007). This chapter attempts to fill this gap by outlining some important methodological concerns for multi-site vulnerability assessments stemming from two idealized, comprehensive methodological frameworks (i.e., the "Eight Steps" and Grounded Vulnerability Assessment approaches) described in Chapter 5. These concerns were translated into two practical approaches for conducting cross-site vulnerability research: Rapid Vulnerability Assessment and Vulnerability Assessment Evaluation. The Rapid Vulnerability Assessment approach is designed for those projects where time and money are in short supply, and the Vulnerability Assessment Evaluation framework is aimed at those projects (not necessarily limited to projects with limited resources) that have produced research findings but have yet to examine how well those results reflect the world the results are intended to represent.

The following two chapters take the next step in this discussion of methods. Chapter 9 presents results from a Rapid Vulnerability Assessment associated with hydroclimatic and socioeconomic variations in the four HERO study sites. Chapter 10 offers a Vulnerability Assessment Evaluation of interview research at those same four sites.

Notes

1. Among many others, Kates (1985), Dow (1992), Böhle *et al.* (1994), Cutter (1996), NRC (1999), Kelly and Adger (2000), Liverman (2001), Turner *et al.* (2003), Adger (2006), Eakin and Luers (2006), Füssel and Klein (2006), Füssel (2007), and Parry *et al.* (2007).
2. See Clark *et al.* (1998), Moss *et al.* (2001), Wu *et al.* (2002), Downing *et al.* (2003), O'Brien *et al.* (2004), Polsky (2004), Schröter *et al.* (2005), Polsky *et al.* (2007), and Eakin and Bojorquez-Tapia (2008); and see Chapter 5 for a more detailed discussion on vulnerability studies and associated methodologies.
3. For a possible exception, see Baxter and Eyles (1997).

References

Adger, W. N., 2006. Vulnerability. *Global Environmental Change* **16**: 268–281.

Adger, W. N., N. Brooks, G. Bentham, M. Agnew, and S. Eriksen, 2004. *New Indicators of Vulnerability and Adaptive Capacity*, Tyndall Centre Technical Report No. 7. Norwich, UK: Tyndall Centre for Climate Change Research.

Barnes, T., and D. Gregory, 1997. Worlding geography: geography as situated knowledge. In *Reading Human Geography: Poetics and Politics of Inquiry*, eds. T. Barnes and D. Gregory, pp. 14–26. London: Arnold Publishers.

Baxter, J., and J. Eyles, 1997. Evaluating qualitative research in social geography: establishing "rigour" in interview anaysis. *Transactions of the Institute of British Geographers* **22**: 505–525.

Böhle, H.-G., T. E. Downing, and M. Watts, 1994. Climate change and social vulnerability: the sociology and geography of food insecurity. *Global Environmental Change* **4**: 37–48.

Chambers, R., 1983. *Rural Development: Putting the Last First.* New York: Longman.

Chambers, R., 1993. *Challenging the Professions: Frontiers for Rural Development.* London: ITDG.

Clark, G. E., S. C. Moser, S. J. Ratick, K. Dow, W. B. Meyer, S. Emani, W. Jin, J. X. Kasperson, R. E. Kasperson, and H. E. Schwarz, 1998. Assessing the vulnerability of coastal communities to extreme storms: the case of Revere, MA, USA. *Mitigation and Adaptation Strategies for Global Change* **3**: 59–82.

Creswell, J. W., 2002. *Research Design.* London: Sage Publications.

Cutter, S., 1996. Vulnerability to environmental hazards. *Progress in Human Geography* **20**: 529–539.

Dow, K., 1992. Exploring differences in our common future(s): the meaning of vulnerability to global environmental change. *Geoforum* **23**: 417–436.

Downing, T. E., A. Patwardhan, R. J. T. Klein, E. Mukhala, L. Stephen, M. Winograd, and G. Ziervogel, 2003. *Vulnerability Assessment for Climate Adaptation*, Technical Paper No. 3, Adaptation Policy Framework. Nairobi: United Nations Development Programme.

Eakin, H., and L. A. Bojorquez-Tapia, 2008. Insights into the composition of household vulnerability from multicriteria decision analysis. *Global Environmental Change* **18**: 112–127.

Eakin, H. and A. Luers, 2006. Assessing the vulnerability of social–environmental systems. *Annual Review of Environment Resources* **31**: 365–394.

England, K., 1994. Getting personal: reflexivity, positionality, and feminist research. *Professional Geographer* **46**: 80–89.

Füssel, H.-M., 2007. Vulnerability: a generally applicable conceptual framework for climate change research. *Global Environmental Change* **17**: 155–167.

Füssel, H.-M., and R. J. T. Klein, 2006. Climate change vulnerability assessments: an evolution of conceptual thinking. *Climatic Change* **75**: 301–329.

Gragson, T. L., and Grove, J. M., 2006. Social science in the context of the Long Term Ecological Research program. *Society and Natural Resources* **93**: 93–100.

Haberl, H., V. Winiwarter, K. Andersson, R. U. Ayres, C. Boone, A. Castillo, G. Cunfer, M. Fischer-Kowalski, W. R. Freudenburg, E. Furman, R. Kaufmann, F. Krausmann, E. Langthaler, H. Lotze-Campen, M. Mirtl, C. L. Redman, A. Reenberg, A. Wardell, B. Warr, and H. Zechmeister, 2006. From LTER to LTSER: conceptualizing the socioeconomic dimension of long-term socioecological research. *Ecology and Society* **11**(2): 13.

Instituto Nacional de Estadística Geografía y Informática (INEGI), 2000. *XII Censo General de Población y Vivienda 2000.* Municipal data. Accessed at www.inegi.org.mx/inegi/.

International Institute for Sustainable Development (IISD), 2008. *Rapid Rural Appraisal (RRA): Participatory Research for Sustainable Livelihoods – A Guide for Field Projects on Adaptive Strategies.* Accessed at www.iisd.org/casl/caslguide/rapidruralappraisal.htm.

Kates, R. W., 1985. The interaction of climate and society. In *Climate Impact Assessment: Studies of the Interaction of Climate and Society*, eds. R. W. Kates, J. H. Ausubel, and M. Berberian, pp. 3–36. New York: John Wiley.

Kelly, P. M., and W. N. Adger, 2000. Theory and practice in assessing vulnerability to climate change and facilitating adaptation. *Climatic Change* **47**: 325–352.

King, G., R. O. Keohane, and S. Verba, 1994. *Designing Social Inquiry: Scientific Inference in Qualitative Research*. Princeton, NJ: Princeton University Press.

Klein, R. J. T., and D. C. MacIver, 1999. Adaptation to climate change and variability: methodological issues. *Mitigation and Adaptation Strategies for Global Change* **4**(3/4): 189–198.

Leathers, N., and L. M. B. Harrington, 2000. Effectiveness of conservation reserves: 'slippage' in southwestern Kansas. *Professional Geographer* **52**(1): 83–93.

Lincoln, Y.S., and E. G. Guba, 1985. *Naturalistic Inquiry*. Beverly Hills, CA: Sage Publications.

Liverman, D., 2001. Vulnerability to global environmental change. *In Global Environmental Risk*, eds. J. X. Kasperson and R. E. Kasperson, pp. 201–216. Tokyo: United Nations University Press.

MASSGIS, 2008. Commonwealth of Massachusetts Office of Geographic and Environmental Information, Boston, Massachusetts. Accessed at www.mass.gov/mgis/massgis.htm.

McCarthy, J. J., O. F. Canziani, N. A. Leary, D. J. Dokken, and K. S. White (eds.) 2001. *Climate Change 2001: Impacts, Adaptation & Vulnerability. Contribution of Working Group II to the Third Assessment Report of the Intergovernmental Panel on Climate Change (IPCC)*. Cambridge: Cambridge University Press.

McDowell, L., 1992. "Doing gender": feminism, feminists and research methods in human geography. *Transactions of the Institute of British Geographers* **17**: 399–416.

Moss, R. H., A. L. Brenkert, and E. L. Malone, 2001. *Vulnerability to Climate Change: A Quantitative Approach*. Richland, WA: Pacific Northwest National Laboratory.

National Research Council (NRC), 1999. *Our Common Journey: A Transition Toward Sustainability*. Washington, D. C.: National Academy Press.

O'Brien, K., R. Leichenko, U. Kelkar, H. Venema, G. Nadal, H. Tompkins, A. Javed, S. Bhadwal, S. Barg, L. Nygaard, and J. West, 2004. Mapping vulnerability to multiple stressors: climate change and globalization in India. *Global Environmental Change* **14**: 1–11.

Oreskes, N., K. Shrader-Frechette, and K. Belitz, 1994. Verification, validation, and confirmation of numerical models in the Earth sciences, *Science* **263**(5147): 641–646.

PA DEP, 2005. Drinking water reporting system: inventory data. Accessed at www.drinkingwater.state.pa.us/dwrs/HTM/DEP_frm.html.

Parry, M. L., O. F. Canziani, J. P. Palutikof, P. J. van der Linden, and C. E. Hanson (eds.), 2007. *Climate Change 2007: Impacts, Adaptation and Vulnerability*, Cambridge: Cambridge University Press.

Polsky, C., 2004. Putting space and time in Ricardian climate change impact studies: agriculture in the U.S. Great Plains, 1969–1992. *Annals of the Association of American Geographers* **94**: 549–564.

Polsky, C., R. Neff, and B. Yarnal, 2007. Building comparable global change vulnerability assessments: the Vulnerability Scoping Diagram. *Global Environmental Change* **17**: 472–485.

Rastetter, E., 1996. Validating models of ecosystem response to global change: how can we best assess models of long-term global change? *BioScience* **46**(3): 190–198.

Redman, C. L., J. M. Grove, and L. H. Kuby, 2004. Integrating social science into the long-term ecological research (LTER) network: social dimensions of ecological change and ecological dimensions of social change. *Ecosystems* **7**: 161–171.

Sayre, R., E. Roca, G. Sedaghatkish, B. Young, S. Keel, R. L. Roca, and S. Sheppard, 2000. *Nature in Focus: Rapid Ecological Assessment.* Washington, D.C.: Island Press.

Schröter, D., C. Polsky, and A. G. Patt, 2005. Assessing vulnerabilities to the effects of global change: an eight step approach. *Mitigation and Adaptation Strategies for Global Change* **10**: 573–596.

Singleton, R. A., and Straits, B. C., 2005. *Approaches to Social Research.* Oxford: Oxford University Press.

Smit, B., I. Burton, R. J. T. Klein, and R. Street, 1999. The science of adaptation: a framework for assessment. *Mitigation and Adaptation Strategies for Global Change* **4** (3/4): 199–213.

Smit, B., I. Burton, R. J. T. Klein, and J. Wandel, 2000. An anatomy of adaptation to climate change and variability. *Climatic Change* **45**(11): 223–251.

Smit, B., O. Pilifosova, I. Burton, B. Challenger, S. Huq, R. J. T. Klein, and G. Yohe, 2001. Adaptation to climate change in the context of sustainable development and equity. In *Climate Change 2001: Impacts, Adaptation, and Vulnerability – Contribution of Working Group II to the Third Assessment Report of the Intergovernmental Panel on Climate Change*, eds. J. J. McCarthy *et al.*, pp. 877–912. Cambridge: Cambridge University Press.

Strauss, A., and J. Corbin, 1998. *Basics of Qualitative Research: Techniques and Procedures for Developing Grounded Theory.* Thousand Oaks, CA: Sage Publications.

Turner, B. L. II, R. E. Kasperson, P. A. Matsone, J. J. McCarthy, R. W. Corell, L. Christensen, N. Eckley, J. X. Kasperson, A. Luerse, M. L. Martello, C. Polsky, A. Pulsipher, and A. Schiller, 2003. A framework for vulnerability analysis in sustainability science. *Proceedings of the National Academy of Sciences of the USA* **100**: 8074–8079.

United States Census Bureau, 1997. American FactFinder, 1997 Economic Census. Accessed at http://factfinder.census.gov/servlet/DatasetMainPageServlet?_program=ECN&_tabId=ECN2&_submenuId=datasets_4&_lang=en&_ts=168466624781.

United States Census Bureau, 2004. 2004 American Community Survey. Accessed at http://factfinder.census.gov/servlet/DatasetMainPageServlet?_program=ACS&_submenuId=datasets_2&_lang=en&_ts=.

United States Department of Agriculture, Economic Research Service (USDA ERS), 2006. *State Fact Sheets: Kansas.* Accessed at www.ers.usda.gov/statefacts/KS.htm.

United States Department of Agriculture, National Agriculture Statistics Service, 2002. Census of Agriculture. Accessed at www.nass.usda.gov/Census_of_Agriculture/.

Wilbanks, T. L., S. M. Kane, P. N. Leiby, R. D. Perlack, C. Settle, J. F. Shogren, and J. B. Smith. 2003. Integrating mitigation and adaptation. *Environment* **45**(5): 28–39.

Wu, S.-Y., B. Yarnal, and A. Fisher, 2002. Vulnerability of coastal communities to sea-level rise: a case study of Cape May County, New Jersey. *Climate Research* **22**: 255–270.

Yohe, G., and R. S. J. Tol, 2002. Indicators for social and economic coping capacity: moving toward a working definition of adaptive capacity. *Global Environmental Change* **12**: 25–40.

9

Rapid Vulnerability Assessments of exposures, sensitivities, and adaptive capacities of the HERO study sites

COLIN POLSKY, ANDREW COMRIE, JESSICA WHITEHEAD, CYNTHIA SORRENSEN, LISA M. BUTLER HARRINGTON, MAX LU, ROB NEFF, AND BRENT YARNAL

Introduction

Vulnerability is a concept that captures the dynamic interactions between complex human systems and complex environmental systems. Thus, a vulnerability assessment that produces a static view of human–environment interactions (i.e., by examining *one place at one time*) will likely provide only limited – and potentially misleading – insight into how the coupled system works. Of course, such static pictures are common in this research domain because it is challenging to establish the temporal evolution of vulnerability (i.e., *one place or many places over time*). Especially in the context of having limited resources to conduct a vulnerability assessment, a solution to this challenge is to ignore variations over time in favor of examining variations over geographic space (i.e., *many places at one time*; see Mendelsohn *et al.* 1994; Carbone 1995; Polsky 2004). We argue that executing a many-places-at-one-time approach requires that all the places adopt a common research protocol; to our knowledge such a networked vulnerability assessment has yet to be reported in the literature. In this chapter, we report results from our effort to examine vulnerabilities – using a rapidly executable and commonly executed methodology – in four distinct study sites in the United States.

As explained in Chapter 1, the HERO project sought to develop infrastructure for studying and monitoring human–environment interactions at individual sites and to enable cross-site comparisons and generalizations. To test how well these concepts and tools work in practice, the project addressed the question, "How does land-use change influence vulnerability to droughts and floods?" and applied a common methodology across the four HEROs. The work presented in this chapter and its companion Chapter 10 were designed to facilitate the same research in the four HERO sites, isolating generalizations while permitting the unique characteristics of each region and their effects on vulnerability to shine through. Chapter 9 presents the results of applying the *Rapid Vulnerability Assessment* methodology, and

Sustainable Communities on a Sustainable Planet: The Human–Environment Regional Observatory Project, eds. Brent Yarnal, Colin Polsky, and James O'Brien. Published by Cambridge University Press.

Chapter 10 the results of applying the *Vulnerability Assessment Evaluation* methodology (Chapters 5 and 8 introduced these methodologies).[1]

Consistent with the definition of vulnerability introduced in Chapter 5, this chapter assesses the three dimensions of vulnerability – exposure, sensitivity, and adaptive capacity – at each of the study sites. The basis of this vulnerability assessment is data collected using the cross-site methodology for rapid vulnerability assessment discussed in Chapter 8.[2] As in Chapter 8, here we restrict the topical focus to the water-dependent sectors of agriculture and public water supply.

Rapid Vulnerability Assessment results: exposure, sensitivity, and adaptive capacity

One of the HERO project's tasks, as described in general terms in Chapter 5 and in operational terms in Chapter 8, was to take a "rapid" approach to its vulnerability assessments. That is, we wanted to learn as much as possible about the vulnerabilities of a set of places and, at the same time, to invest as few resources as possible in the research. This approach allows our methods to be applied by what we view to be the majority of potential future producers of vulnerability assessments – namely people who wish to generate insight about place-based vulnerability but who do not have hundreds of thousands of dollars and multiple years to conduct the research.

The general approach to conducting a Rapid Vulnerability Assessment, as outlined in Chapter 8, is to follow the nine steps in Table 8.1. These steps include selecting which sector(s) to investigate in each site, conducting a literature search, developing a framework for assessment, generating a list of possible indicators, discussing and prioritizing indicators, gathering data, evaluating individual site matrices, cross-evaluating all site matrices, and drawing conclusions for each site based on matrix sets.

Rapid Vulnerability Assessment results: exposure

Exposure is one of three fundamental dimensions of vulnerability (see Chapter 5). It denotes the environmental hazard (in the HERO project, the hazard is hydroclimatic variability, manifest as droughts and floods) to which an exposure unit (such as a family, neighborhood, county, watershed, resource, or economic sector) is put at risk. This section describes the drought phenomenon in each HERO site by describing the four climates over recent decades.

The climate data comprise monthly climate division data from the National Climatic Data Center, 1895 to 2002, for the climate division in which each study site is located (i.e., Massachusetts climate division 02, Pennsylvania climate division 07, Kansas climate division 07, and Arizona climate division 07). For the

Sonoran Desert Border Region HERO, these data do not include the neighboring part of the study area in Mexico; given the large size of the climate division, we believe the United States data are representative of the larger HERO region. The Standardized Precipitation Index (SPI) was computed for the different averaging periods using software from the National Drought Mitigation Center. In evaluating exposure to drought, various seasons or months may have a different relative importance at each site according to the subject of the drought impact. For example, late spring and early summer precipitation strongly influences water supply in Pennsylvania, but July precipitation is critical for corn growth (Dilley 1992). Similarly, ranching activities depend on both winter and summer precipitation in Arizona, whereas high-elevation winter snowpack largely determines surface water supply (Eakin and Conley 2002; Pagano *et al.* 2002). We therefore acknowledge that scientists must account for the differential importance of particular monthly or seasonal precipitation in detailed site analyses, but for conciseness and clarity in this overview, we characterize drought exposure more generally without particular emphasis on different times of year. The basis of the analysis is a long record of simple climate data and its conversion to a well-established precipitation index.

The environmental hazard: drought

General precipitation climatology and seasonality

Mean annual precipitation at the four climate divisions [with coefficients of variation in brackets] are Central Massachusetts HERO 42.6 inches (1082 mm) [0.16], Central Pennsylvania HERO 38.1 inches (968 mm) [0.15], High Plains–Ogallala HERO 18.8 inches (478 mm) [0.22], and Sonoran Desert Border Region HERO 14.3 inches (363 mm) [0.23]. The two western sites are clearly drier and experience roughly 50% more annual variability than the two eastern sites. Annual means obscure the important seasonal cycle of precipitation for all sites except the Central Massachusetts HERO (Figure 9.1a). This site has relatively uniform year-round precipitation, which is generated by the midlatitude synoptic-scale circulation (Keim and Rock 2001). The Central Pennsylvania HERO has a spring–summer maximum, which is also controlled by synoptic circulation patterns (Yarnal 1993, 1995). The High Plains–Ogallala HERO has a pronounced summer precipitation peak associated with the synoptic circulation of moist subtropical air masses from the Gulf of Mexico (Rosenberg 1987; Bark and Sunderman 1990; Englehart and Douglas 2002; Ojima and Lackett 2002). Finally, the Sonoran Desert Border Region HERO has a bimodal winter–summer precipitation pattern resulting from midlatitude synoptic frontal systems in winter and from thunderstorms within the regional North American monsoon circulation in summer (Adams and Comrie 1997; Sheppard *et al.* 2002). The coefficients of variation for monthly precipitation in

Figure 9.1a average 0.49, 0.43, 0.73, and 0.86 across the year for the Massachusetts, Pennsylvania, Kansas, and Arizona study sites, respectively. Variation in monthly precipitation at the western sites is thus relatively large, meaning that it is not unusual for these areas to receive anywhere between zero and double the average monthly total, while for the eastern sites the typical variation is plus or minus about half of the average.

Drought history, intensity, and duration

Any study of drought needs to select one of the several drought indices commonly used in the literature. We selected the Standardized Precipitation Index (SPI) because it compares precipitation relative to historical averages, which permits

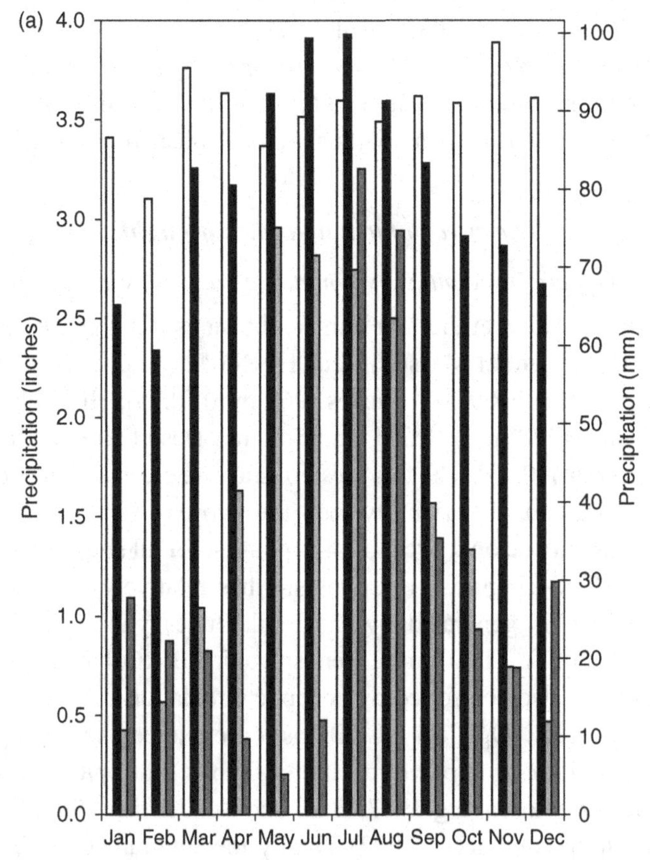

Figure 9.1. (a) Mean monthly precipitation and (b) frequency distributions of five-year Standardized Precipitation Index (SPI). Both graphs represent the period 1895–2002 for the four climate divisions. The four HERO regions are presented by month left to right Massachusetts (white bar), Pennsylvania (black bar), Kansas (light gray bar), Arizona (dark gray bar).

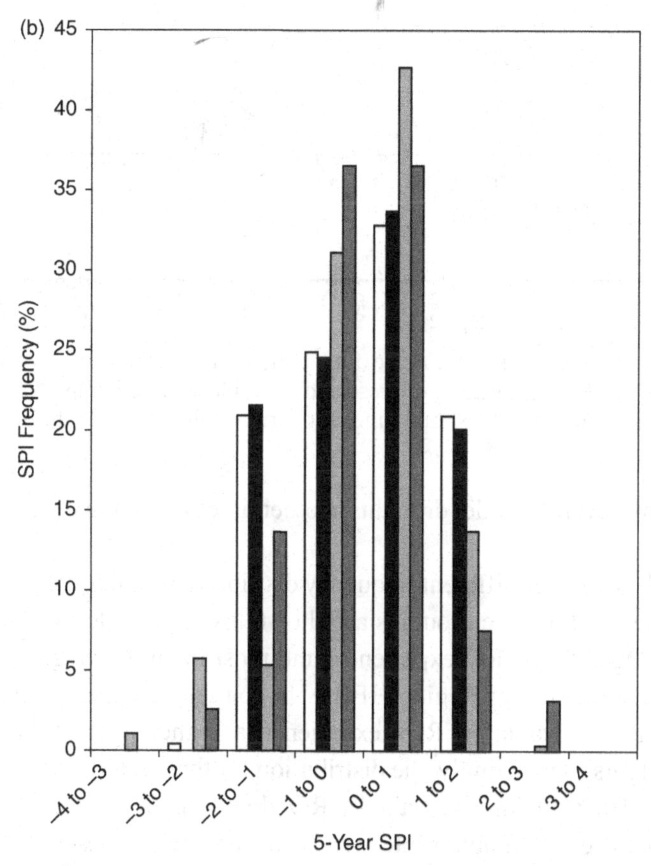

Figure 9.1. (cont.)

direct comparisons across sites. SPI also is more straightforward to compute than other indices: it is calculated as the standardized value of precipitation for a given averaging period expressed in standard deviation units (i.e., *Z*-scores). We present specific drought histories at the four HERO sites using five-year (60-month) SPIs to highlight major multi-year droughts (Figure 9.2).

The Central Massachusetts and Central Pennsylvania HEROs have similar drought histories that are relatively less variable than the western HEROs, and the ranges are therefore smaller. Both eastern HEROs experienced an extended dry period for the first few decades of the twentieth century, followed by a wetter phase for the remainder of the record punctuated by a drought in the 1960s. The High Plains–Ogallala HERO record is highly variable and has a large range too, but the major droughts took place in the 1930s and in the late 1950s. The Sonoran Desert Border Region HERO SPI is even more variable with a large range (−3 to +3 standard deviations). Major droughts occurred in the late 1890s/early 1900s and in

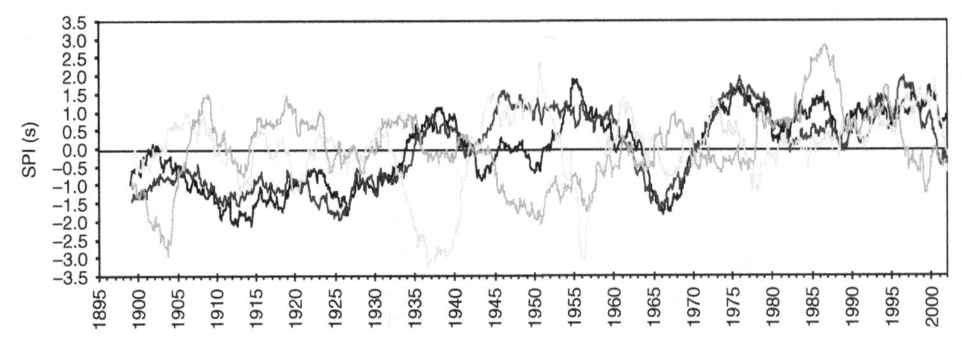

Figure 9.2. Drought history for the four climate divisions using five-year (60-month) SPIs. Massachusetts is represented by a black line, Pennsylvania a dark gray line, Kansas the lightest gray line, and Arizona the second lightest gray line.

the 1950s, with several smaller droughts at other times including the most recent few years.

Figure 9.1b shows the different frequency distributions of five-year SPI across the four sites. The most intense droughts are shown towards the left of the figure. The High Plains–Ogallala HERO experiences the most intense droughts, followed by the Sonoran Desert Border Region HERO. In contrast, the Central Massachusetts and Central Pennsylvania HEROs experience a higher proportion of moderate intensity droughts. Interestingly, the distribution for the western sites is not symmetrical, as the High Plains–Ogallala HERO does not experience matching wet extremes. Defined as uninterrupted runs of negative five-year SPI, Central Massachusetts had five, Central Pennsylvania had seven, High Plains–Ogallala had 11, and Sonoran Desert Border Region had eight droughts over the instrumental period of record (not shown). The High Plains–Ogallala and Sonoran Desert Border Region HEROs have longer droughts, and have them more frequently, than do the Central Massachusetts and Central Pennsylvania HEROs (with the exception of the multi-decadal drought of the early twentieth century). The Sonoran Desert Border Region experiences a higher proportion of longer droughts than any other site (not shown). Central Pennsylvania and High Plains–Ogallala have a relatively even balance of drought durations, whereas Central Massachusetts has a dominance of short-term (less than five years) droughts.

Recapitulation

This section on exposure demonstrated that each of the HERO study sites has different precipitation and drought regimes, as well as differing variations in those regimes. For example, the Central Massachusetts HERO normally experiences plentiful, reliable, and evenly distributed year-round rainfall, whereas the High Plains–Ogallala HERO typically has considerably less and more variable rainfall, with a distinct summertime peak; droughts in Central Massachusetts are infrequent and of moderate intensity, whereas droughts in the High Plains–Ogallala region are twice as frequent and can be much more intense. These dissimilarities, in conjunction with the biophysical and socioeconomic contexts spelled out in Chapter 8, mean that the nature of exposure to drought is different in each area. Note that although the four are different, they cluster into two sets of fairly similar pairs, i.e., the humid eastern HEROs (Central Massachusetts and Central Pennsylvania) and the arid western HEROs (High Plains–Ogallala and Sonoran Desert Border Region).

Rapid Vulnerability Assessment results: sensitivity

The second of the three vulnerability dimensions is sensitivity. Sensitivity characterizes the factors that modulate the impacts on an exposure unit associated with a given exposure event. This section explores the sensitivity of two specific exposure units – water resources and agriculture – to drought, starting with water resources.

The data for the following analysis came from varying sources. For the Central Massachusetts HERO, the analysis used data from nine reservoirs supplying the Worcester metropolitan area. For the Central Pennsylvania HERO, we judged the Harrisburg station to have the best quality streamflow data near Centre County. We obtained groundwater data from the United States Geological Survey (USGS) for Central Massachusetts, Central Pennsylvania, and the Arizona portion of the Sonoran Desert Border Region HERO, and from Kansas Geological Survey for the High Plains–Ogallala HERO. For the two western HEROs, we used annual groundwater data because monthly records were too sparse. We standardized all well data and surface data records into *Z*-scores and then averaged those scores to obtain a single summary series for each site. For groundwater, we changed the algebraic sign in the dataset so that the data represent water table height for ease of comparison with surface water records. We compared water supply data with six-month SPI (smoothed with a 12-month moving average for clarity) to analyze seasonal variations. For Central Pennsylvania surface water, we used precipitation rather than SPI because streamflow was more sensitive to short-term variations in precipitation than other types of water resources. As the results presented below demonstrate, human management often mitigates the effects of variations in precipitation on reservoirs, whereas natural buffers mitigate the effects of precipitation on groundwater resources.

Water resource exposure units

As noted earlier, there are many possible exposure units that might make sense for a Rapid Vulnerability Assessment to examine where the hazard is hydroclimatic variability. Examples include natural areas (e.g., watersheds), human-defined areas (e.g., HEROs or counties), natural resource sectors (e.g., water resources), and economic sectors (e.g., agriculture). In this subsection, we explore the sensitivity of one particular exposure unit: water resources. We divide this unit into its two source components: surface water and groundwater. In the subsequent subsections, we investigate two water resource sectors that are crucial to the four HERO regions: public water systems and agriculture.

Surface water

Surface water is a significant source of supply only at the Central Massachusetts and Central Pennsylvania HERO sites. Visual examination of these data shows a clear link between climate variability and water supply. Figure 9.3a shows time series of

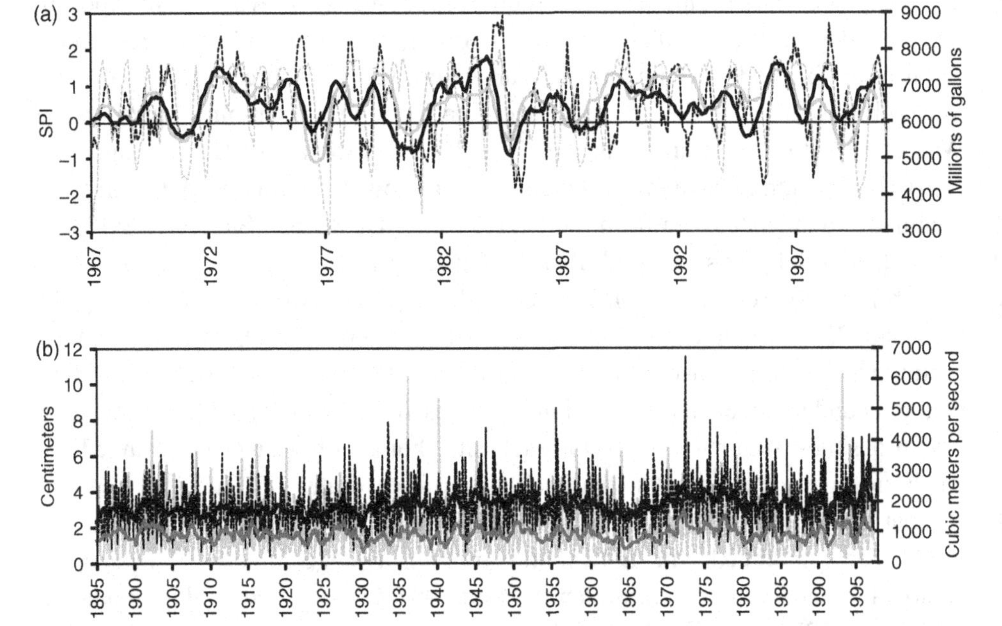

Figure 9.3. Relationships between precipitation indicators and surface-water indicators for the (a) Massachusetts and (b) Pennsylvania study sites. Correlations are significant at the 0.99 confidence level for both sites: $R^2 = 0.33$ for the (a) Massachusetts site and 0.59 for the (b) Pennsylvania site. For (a) the solid black line represents the six-month SPI (smoothed), the dashed black line represents the six-month SPI (unsmoothed), the solid gray line represents the monthly average reservoir level (smoothed), and the dashed gray line represents the monthly average reservoir level (unsmoothed). For (b) the solid black line represents monthly precipitation (smoothed), the dashed dark gray line represents the monthly precipitation (unsmoothed), the solid gray line represents the monthly average streamflow (smoothed), and the dashed light gray line represents the monthly average streamflow (unsmoothed).

reservoir levels and SPI for Central Massachusetts, where SPI explains about one-third of the variance in reservoir levels over time. Using the smoothed values in Figure 9.3b, precipitation variability explains about 60% of the variance in stream-flow at the Central Pennsylvania HERO. Similarly, Figure 9.3b shows close links between surface water supply and climate at the Central Pennsylvania site.

Groundwater

Figure 9.4 shows time series of SPI and groundwater at the four sites. SPI explains 40% of the variance in the Central Massachusetts HERO groundwater levels. There is a possible seasonal lag between precipitation and subsequent groundwater in central Massachusetts (Figure 9.4a), although the effects of greater pumping in drier years may also produce this effect. At the multi-year scale, there are clear

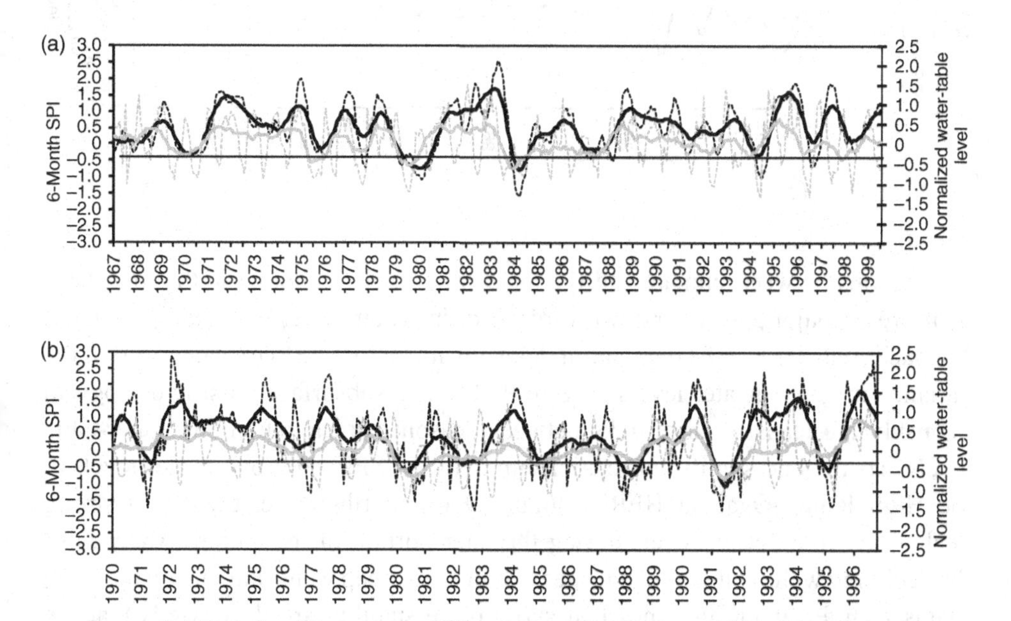

Figure 9.4. Relationship between SPI and normalized water table height for the (a) Massachusetts, (b) Pennsylvania, (c) Kansas, and (d) Arizona study sites. The solid black line represents the six-month SPI (smoothed), the dashed black line represents the six-month SPI (unsmoothed), the solid gray line represents the normalized water-table height (smoothed), and the dashed gray line represents the normalized water-table height (unsmoothed). In the chart for the Arizona–Mexico (d) site, the lightest line represents water-table height in the Santa Cruz Watershed, while the next darkest line represents water-table height in the Cochise Watershed. As only annual data for water-table height are available for the Kansas (c) and Arizona (d) sites, only smoothed data are presented in those charts. Correlations are significant for the Massachusetts ($R^2 = 0.3982$) and Pennsylvania ($R^2 = 0.5562$) sites, and the Santa Cruz Watershed ($R^2 = 0.2732$) in the Arizona study area.

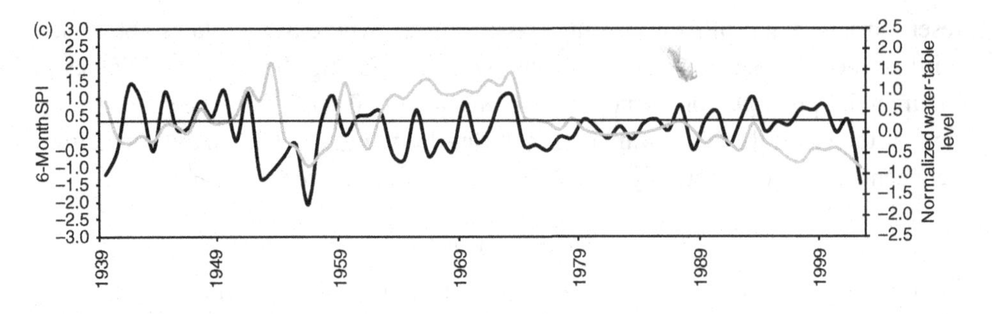

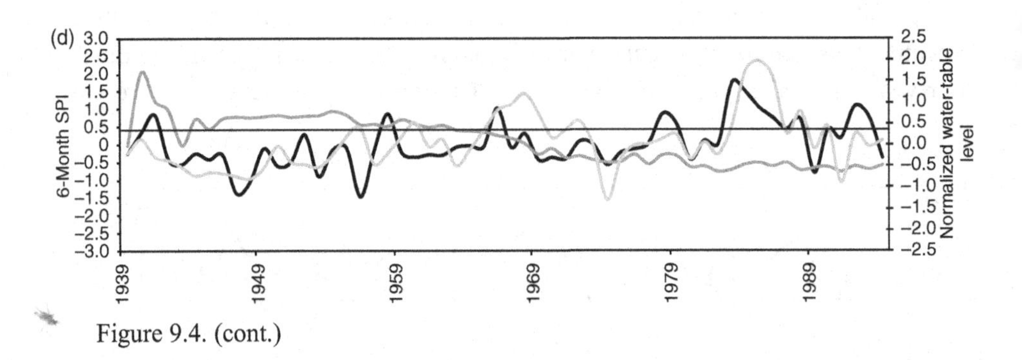

Figure 9.4. (cont.)

associations between climate and the local water table, although in the long term there are no major upward or downward trends in groundwater levels for the Central Massachusetts HERO. At the Central Pennsylvania HERO, SPI explains 56% of the variance in groundwater levels (Figure 9.4b). The subsurface karst hydrology in Pennsylvania leads to close links between seasonal precipitation and water-table levels. There is no significant relationship between SPI and groundwater levels at the High Plains–Ogallala HERO (Figure 9.4c). Furthermore, there is a strong decline in groundwater levels during the later portion of the record. Analysis of the declining wells at this site indicates that they are deep wells drawing on the High Plains Aquifer, while the ones that show more stability are those wells that are shallow and draw on alluvial groundwater. These data therefore suggest that over time there has been a switch from alluvial groundwater to deeper fossil water. The Sonoran Desert Border Region HERO (Figure 9.4d) has a relatively shallow aquifer in the Nogales area (Santa Cruz County) and a relatively deep aquifer further east (Cochise County). SPI explains about 26% of the variation in the Santa Cruz groundwater levels, but there is no significant relationship for Cochise County. The two groundwater curves have very different longer-term trends. Santa Cruz County groundwater levels are relatively variable and track well with decadal precipitation changes, although withdrawals may have increased in the latter part of the record. In contrast, the deeper wells in Cochise County show a steady decrease in long-term groundwater levels so that, similar to the High Plains–Ogallala HERO, the trends in deeper groundwater records do not link to climate.

Public water systems

The public water systems (PWSs) of the four study areas have three distinct profiles for the sources of their water (Figure 9.5). For the Central Massachusetts HERO (Figure 9.5a), although the source contributions changed rapidly during the 1980s and 1990s, by 1995, about half of all PWSs drew their water from surface water sources; public and private well water made up the other half. The sources of Central Pennsylvania HERO water (Figure 9.5b) also included surface water and private

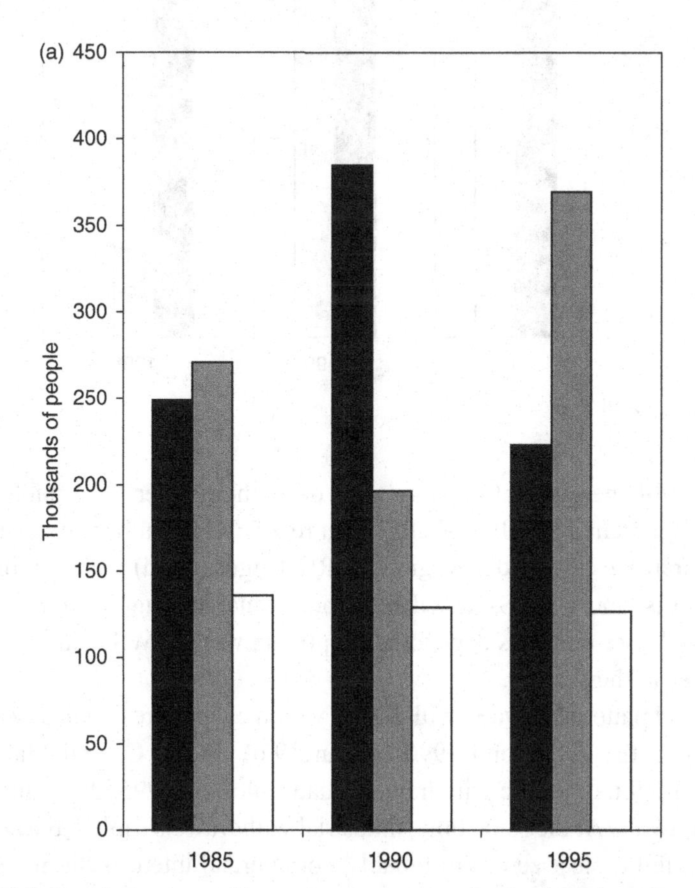

Figure 9.5. Population served by water-source type for the (a) Massachusetts, (b) Pennsylvania, (c) Kansas, and (d) Arizona portion of the Arizona study sites (Solley *et al.* 1985, 1990, 1995). The black bar represents public groundwater, the gray bar represents public surface water, and the white bar represents self-supplied water.

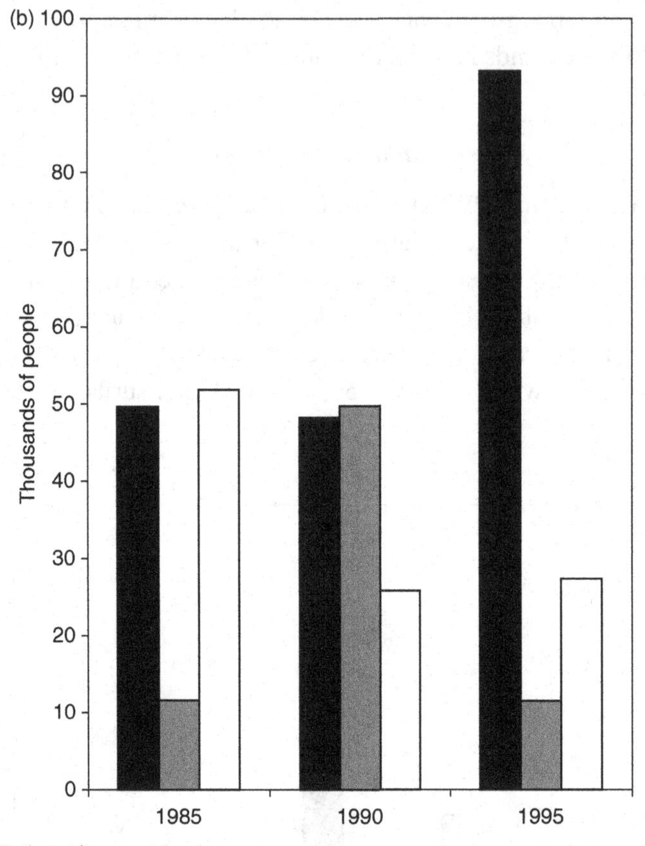

Figure 9.5. (cont.)

wells during this period, but most systems drew their water from public wells by 1995. The High Plains–Ogallala HERO (Figure 9.5c) and southern Arizona portions of the Sonoran Desert Border Region HERO (Figure 9.5d) were similar in their source patterns: most PWSs relied on groundwater throughout the period with considerably lesser numbers depending on private wells. PWSs did not draw from surface water in these areas.

The sectoral patterns of water withdrawal varied even more drastically among the four sites over the 1980s and 1990s (Figure 9.6). In the Central Massachusetts HERO (Figure 9.6a), public withdrawals quadrupled by 1995, dwarfing the area's other withdrawals. At the same time, industrial withdrawals tripled, but commercial withdrawals fell to near zero. Presumably, commercial interests did not stop using water, but started drawing it from public sources. The picture was more stable in the Central Pennsylvania HERO (Figure 9.6b), where public withdrawals also dominated other sectors. The greatest changes occurred in industrial withdrawals, which essentially ceased, and mining, which accounted for nearly one-third of the

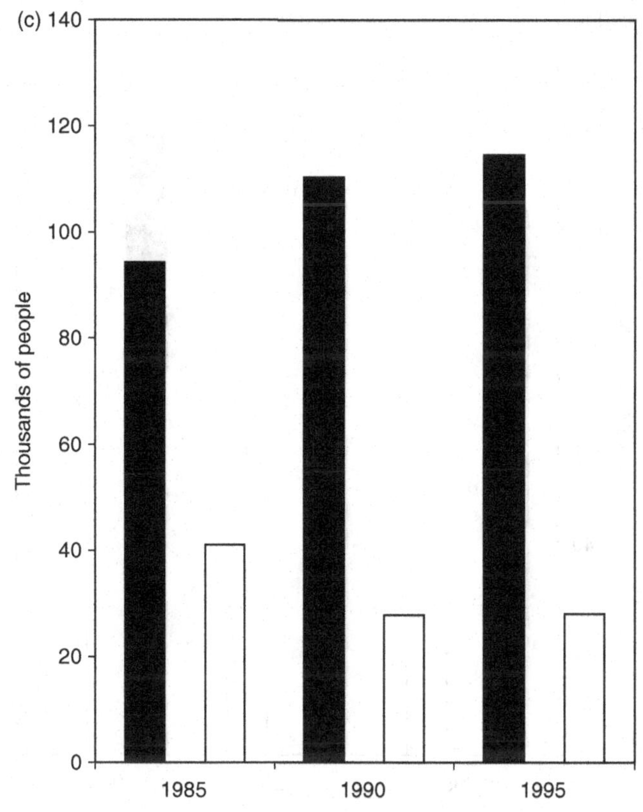

Figure 9.5. (cont.)

withdrawals in 1995. Irrigation towered above other withdrawals in the High Plains–Ogallala HERO in southwestern Kansas (Figure 9.6c), but declined by over 10% from 1985 to 1995. Leaving aside irrigation, agriculture still overshadowed other withdrawals, with 40% of remaining 1995 withdrawals coming from the livestock sector. Public and industrial withdrawals made up much of the remainder with 30% and 15% of withdrawals, respectively. In the Sonoran Desert Border Region HERO (Figure 9.6d), Southern Arizona withdrawals were again similar to those of the High Plains–Ogallala HERO, with irrigation having the greatest withdrawals (but one order of magnitude less than those of southwestern Kansas). Public supplies comprised 55% of remaining withdrawals.

The above patterns result from seemingly different water resource stories at each study site. The Central Massachusetts HERO would appear to be insensitive to all but the most significant variations in precipitation because the region receives abundant, reliable precipitation. Yet, drought is a prominent concern. The state Water Resources Commission (2001) issued a report identifying significant portions

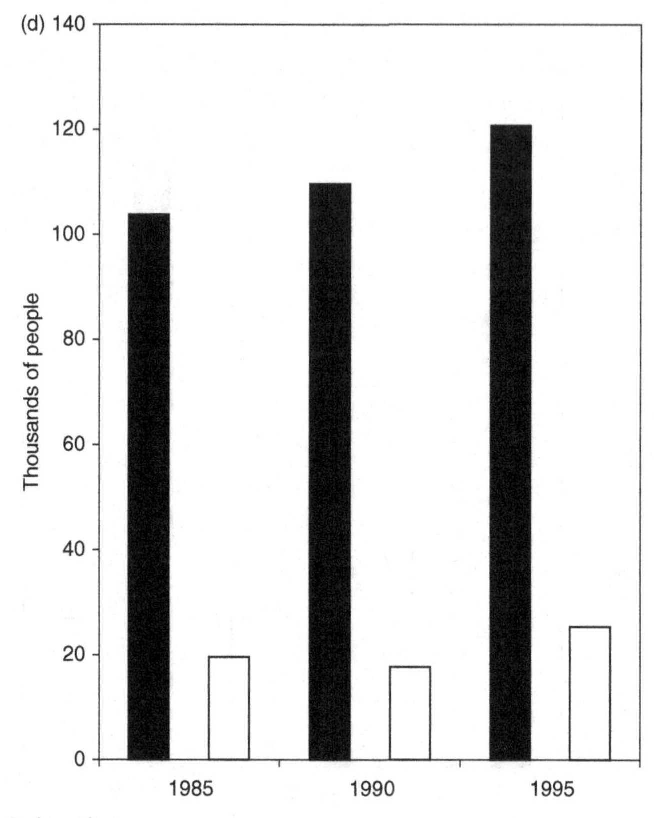

Figure 9.5. (cont.)

of Central Massachusetts as stressed in terms of local abilities to meet water demand. This stress comes from population growth, estimated at 15% for the period 1985–2000. Projected water needs will more than double by 2030. Average demand should rise from the current level of 242 000 m^3d^{-1} to about 613 000 m^3d^{-1}, which, when coupled with remaining capacity of 26 000 m^3d^{-1}, translates into a future supply deficit of 344 000 m^3d^{-1}. Addressing this supply–demand imbalance must involve both increasing supply and reducing losses from aging infrastructure.

These water sources are, in fact, sensitive to precipitation variations. Approximately 33% of the 1967–2000 variation in reservoir levels for the city of Worcester can be explained by a six-month running average of regional SPI values (Figure 9.3a). Given this sensitivity, the central issue defining vulnerability to drought is whether the rapidly growing region can adapt. One possible adaptation option is for local communities to contract with the Massachusetts Water Resources Authority (MWRA) to tap two enormous reservoirs located in Central Massachusetts. These reservoirs possess an active capacity of 1.8 billion m^3, a safe

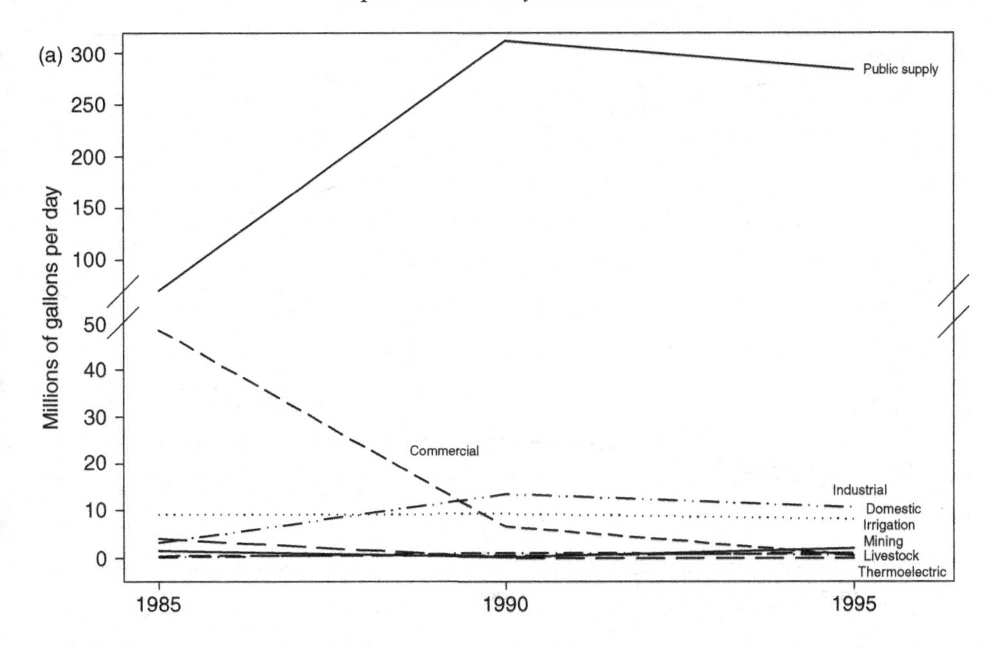

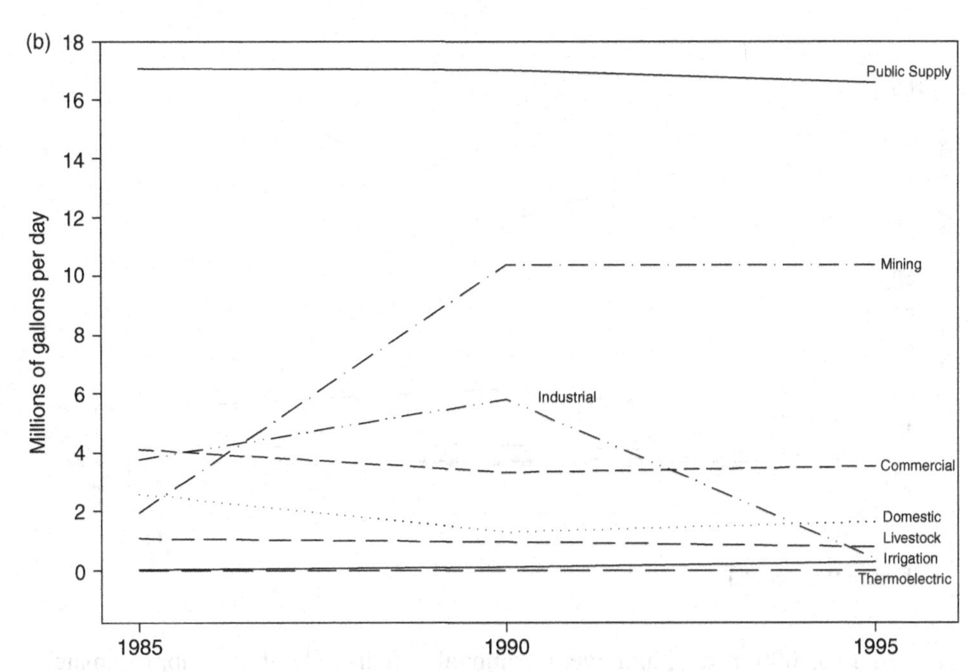

Figure 9.6. Water withdrawals by sector for the (a) Massachusetts, (b) Pennsylvania, (c) Kansas, and (d) Arizona study sites (USGS 1985, 1990, 1995).

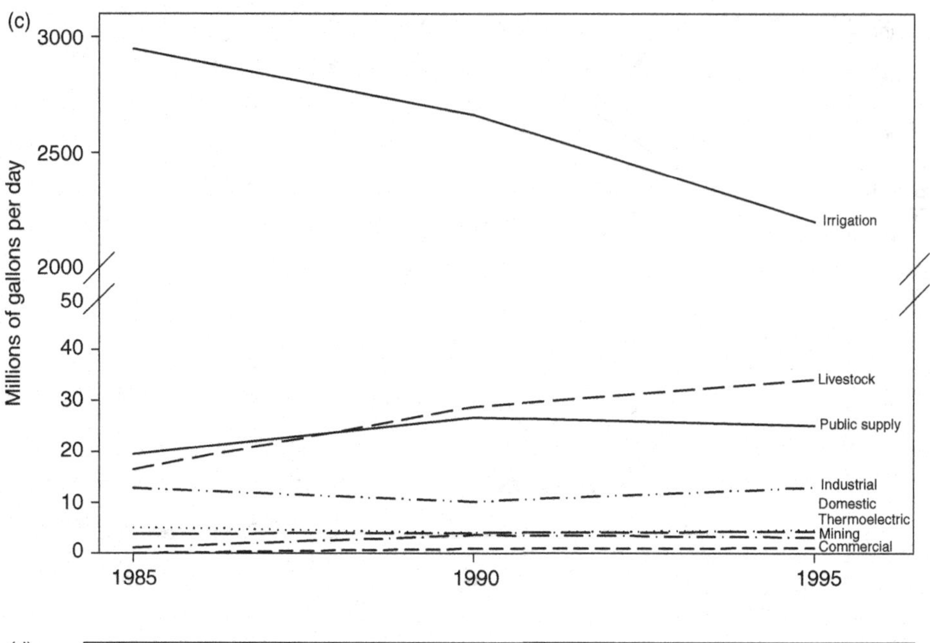

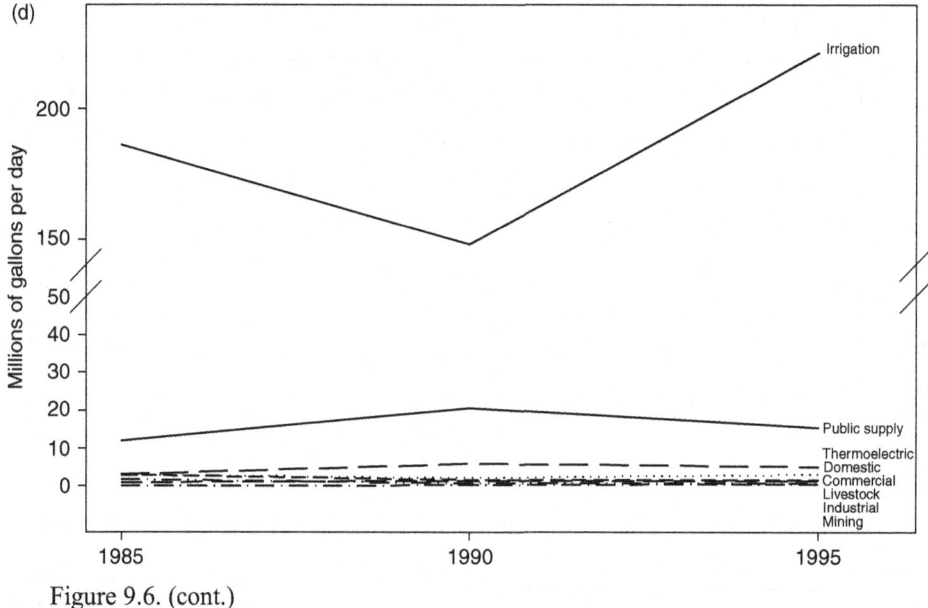

Figure 9.6. (cont.)

yield of 1 135 000 m^3d^{-1}, and average annual withdrawals of only approximately 984 000 m^3d^{-1}. Yet, this water is reserved for the Boston metropolitan area – and any communities the MWRA elects to enroll in its service list (as of 2007, the MWRA was actively courting towns to join its service list). Contingency plans for central Massachusetts communities to purchase MWRA water in times of stress are currently in force, but have not been tested (Hill and Polsky 2005). The nightmare scenario for local water planners is a severe drought that affects the rapidly growing Boston and central Massachusetts areas simultaneously. In short, although central Massachusetts's residents seem to be sufficiently resilient to current demand/drought shortfalls, they also appear to be increasingly exposed to drought impacts in coming years (Hill and Polsky 2007).

The Central Pennsylvania HERO is also a well-watered place, but the source used to supply water demand primarily determines vulnerability to drought. By 1995, most of the county's public systems shifted to groundwater in response to the Surface Treatment Rule of the 1986 amendments to the Safe Drinking Water Act (O'Connor *et al.* 1999). At that time, 71% of the county's 132 000 residents relied on public groundwater systems, while only 8% received their drinking water from surface PWSs. Twenty-one percent still supplied their own water from wells. Several researchers have hypothesized that the shift from surface water to groundwater made regional PWSs less vulnerable to climate variation and change, including droughts (e.g., O'Connor *et al.* 1999; Neff *et al.* 2000).

Groundwater levels are much less sensitive to droughts than surface water flows in the Central Pennsylvania HERO site. Groundwater is a renewable resource in the region due to the karst topography, which allows considerable direct interaction between surface and groundwater. More so than other groundwater-fed regions, water tables decline quickly in response to below-normal precipitation, but recharge quickly in response to increased precipitation. A smoothed six-month SPI explains as much as 56% of the variation in water-table height.

During dry periods, the Central Pennsylvania HERO's smaller groundwater-based systems run short of water of because their infrastructure is aging and their tax-base is too small to repair or replace it. Exacerbating this situation, growth from the larger population centers is spilling into small towns and rural areas, further straining the infrastructure of the smaller systems. Those smaller systems that can tie into the larger systems do join those systems, whereas other smaller systems consolidate with other smaller systems. For reasons of preserving local sovereignty, however, some smaller systems have refused to join other systems and are much more sensitive to drought than neighboring systems are. In sum, smaller, poorer systems are sensitive to drought, even in this landscape of water-rich groundwater-fed systems.

The High Plains–Ogallala HERO has a semi-arid climate and limited surface water. The PWSs in the study area completely rely on groundwater. The major source of that groundwater is the Ogallala aquifer, a fossil aquifer underlying much of the High Plains (Kromm and White 1992). The exclusive use of groundwater and negligible recharge to the aquifer mean that public water supply in the region is not directly sensitive to climate variability; the impact of drought on public water supply is likely to be indirect, through competition for water from crop irrigation, livestock production, and other uses.

In the High Plains–Ogallala HERO study area, groundwater withdrawal was a staggering 2.2 billion gallons per day (8.6 million m^3d^{-1}) in 1995 (Figure 9.6c), with irrigation usually consuming more than 96% of reported annual groundwater withdrawn (White 1994; Alley *et al.* 1999). In the event of a severe, protracted drought, demand for water from crop irrigation and livestock production increases sharply. The resultant drawdown in the groundwater table may affect the yield of PWS wells. Some shallow wells may even run dry during an extended drought. Compounding this problem is the gradual dwindling of the local Ogallala aquifer caused by continuous large-volume pumping since the 1960s (Kromm and White 1992). The decline during 1980–1999 alone was more than 20 feet in many areas (McGuire 2001).

Nevertheless, competition for water from irrigation and other uses is only one factor that may affect PWSs during a drought. To meet the high demand for water in this dry area and to maintain water level in storage tanks, public water facilities often must run at maximum capacity, which increases the probability of mechanical problems, especially in older systems with aging infrastructure.

A recent change that may have reduced sensitivity to drought is the switch from self-supplied water to public water. In 1985, more than 41 000 people relied on self-supplied water in the High Plains–Ogallala HERO. This number reduced to a little over 28 000 in 1995 even as the region's total population increased from 135 400 to 142 620.

The water supply for the arid Sonoran Desert Border Region HERO largely draws on groundwater located in the region's alluvial and low-yielding bedrock aquifers. On the Arizona side of the United States–Mexico border, regulation tempers vulnerability to drought. Arizona initiated the Groundwater Management Act in 1980 to reduce overdraft and sustain water resources. In areas where overdraft was most severe, the state set up Active Management Areas (AMAs) to reduce overdraft, augment supply through increased infrastructure, and distribute water resources equably and efficiently. The state also established Irrigation Non-Expansion Areas (INAs), which prohibit increased irrigation but does not restrict or monitor other water demands. The Santa Cruz County AMA and the Douglas County INA are located within the study region. The act works to insure long-term use by mandating secure water supplies for 100 years and thereby to buffer residents from drought.

After irrigation, drinking water demands from public systems and private wells use the largest proportion of water resources on the United States side of the Sonoran Desert Border Region HERO. Between 19% and 25% of county residents have relied on private wells for drinking water over the past 20 years (Figure 9.5d); these self-suppliers are potentially more vulnerable to drought. Law requires PWSs to have backup emergency plans to supply water when their local sources become unreliable; households that rely on water from private wells do not share this benefit. Consequently, in Santa Cruz County, there has been a trend towards conversion to public systems, with private well owners asking to connect to public systems. Remoteness of many other households makes hook-up costs prohibitive.

Southern Arizona is a desirable spot for retirement and retirement-driven development is occurring in rural areas. However, this development occurs on dry-lot subdivisions and requires private wells. Most private wells associated with this type of development are too small for the Douglas County AMA and Cochise County INA to regulate them. These private wells have had minimal impact on the water table, but concentration of these wells in the future may cause overdraft problems, thus making these elderly individuals vulnerable to continued development and to climate variation and change.

Within the major cross-border urban area of Ambos Nogales, access to drinking water on the Sonora side is a concern, especially during drought. Over half of the water resources of Nogales, Sonora come from pumping galleries that draw water from shallow aquifers in the Santa Cruz basin, east of the city. It is common for wells to run dry in summer. Further north on the Arizona side, the aquifers are deeper and less volatile to short-term fluctuations in surface flows. In addition, over half of all Nogales, Sonora residents do not have water 24 hours per day, and a substantial portion of that population is not hooked up to any water infrastructure and relies on weekly truck distribution of water. Existing water infrastructure suffers a high percentage of system loss, adding to the inefficiency of the system. Although Nogales, Arizona largely relies on groundwater for its drinking water and thus is not dependent on fluctuations in surface water flows coming from the Sonora side, the sister city complex remains linked in terms of water resources. Often in the hottest summer months, the mayor of Nogales, Sonora calls the mayor of Nogales, Arizona to request additional water. With authorization from the Governor of Arizona, taps located at the border fill Sonora water trucks to distribute water to Sonora residents. Thus, despite the efforts of active management on the Arizona side, Ambos Nogales faces water shortages during drought years. Especially hard hit are populations that lack adequate infrastructure. Population growth is one of the major challenges for the two countries to insure sustainable water resources in this region.

Although the PWS water resources pictures appear to be radically different in the four study sites, there are three parallel themes running through them. First, growth is a problem in each area. Although the magnitude of growth is far greater in central Massachusetts than in the other areas, the relative growth is considerable in each place, straining existing infrastructure and resources while necessitating new infrastructure and resources. Second, aging infrastructure is a problem in each area. For places struggling to keep up with growth, replacing existing infrastructure is a lower priority and repairing leaky systems a daily necessity. Third, the inability to cooperate with neighbors is a problem in each area. Whether it is neighboring Boston, neighboring water systems in central Pennsylvania, neighboring farms and industry in southwestern Kansas, or neighboring communities, ranches, and industries in the Arizona–Sonora border region, conflicts over water resources contribute to each region's sensitivity to drought.

Agriculture

The four HERO sites vary greatly in the nature and importance of agriculture, although we use similar crops here for comparisons. Likewise, irrigation use of water is highly variable among the study areas. Farms in the Central Massachusetts HERO tend to be small and family-owned and -operated, with about half of farm operators reporting their principal occupations as "off the farm." For this study area, agricultural activities operate on a small scale relative to the other HEROs and contribute little to the regional economy. Irrigation is a limited water use (Figure 9.6a): in 1997, only 1% of the total farmland in the county was irrigated, despite an overall increase in amount of irrigated farmland since 1969. Despite this low profile, Worcester County ranks among the top counties in the United States in the value of direct agricultural sales through such activities as roadside stands, farmers' markets, pick-your-own produce, and subscription farming. The major agricultural commodity group in Worcester County by sales is nursery and greenhouse crops. These crops often require supplemental water and constitute the fastest-growing agricultural sector in the county, with their share of sales more than quadrupling from 1974 to 1997. It is possible that agriculture in Worcester County is subject to too much water rather than not enough: major climatic events in Massachusetts from 1927 to 1988 include over twice as many floods (often connected with hurricanes) as droughts (Hurd *et al.* 1999). Overall, there appears to be relatively little agricultural sensitivity to drought because of the low reliance on crop production compared to other activities.

For the Central Pennsylvania HERO, agricultural production is a conspicuous local land use, but there is little irrigation here, as well. Similar to Worcester County, less than 1% of total water use is attributable to irrigation, and periods of too-wet conditions may occur more frequently and with greater attendant problems than drought. Total farmland has been decreasing, although there has been a slight increase in irrigated acreage (Figure 9.6b). The production trend indicates a slight increase over time, with little apparent trend in crop mix (Figure 9.7a). For Centre County, there is a correlation between total corn production and SPI, with SPI accounting for 22% of the total variation in corn production. The correlation is

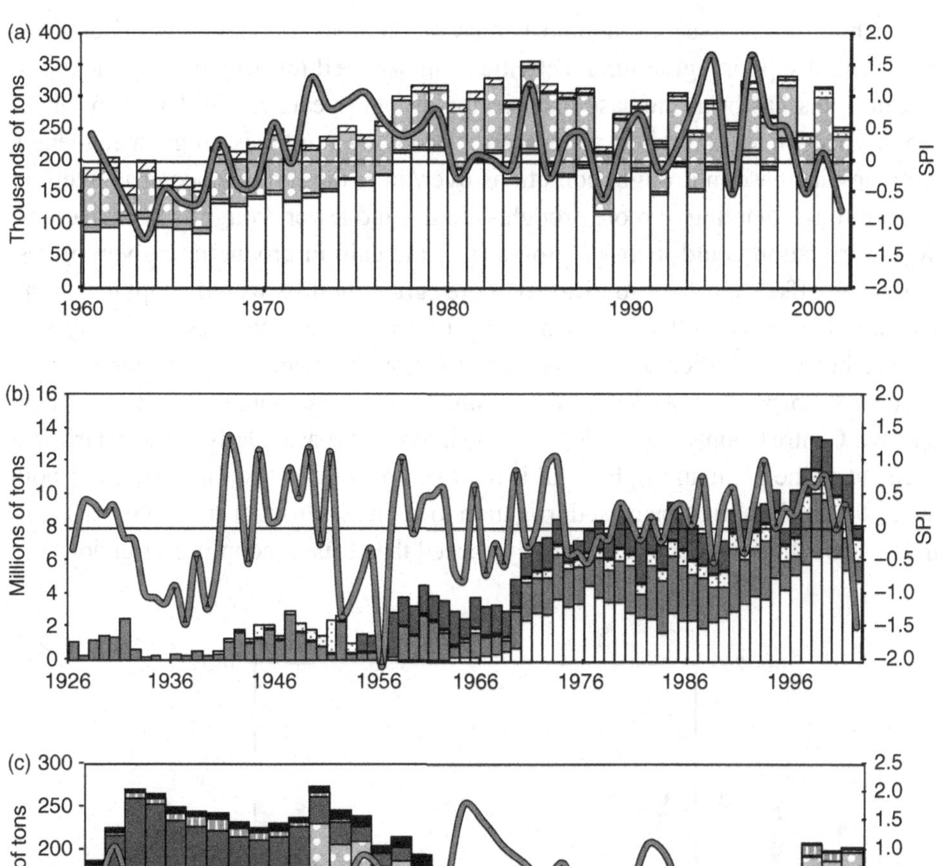

Figure 9.7. Relationship between production of regionally important crops and SPI for (a) Pennsylvania, (b) Kansas, and (c) Arizona study sites. The crops are arranged in stacked bars with the white bar representing corn, the light gray bar wheat, the polka-dot pattern hay, the dark gray sorghum (b and c only), grey and white vertical stripe barley (c only), white with gray checks soybeans (a and b only), diagonal hatching oats (a only), and black cotton (c only).The smoothed gray line is the SPI.

positive, demonstrating that in years of drought, corn production drops. As might be expected, the percentage of total crop production provided by wheat is negatively correlated with SPI; growing this drought-tolerant crop during dry years may help farmers insulate themselves from the negative effects of drought, depending on the balance of market prices.

For both the Massachusetts and Pennsylvania areas, precipitation generally is sufficient to support agricultural activities with no need for irrigation supplements. Because this has been the case historically, drought years can be harder on these regions than on regions where water scarcity from low rainfall is a regular challenge. With any increase in drought conditions over time, the farming sector is likely to adapt through planting of more drought-tolerant and fewer drought-sensitive crops. Where irrigation could increase profits (i.e., the gain in production, given prices, more than offsets the costs to irrigate), more agriculturalists would adopt irrigation, which would serve to decrease sensitivity to drought. For Worcester County as a whole, however, agricultural sensitivity to drought would be a minor concern because crop production plays such a small role in economic activity and land use. For Centre County, agriculture is more important (particularly to the Anabaptist segment of the population), but still is not the dominant economic sector. For both areas, the likelihood of increased irrigation to decrease drought sensitivity is small, unless drought costs begin to regularly exceed the costs of adopting irrigation.

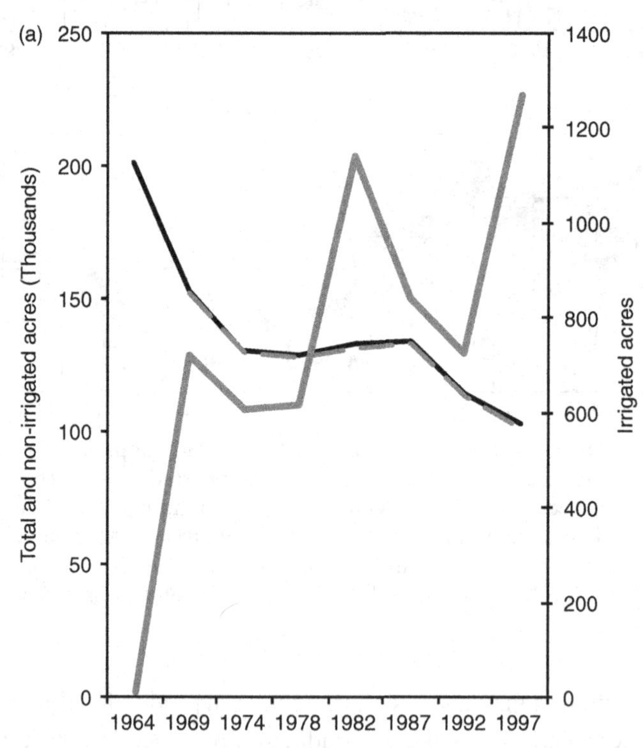

Figure 9.8. Trends in farmland area and area irrigated for the (a) Massachusetts, (b) Pennsylvania, (c) Kansas, and (d) Arizona study sites. The solid black line represents total farmland, the dashed gray line non-irrigated farmland, and the solid gray line irrigated farmland.

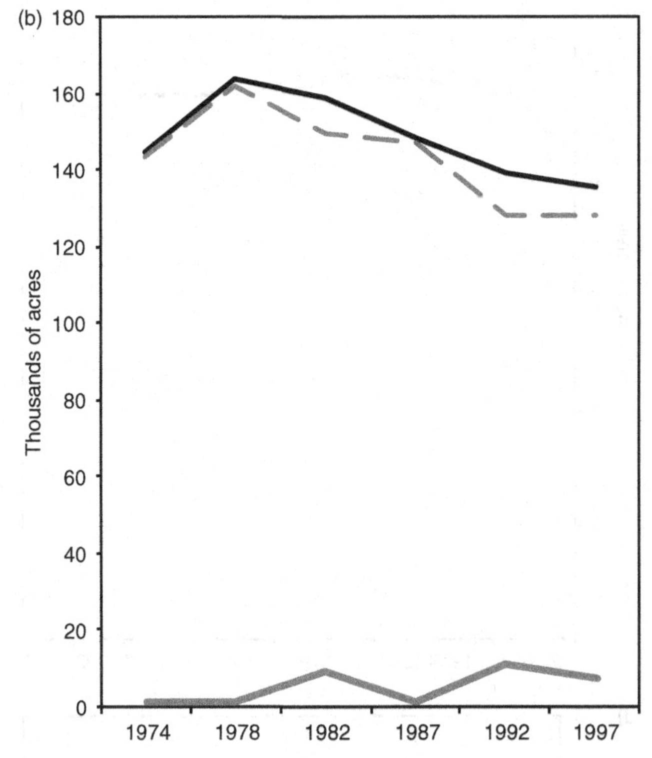

Figure 9.8. (cont.)

The agricultural and irrigation situations are much different for southwestern Kansas and the desert borderlands between Arizona and Sonora. Agriculture constitutes the economic base for the High Plains–Ogallala HERO, both directly through production and indirectly through support industries (Kromm and White 1992). The dominance of irrigation in overall water use in southwestern Kansas is dramatic: more than 95% of freshwater withdrawals are used for irrigation (Figure 9.8c). Virtually all available freshwater in the region is groundwater-based; there are a few perennial streams, and even the most important (the Arkansas River) has been dry in some years – largely due to irrigation withdrawals in Colorado.

For the High Plains–Ogallala HERO, rapid expansion of groundwater for irrigation in the 1950s to 1970s followed the disastrous Dust Bowl of the 1930s (Figure 9.6c). Much of the crop production in the region is now dependent on the use of fossil water from the High Plains–Ogallala aquifer system. In parallel, the mix of crops grown in the High Plains–Ogallala HERO has changed dramatically following the Dust Bowl and expansion of irrigation (Figure 9.7b). High water-use plants include those for cattle feed (alfalfa and corn for silage) in support of beef production and the dairy industry in the region.

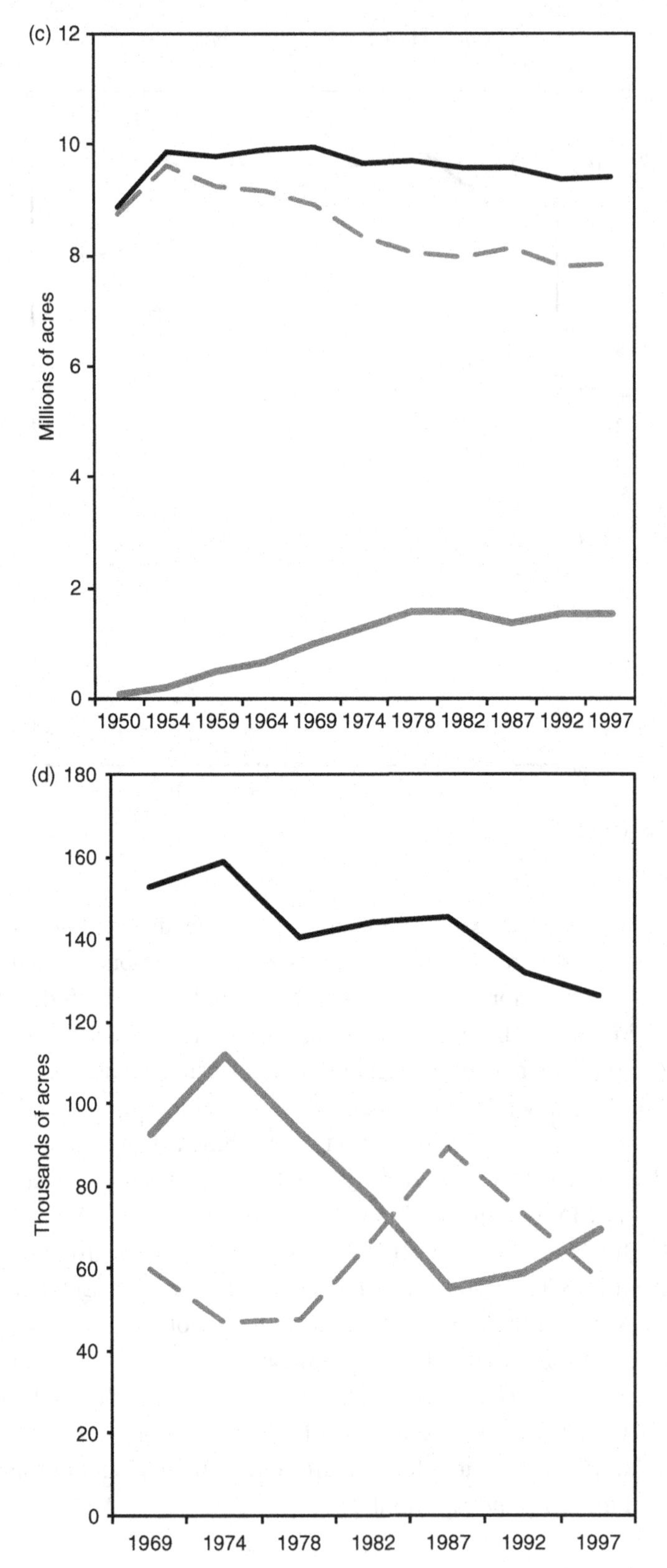

Figure 9.8. (cont.)

An abiding hypothesis in the literature is that this dramatic increased reliance on fossil water-based irrigation has decoupled agriculture from weather conditions (Polsky and Cash 2005). It is clear that for the High Plains–Ogallala HERO, use of groundwater has been an adaptation to unreliable rainfall and has worked well for farmers with access to this resource. Inhabitants of the region know that the groundwater resource is finite, however, even though its use has become more efficient and rates of decline have slowed in some portions of the Ogallala Aquifer over the last three decades (McGuire *et al.* 2003).[3] In some locations, irrigated land has decreased already, with moves toward less water-intensive crop types and varieties. Even with shifts in crop types and continued increases in dryland acreage as compared to irrigated acreage, an increase in vulnerability to drought is likely for southwestern Kansas. Irrigation is an adaptation that has decreased the region's overall sensitivity to drought, but the gains through irrigation have spatial and temporal limits. Much of the agriculture in High Plains–Ogallala HERO remains dryland. For dryland farmers, mitigation of drought conditions is difficult and must take the form of crop choice, enrollment in federal programs (e.g., the Conservation Reserve Program), and crop insurance. Drought sensitivity and vulnerability will likely increase for much of the area as the aquifer depletes and climate changes, although farmers in the region have shown themselves to have great adaptive capacity in the past.

The Sonoran Desert Border Region HERO is similar to the High Plains–Ogallala HERO in having water primarily distributed to agriculture, but is not as extreme in the dominance of irrigation among all water uses (Figure 9.6d). Nevertheless, the agricultural sector is by far the greatest user of water resources on the Arizona side of the study area, with irrigation using over 70% of all groundwater withdrawn in the two counties combined.[4] Unlike any of the other study areas, much of the crop production in the Arizona portion of the Sonoran Desert Border Region HERO is from irrigated land.

Access to irrigation water appears to have decoupled agriculture from precipitation: SPI and production show little relationship over the period 1965–2000 in the Sonoran Desert Border Region HERO (Figure 9.7c). The two main crops cultivated in the two Arizona counties, pecans and alfalfa, are relatively high water-demanding crops, requiring the most water per acre for production.[5] These crops are also favored in the two counties with production of both increasing between 1992 and 1997 (NASS 1994, 1999). During drought years, these crops will be more expensive to maintain because of their high water demands. On the one hand, one year's loss of alfalfa during drought may be substantial, but it is possible to grow new crops in subsequent years. On the other hand, pecans represent a substantial long-term investment. Damage to these crops would signify not only loss of production in the drought year, but also loss in the following years. In other words, farmers specializing in annual or herbaceous crops are apt to be less vulnerable than orchardists are.

Rapid Vulnerability Assessment results: adaptive capacity

The third and final dimension of vulnerability to be assessed is adaptive capacity, which refers in this case to the ability of the exposure units to respond to the effects of drought in the HERO sites. This ability to respond can be manifest as an observed or theoretical ability and can be executed in reactive and anticipatory modes. For this rapid assessment of adaptive capacity associated with the effects of hydroclimatic variability (principally drought) in the HERO sites, we restricted our inquiry to readily available secondary data applicable to the agricultural and domestic water supply sectors.

As described in Chapter 8, a team composed of researchers from all four sites developed a bank of possible qualitative and quantitative indicators of adaptive capacity. Based upon the available secondary data for counties in each site, we selected Worcester County, MA, Centre County, PA, Ness County, KS, and Santa Cruz County, AZ for our study areas. We used or developed indices to compare biophysical and socioeconomic conditions in these counties. For example, to analyze economic diversity, we used a Shannon–Weaver index based on the United States Economic Census data covering the number of people employed by each industry categorized by the US Census. This index uses a scale from 0 to 1, with 0 indicating employment concentrated in a single industry and 1 indicating employment evenly divided among multiple industries. We placed the data or indices for each county into a matrix and evaluated each variable's contribution to local adaptive capacity: positive (enhancing), negative (reducing), or neutral.

To the best possible extent, we attempted to reduce researchers' biases (from familiarity with their own sites) by having each county matrix evaluated by at least one team member from outside that study area. One drawback of relying on additional outside expert judgment was the increased time commitment required from each researcher. We obtained multiple site evaluations for Worcester County, Ness County, and Santa Cruz County, but because of researchers' other commitments only one site evaluation exists for Centre County. We used the site-specific matrix sets to compose narratives describing adaptive capacity across the four sites. We highlight some of the Rapid Vulnerability Assessment indicator rankings for which we achieved agreement across sets.

Worcester County, Massachusetts

It was difficult to reach agreement on several of the factors reducing Worcester County's adaptive capacity. Clearly the poverty rate in 1999 represents a significant lack of financial resources for roughly 9% of the population. Population density and the population growth rate, however, were difficult to interpret in the

Massachusetts context. The density of 192 people per square kilometer can be construed as a negative contribution reducing adaptive capacity because more people will need to adapt, requiring greater resources and coordination. Yet, it can also be true that in urban areas a dense population facilitates resource distribution and enhances accessibility to assistance for adaptation. Interpretations of the population growth rate of 5.8% from 1990 to 2000 demonstrate the value of familiarity with a place. On the one hand, for the researchers outside Worcester County this population growth rate initially did not seem to be large, and the benefits of an influx of new resources appeared to outweigh the costs of expansion. On the other hand, the local researchers noted that the growth primarily occurs in areas on the fringe of the pre-existing urban core, resulting in a limited ability to expand the core infrastructure and to acquire new water resources. Thus, on balance we conclude that the low population growth rate appears to constitute a reduction of adaptive capacity in this study area.

Several factors stand out as contributors to adaptive capacity in Worcester County. At $47 874 in 1999, it has the advantage of the highest median income of our four study counties, indicating greater economic resources even if the cost of living is higher. The percentage of people speaking a language other than English at home was less than 10% in 1999. Employment is also diverse. Regional planning activities exist, as do some government program financial contributions and social services; again, our data do not allow us to determine how effective these programs are. In addition, some towns in the county do have emergency plans that strengthen resilience as long as the plans are followed. This finding shows evidence of some efforts to plan ahead, which may increase likelihood that individuals and municipalities perceive the need to adapt.

Unsurprisingly, agricultural diversity is limited in Worcester County, reducing the adaptive capacity of the agricultural sector. A farm expense-to-income ratio of 0.85 suggests that funding may be available for reinvestment in adaptation to floods and droughts. Reliance on irrigation has increased, further enhancing the sector's drought adaptive capacity. The PWS sector also exhibits clear contributions to adaptive capacity. Of the many community organizations in the county, at least seven address water-related issues. Water resource legislation exists, and as long as it is prudent, such regulations will continue to drive local water providers to adapt in order to preserve compliance with the law.

Centre County, Pennsylvania

Because of individual researchers' time constraints, Centre County is the only county with an analysis based on a single evaluation from researchers, which reduces the confidence in its comparative value in a cross-site assessment. Still, the Centre County analysis demonstrates the value of Rapid Vulnerability Assessment in support of more in-depth analyses of local adaptive capacity.

Centre County is unusual because of the contrast between the urban areas surrounding Penn State University and the region's agricultural society. For example, the 1999 poverty rate of 18.8% is misleading. The poverty rate in the borough surrounding Penn State was about 46.9%, with the large number of university students skewing the county-wide poverty rate.[6] In another instance, the 8.8% growth in population from 1990 to 2000 put stress on both agricultural land and water supplies because much of this growth was in previously rural areas. Private wells in new developments draw from the same groundwater supplies as farms and PWSs, so in times of drought, the increasing number of users may decrease the availability of water that farmers could use to adapt by increasing irrigation. Moreover, any increased runoff due to development increases both agricultural and PWS sensitivity to flood because farmers and water managers would need additional financial resources to develop preventive flood measures.

Centre County has many indicators of positive adaptive capacity. Occupational employment diversity is high. Centre County also has a diverse economy and extensive regional planning activities. Centre County stands out in its positive contributions to adaptive capacity from high educational attainment: 88% of the population age 25 and over have high school diplomas and nearly 42% have college degrees. Although local knowledge does not require advanced education, the high educational level may contribute to adaptive capacity by increasing the likelihood that people perceive the need to adapt, by providing further skills enabling them to adapt, or by enhancing their ability to learn about adaptations.

The picture for Centre County agricultural adaptive capacity is mixed. Agricultural diversity is low because the high proportion of dairy production skews livestock diversity. In contrast, the county has highly diverse sources of crop-related income. The Pennsylvania site developed a site-specific indicator based on the existence of agricultural preservation programs. These programs helped place over 80 000 acres of farmland in Agricultural Security Areas (Centre County Planning Office 2004), representing nearly 50% of the county's total farmland acreage in 2002. This designation shields farms from "unreasonable" agriculturally restrictive municipal ordinances, limits the ability of the state to condemn the land, and makes the land eligible for conservation easement purchases. The increased flexibility these measures give to farmers enhances their adaptive capacity.

There are 31 water-related, community-based organizations in Centre County; of these, 13 are local (POWR 2006). These organizations may not take an active interest in public drinking water protection, but local organization activities like erosion prevention and watershed restoration are adaptations to floods. Additionally, the low number of surface water sources compared to groundwater sources indicates that more groundwater is available in some areas for switches to less drought-sensitive supplies. Finally, historical adaptations in the form of the production of water system emergency plans have been limited – in 2005, only 19 of the county's 45 active water systems had emergency plans (Pennsylvania DEP 2005). Without formal plans to reduce drought and flood damages, water systems are less adaptable and therefore more sensitive.

Ness County, Kansas

Ness County is a sparsely populated county in the northeastern corner of the HERO study area in Kansas. Its population decreased by about 14% between 1990 and 2000, suggesting an economic or resource-related stress that made the county a less desirable place to live compared to other locations. Adaptive capacity is reduced by the low economic diversity. The poverty rate of 8.7% in 1999 is small in comparison to other places, but still means that a sizable portion of this county lacks resources and concerns itself with necessities. One alarming indicator notes that there are no regional planning activities; the county has only one part-time employee in planning. This lack of foresight raises concerns that the need to adapt may not become apparent until a drought or flood occurs. Even if the need to adapt is recognized, the lack of planning infrastructure eliminates a forum for inter-municipality communication. With only a part-time staffer available, it will be difficult for planning activities to be initiated should the county realize that they are necessary.

Despite these negatives, many factors contribute positively to adaptive capacity in Ness County. Occupational employment options are diverse and current water use is somewhat conservative. The population speaking a language other than English is very low, suggesting that the number of people without English skills may be negligible, facilitating communication. Several government programs and social services exist, but the impact of such programs depends on access and disbursement of funds and resources. Finally, the county received substantial disaster assistance in the past, setting the precedent that such funds would be available in the future to offset coping costs and to free more personal resources for adaptation.

The data suggest that the adaptive capacity of the agricultural sector in Ness County may be low, which is a chilling prospect for this agricultural region. Agricultural diversity is very limited, and costs nearly equal expenses. The agricultural extension agents can assist farmers in a variety of ways through difficult times, but there are few agents relative to the number of farmers. The adaptive capacity of Public Water Systems may be affected by the population decline and the low population density (about 8 people per square kilometer). As customers leave, these systems find themselves operating with less financial capital, making it more difficult to maintain their infrastructure. An estimated 39% of the county's population uses private wells, which could reduce the pressure PWSs face. However, private wells do not necessarily provide reliable quality and quantity of water when compared to community systems.

Santa Cruz County, Arizona

Several factors act to reduce adaptive capacity of individuals in Santa Cruz County. According to the United States Census Bureau (US Census Bureau 2000), the population in Santa Cruz County grew by roughly 29% between 1990 and 2000. A large proportion of the population speaks a language other than English at home, suggesting that some portion may not speak English well or at all. Water supply infrastructure may be stressed, and authorities must make an effort to communicate information about droughts and floods and adaptation options in multiple languages. The median income was only $29 710 in 2000, and the poverty rate at the time was a staggering 24.5%. Many individuals are left with few economic resources and are forced to use existing income to supply immediate necessities instead of implementing adaptation.

The Shannon–Weaver diversity index of 0.35 for Santa Cruz County suggests that the area's industry is fairly homogeneous, making it difficult for employees to transfer their jobs to other industries in times of stress. Thus, in a case in which one industry is harmed by a drought or a flood, the blow to the local economy may be disproportionately large because of the relatively limited ability of workers to respond. The picture for Santa Cruz County's adaptive capacity is not entirely negative. For instance, there are regional-level planning activities that could be used in the future to address adaptation to floods and droughts (Sorrensen 2005).

The agricultural sector in the Arizona HERO is limited by a lack of diversity (US Department of Agriculture, National Agriculture Statistics Service 2002). If one crop is overwhelmingly affected by drought or flood, economic resources that could have been directed toward adaptation may be reduced in the short term. Without existing structures, it may be difficult for farmers to adapt by switching to other types of crops or livestock. Nonetheless, farmers have some capacity to adapt to drought as demonstrated by recent shifts away from irrigated agriculture to more sustainable, less sensitive farming practices. On average, farm income exceeds expenses by a comfortable margin, increasing the likelihood that farmers will have funds available for adaptive measures.

Information specific to the capacity of the domestic water supply sector proved more difficult to locate. An estimated 20% of the population uses private wells for their water supply. Such individuals may find drought adaptations, such as drilling new wells or deepening existing wells, burdensome. They may also be ill-equipped to enact measures that limit water quality degradation caused by floods.

Challenges to Rapid Vulnerability Assessment implementation

In general, our team agreed often about the ranks of adaptive capacity indicators for the agricultural and public water supply sectors. Several of our disagreements arose from different perspectives on indicator interpretation that only became apparent after we began analysis. For example, in Santa Cruz County, 60.7% of the population age 25 and over had a high school diploma in 1999, and 15.2% had at least a college or university bachelor degree. These percentages are lower compared to the other counties in our study and to the national averages of 80.5% for high school diplomas and 24.4% with at least a bachelor's degree (US Census Bureau 2000). As a result, some of our team members ranked educational attainment in Santa Cruz County as a negative contribution to adaptive capacity because the high proportion of people with lower education could be at a disadvantage when interpreting communication about adaptation needs and when researching viable adaptations. However, others interpreted educational attainment as a positive contribution in Santa Cruz County because sufficient people have enough education to help find alternative employment if necessary, limiting the economic damage as people fall behind in loan and bill payments or move outside of the area.

In other situations, the issue became one of determining thresholds at which an indicator's value shifts from a negative to a positive adaptive capacity contribution or vice versa. Economic diversity indices in Santa Cruz County and Ness County were clearly low, but these indices for Worcester County and Centre County were 0.65 and 0.67, respectively. Some of our team members thought that economic diversity in Massachusetts and Pennsylvania contributed positively to adaptive capacity, while some thought that diversity could still be improved, constituting a negative contribution to adaptive capacity.

Conclusions

Until the twentieth century, absolute water availability does not appear to have been a problem in any of the four HERO research sites because the natural environments and early inhabitants of the regions were adapted to the hydroclimatic environments. This chapter used the results of a Rapid Vulnerability Assessment to establish that those adaptations were overwhelmed by recent human activities and the underlying forces driving those activities, including population growth, technological advancements, economic growth, institutional change, and human choice. The Rapid Vulnerability Assessment results suggest that each of the four HERO regions would be able to deal better with floods and droughts if it were not for social and human dynamics – e.g., population and socioeconomic growth, decaying infrastructure, and conflict over water in the case of water resources. In short, in all four places, vulnerability to floods and droughts is more about demographics, money, politics, and other human dimensions of environmental change than it is about hydroclimatology.

The following chapter presents the final step in our methodological development for assessing place-based vulnerabilities in a coordinated and networked multi-site research environment. This final step is to develop a means for validating – i.e., establishing the accuracy of – a vulnerability assessment.

Notes

1. We do not refer to these common methodologies as protocols because they evolved during the course of the research. We believe that it is essential to fix protocols at the beginning of a research project, so – by this requirement – these methodologies were not protocols. The common methodologies, however, were an important step towards developing vulnerability assessment protocols for subsequent research.
2. In Chapter 8, the "nine steps" methodology is only illustrated in terms of adaptive capacity. Similarly, in this chapter, the "nine steps" methodology is only operationalized in terms of adaptive capacity – not exposure or sensitivity. As explained in endnote 1, the methodologies evolved with the project, and in this case, we conducted the exposure and sensitivity research before we developed the "nine steps" methodology. In theory, we could apply the "nine steps" to exposure and sensitivity, but in practice, we did not because of timing.
3. Pumping for agricultural use will not completely deplete the Ogallala Aquifer of its water. Instead, the cost of pumping the water to the surface will increase as the water table falls. The increased costs will eventually make use of groundwater uneconomical over most of the region.
4. Groundwater data for Mexico are not available.
5. Alfalfa and pecans are not shown in Figure 9.7 because they are not grown in the other HERO regions.
6. The University Park campus of Penn State University has approximately 42 000 students, with most of those students counted in the census (i.e., students who live off-campus and claim State College as their legal place of residence) having incomes below the poverty line. These students tend to be upwardly mobile and above the poverty line, if not affluent, within a few years of the census count. Mixing this anomalous population with the non-student county population masks any underlying poverty that might exist.

References

Adams, D. K., and A. C. Comrie, 1997. North American Monsoon. *Bulletin of the American Meteorological Society* **78**(10): 2197–2213.

Alley, W. M., T. E. Reilly, and O. L. Franke, 1999. *Sustainability of Groundwater Resources*, USGS Circular No. 1186. Denver, CO: US Geological Survey.

Bark, L. D., and H. D. Sunderman, 1990. *Climate of Northwestern Kansas*, Report of Progress No. 594. Manhattan, KS: Agricultural Experiment Station, Kansas State University.

Carbone, G. J., 1995. Issues of spatial and temporal variability in climate impact studies. *Professional Geographer* **47**: 30–40.

Centre County Planning Office, 2004. *Centre County Fact Sheet*. Accessed at www.co.centre.pa.us/planning/data.asp.

Dilley, F. B., 1992. The statistical relationship between weather-type frequencies and corn (maize) yields in southwestern Pennsylvania, USA. *Agricultural and Forest Meteorology* **59**(3/4): 149–164.

Eakin, H., and J. Conley, 2002. Climate variability and the vulnerability of ranching in southeastern Arizona: a pilot study. *Climate Research* **21**(3): 271–281.

Englehart, P. J., and A. V. Douglas, 2002. On some characteristic variations in warm season precipitation over the central United States (1910–2000). *Journal of Geophysical Research – Atmospheres* **107**(D16).

Hill, T., and C. Polsky, 2005. Adaptation to drought in the context of suburban sprawl and abundant rainfall. *Geographical Bulletin* **47**(2): 85–100.

Hill, T., and C. Polsky, 2007. Development and drought in suburbia: a mixed methods rapid assessment of vulnerability to drought in rainy Massachusetts. *Global Environmental Change B – Environmental Hazards* **7**: 291–301.

Hurd, B., N. A. Leary, R. Jones, and J. Smith, 1999. Relative regional vulnerability of water resources to climate change. *Journal of the American Water Resources Association* **35** (6): 1399–1409.

Keim, B., and B. Rock, 2001. New England region's changing climate. In *Preparing for a Changing Climate: The Potential Consequences of Climate Variability and Change* eds., N. E. R. A. Group, pp. 8–17. Durham, NH: University of New Hampshire.

Kromm, D. E., and S. E. White (eds.), 1992. *Groundwater Exploitation in the High Plains*. Lawrence, KS: University of Kansas Press.

McGuire, V. L., 2001, *Water-Level Changes in the High Plains Aquifer, 1980–1999*, US Geological Survey Fact Sheet FS-0029–01. Denver, CO: VS Geological Survey.

McGuire, V. L., M. R. Johnson, R. L. Schieffer, J. S. Stanton, S. K. Sebree, and I. M. Verstraeten, 2003. *Water in Storage and Approaches to Groundwater Management, High Plains Aquifer, 2000*, US Geological Survey Circular No. 1243. Denver, CO: US Geological Survey.

Mendelsohn, R., W. D. Nordhaus, and D. Shaw, 1994. The impact of global warming on agriculture: a Ricardian approach. *American Economic Review* **84**(4): 753–771.

Neff, R., H. J. Chang, C. G. Knight, R. G. Najjar, B. Yarnal, and H. A. Walker, 2000. Impact of climate variation and change on Mid-Atlantic Region hydrology and water resources. *Climate Research* **14**(3): 207–218.

O'Connor, R. E., B. Yarnal, R. Neff, R. Bord, N. Wiefek, C. Reenock, R. Shudak, C. L. Jocoy, P. Pascale, and C. G. Knight, 1999. Weather and climate extremes, climate change, and planning: views of community water system managers in Pennsylvania's Susquehanna River Basin. *Journal of the American Water Resources Association* **35** (6): 1411–1419.

Ojima, D. S., and J. M. Lackett, 2002. *Preparing for a Changing Climate: The Potential Consequences of Climate Variability and Change – Central Great Plains*. Fort Collins, CO: Colorado State University.

Pagano, T. C., H. C. Hartmann, and S. Sorooshian, 2002. Factors affecting seasonal forecast use in Arizona water management: a case study of the 1997–98 El Niño. *Climate Research* **21**(3): 259–269.

Pennsylvania Department of Environmental Protection (PA DEP), 2005. *Drinking Water Reporting System: Inventory Data*. Accessed at www.drinkingwater.state.pa.us/dwrs/HTM/DEP_frm.html.

Pennsylvania Organization for Watersheds and Rivers (POWR), 2006. Home page Accessed at www.pawatersheds.org.

Polsky, C., 2004. Putting space and time in Ricardian climate change impact studies: agriculture in the U. S. Great Plains, 1969–1992. *Annals of the Association of American Geographers* **94**: 549–564.

Polsky, C., and D. Cash, 2005. Reducing vulnerability to the effects of global change: drought management in a multi-scale, multi-stressor world. In *Drought and Water Crises: Science, Technology, and Management Issues*, ed. D. Wilhite, pp. 215–245. Amsterdam: Marcel Dekker.

Rosenberg, N. J., 1987. Climate of the Great Plains Region of the United States. *Great Plains Quarterly* **7**(1): 22–32.
Sheppard, P. R., A. C. Comrie, G. D. Packin, K. Angersbach, and M. K. Hughes, 2002. The climate of the US Southwest. *Climate Research* **21**(3): 219–238.
Solley, W. B., C. F. Merk, and R. P. Pierce, 1988. *Estimated Use of Water in the United States in 1985*, US Geological Survey Circular No. 1004.
Solley, W. B., R. R. Pierce, and H. A. Perlman, 1993. *Estimated Use of Water in the United States in 1990*, US Geological Survey Circular No. 1081.
Solley, W. B., R. R. Pierce, and H. A. Perlman, 1998. *Estimated Use of Water in the United States in 1995*, US Geological Survey Circular No. 1200.
Sorrensen, C., 2005. Adapting to drought and floods in the semi arid landscape of Santa Cruz County, Arizona. The Geographical Bulletin **47**(2): 101–118.
United States Census Bureau, 2000. *The US Census Bureau*. Accessed at http://factfinder.census.gov.
United States Department of Agriculture National Agricultural Statistics Service (NASS), 1994. *1992 Census of Agriculture*. Accessed at www.nass.usda.gov.
United States Department of Agriculture National Agricultural Statistics Service (NASS), 1999. *1997 Census of Agriculture*. Accessed at www.nass.usda.gov.
United States Department of Agriculture National Agricultural Statistics Service (NASS), 2002. *2002 Census of Agriculture*. Accessed at www.nass.usda.gov/Census_of_Agriculture/index.asp.
White, S. E., 1994. Ogallala oases: water use, population redistribution, and policy implications in the High Plains of western Kansas, 1980–1990. *Annals of the Association of American Geographers* **84**(1): 29–45.
Water Resources Commission (WRC), 2001. *Stressed Basins in Massachusetts*. Boston, MA: The Commonwealth of Massachusetts.
Yarnal, B., 1993. *Synoptic Climatology in Environmental Analysis*. London: Belhaven Press.
Yarnal, B., 1995. Climate. In *A Geography of Pennsylvania*, ed. E. W. Miller, pp. 44–55. University Park, PA: Pennsylvania State University Press.

10

Evaluating vulnerability assessments of the HERO study sites

COLIN POLSKY, CYNTHIA SORRENSEN, JESSICA WHITEHEAD, LISA M. BUTLER HARRINGTON, MAX LU, ROB NEFF, AND BRENT YARNAL

Introduction

As described in preceding chapters, one of the overarching HERO research activities was to establish a set of methodological protocols for vulnerability assessments. Within the methodological research plan were two principal activities, as introduced in Chapter 8: the development and testing of a Rapid Vulnerability Assessment methodology, and the development and testing of an Vulnerability Assessment Evaluation methodology, designed to validate, or assess the accuracy, of our work. Chapter 9 summarized our efforts on the former task; this chapter presents the results of the latter task. Accordingly, at the end of this chapter, we will be in a position to posit some synthetic conclusions about vulnerability in and across the four HERO sites.

Accuracy assessments are difficult to conduct in many research domains, but they are particularly challenging in the domain of vulnerability because the multidimensional nature of this concept makes it difficult for an individual researcher to observe and measure the principal variable of interest. The multi-site context, where the number of places and researchers is larger, amplifies this challenge. Clearly, then, conducting vulnerability research in a networked environment will present particular challenges to validating the research. In this light, a methodology for validating – or what we term evaluating – our research findings is needed.

Methods and data

There are two datasets involved in the application of our Vulnerability Assessment Evaluation methodology: the reference dataset and the validation dataset. The reference dataset is derived from a set of in-depth, semi-structured interviews conducted by HERO students and faculty in each study site in summers 2003 and 2004 (detailed below). The validation dataset is derived from a set of secondary interviews, conducted in summer 2005, with as many as possible of the 2003 and

Sustainable Communities on a Sustainable Planet: The Human–Environment Regional Observatory Project, eds. Brent Yarnal, Colin Polsky, and James O'Brien. Published by Cambridge University Press.

Table 10.1 *Description of interview participants: 2003 and 2004*

	Worcester County	Centre County	High Plains–Ogallala	Southeast Arizona
Agricultural Extension agents			1	
City/County/State public officials		6	5	8
Community Water System managers/spokespersons	14	13	10	15
Conservation advocate/outreach		3		1
Developers		4	5	
Federal Conservation Program officials	3			
Local scholars (university researchers, historians, scientists)	7	1	3	1
Miscellaneous land-owners and water consumers	1			1
Ranchers/farmers/feedlot owners	2		3	
Public resource managers (land/water/forest)	6		2	3
Soliciting method	Individual interviews	Individual interviews, focus groups	Individual interviews	Individual interviews, focus groups
Total sample size	33	27	29	29

2004 interview participants. For both datasets, the focus of the data collection and analysis was on two of the three vulnerability dimensions: sensitivity and adaptive capacity. Thus, the illustration and discussion in this chapter address these two dimensions only, although the evaluation methodology could also apply to data collected and analyzed for the exposure dimension.

The reference data from 2003 and 2004 consisted of archival research on recent climate impacts, extensive interviews – with local stakeholder groups, public officials, and public resource managers – and qualitative analysis of all information collected (see Table 10.1 for a description of the sample). Interviews were conducted by select undergraduate students who participated in the HERO REU (Research for Undergraduate) program in 2003 and 2004 (Sorrensen *et al.* 2005). In both years, the general focus was on local sensitivities and adaptive capacities associated with hydroclimatic variability. In 2003, the specific focus was on Community Water Systems to understand the *sensitivity* of these systems to exposure to hydroclimatic events, in particular droughts and extreme precipitation. As such, interviews were conducted with water system managers or spokespersons in each of the study sites. In 2004, the specific focus was on land-use change issues specific to each study site and the impact of these changes on the *adaptive capacity* of each region to the effects of hydroclimatic events. Thus, the population interviewed in 2004 was more expansive, including public officials, private landowners, scholars, and environmental advocates. Interviews at all sites were based on a common interview script, then transcribed and coded to allow for both unique content analysis and comparison across sites.

Then, in 2005 and as part of executing the Vulnerability Assessment Evaluation, the HERO team revisited as many of these interview participants as possible to learn from them how well our interpretation of the data corresponded to their views. Before describing the results of the multi-part evaluation process that was outlined in general form in Chapter 8, we turn to a summary of the 2003 and 2004 interviews that served as the reference dataset, i.e., the focus of the subsequent Vulnerability Assessment Evaluation.

Themes and result statements from 2003 and 2004 interviews

The 2003 and 2004 interviews produced a set of insights that we aggregated into larger themes surrounding the dimensions of sensitivity and adaptive capacity (Table 10.2). The first major theme to emerge from the student interviews of local stakeholders addresses the potential impacts of population on the sensitivity of drinking water systems exposed to hydroclimatic events, especially drought and extreme precipitation (Theme 1: Result Statement 1). At the four sites, population issues consistently appear in the responses of interview participants, yet each site faces different challenges with population change. The most obvious concern is that population growth increases demand for water resources, further stresses drinking water systems, and heightens sensitivity under drought circumstances. Respondents at all sites commented on the association between population numbers and water demand to some extent, but the Sonoran Desert Border Region HERO and Central Massachusetts HERO stakeholders saw this as a greater concern, whereas the Central Pennsylvania HERO stakeholders believed that proper planning could mitigate any increased sensitivities. Considering the relative aridity of the Sonoran Desert Border Region HERO region and its scarcity of water resources, it came as little surprise that residents saw a clear relationship between population growth and water scarcity. Understanding why the Central Massachusetts HERO stakeholders, where climate conditions are significantly wetter, held a similar view requires further understanding. In this area, state regulations controlling water supply may have limited the ability of towns to keep pace with growing populations – despite the relatively wet climate. In contrast to the other three sites'

Table 10.2 *List of results statements from student interviews of local stakeholders: focus on sensitivity and adaptive capacity (abridged from the version shown to stakeholders)*

Theme 1: Population issues and sensitivity to drought

1. Population change can affect water system sensitivity to drought in several ways:
   (a) Population growth increases water demand and can stress water supplies, making public water systems more sensitive to drought
   (b) Population growth increases the revenue base available for infrastructural maintenance and development, both of which can reduce problems when there is a drought
   (c) Population decreases reduce revenues, making it more difficult for water systems to maintain infrastructure and making them more sensitive to drought.
2. Population-driven changes to more intensive land uses (e.g., golf courses, suburban sprawl into farmland) increase water use and affect sensitivities to drought and other climate variations. Some of these changes decrease permeable surfaces, which reduces groundwater recharge and results in less water available during droughts. These changes also increase runoff, which affects water quality.

Theme 2: Institutional presence and regulation

3. Mandated water system contingency plans can help reduce sensitivity to drought. Emergency plans are the most common forms of these contingency plans, but plans that include interbasin water agreements also exist and may help reduce water system sensitivity to drought.
4. Areas where water rights remain ambiguous and conflicts over water exist are more sensitive to drought than areas where water rights are clearly defined and there are no conflicts.
5. Areas under some form of water regulation are less sensitive to drought and other climate variations, making stakeholders feel more secure about their resources. However, regulation can also place constraints on stakeholders and make it more difficult to adapt to climate variation and change.
6. Local, state, and federal government can play an important role in helping people adapt to climate variation and change. Important factors that affect the usefulness of these institutions include:
   (a) Access to resources provided by these institutions
   (b) Communication and coordination among these institutions
   (c) Development and enforcement of regulations
   (d) Use of different approaches in responding to emergencies related to weather or climate
   (e) Dissemination of information.

Theme 3: Agency

7. Individuals and groups of stakeholders can work to increase their own adaptability to climate variations and change. This agency takes the form of increased public awareness and participation in planning and management decisions, their individual willingness to be proactive and take risks, and the political influence they have in their own stakeholder groups.

Theme 4: Physical geography

8. The physical geographical context of the water resource affects sensitivity to weather and climate.
9. Adaptations to climate variation and change in the water resource sector have happened in the past and continue to occur. Physical manifestations of these adaptations (e.g., dams, switches to dryland agriculture) are more apparent than the human dimensions of adaptations.

relatively straightforward views, the High Plains–Ogallala HERO reported assorted levels of concern with population change and water supply. Population change among the region's counties has ranged from sharp decline to significant growth in recent decades. Consequently, stakeholder assessment of population change's role in water supply challenges is mixed.

Other outcomes of population change hold both positive and negative implications for sensitivity to climate variation and change. A potentially larger pool of revenue accompanies increases in population, which could be utilized to upgrade and maintain water systems and reduce their sensitivities, whereas stagnant or decreasing population and revenue pools make it more difficult for water system managers to finance the necessary upgrades, therefore increasing the sensitivity of their systems to drought. Across the sites, the relationship between population and potential revenue sources was particularly important to managers of smaller water systems who could not easily make upgrades based on current revenues. Other factors influence the potential reliance of water system management on population-based revenue sources. External funding sources are an obvious resolution for many water managers, but access to these resources varied across the sites. In general, larger systems seem to have the personnel, resources, and will to approach external funding agencies. Nonetheless, there appears to be some site-to-site variation on this matter. For example, although the Arizona Groundwater Act provides additional provisions for resource acquisition in the Sonoran Desert Border Region HERO, state restrictions on expanding water supply make it difficult for towns in central Massachusetts to keep up with suburban development regardless of the increasing revenue pools that accompany that growth. When water supply gets stressed, managers of water systems find themselves in a "catch-22" situation. They enforce water restrictions in order to maintain the supply for overall new growth, but at the same time lose revenue for necessary upgrades because restrictions lower overall water use. In summary, even though the relationships among population, water management and demand, and consequent sensitivity of water systems to drought seem straightforward, the local specifics of these relationships suggest additional complexity.

A second finding associated with the population theme concerns land-use/land-cover changes resulting from human choices, preferences for development, and overall population demands upon the landscape (Theme 1: Result Statement 2). The most clear-cut worry relates to land-use change with water-intensive activities, such as conversions of open lands to golf courses or suburban developments, which increase water demands, which can in turn affect water system sensitivities to droughts and other climatic variations. Stakeholders in all four sites found that population-driven land-use/land-cover changes directly affect their sensitivity to drought. There is a broader range of water sensitivities to land-use/land-cover changes than simple direct impacts on water demand and availability. For instance, increases in impermeable land cover, such as paved roads or buildings, reduce surface recharge and result in less available water during droughts. Impervious surfaces also boost channel runoff, which can affect water quality particularly during extreme precipitation events.

These relationships between land-use/land-cover change, population growth, and diminishing water supplies vary with HERO region. The Central Massachusetts HERO saw perhaps the most direct of these relationships as it faced rapid development from exurban growth in the 1990s. In the Central Pennsylvania HERO, until amendments to the Safe Water Drinking Act took hold in the early 1990s, water managers operated systems that relied more on surface water than they do today. Although they now primarily manage groundwater systems, those managers still showed particular concern with runoff impacts of storm events. In the High Plains–Ogallala HERO, the issues of land-use/land-cover change relate more to competition for water from irrigated crop production and to recent shifts to dryland agriculture. In the event of drought, the effects of irrigation on public water supplies can be particularly great because more groundwater is needed for irrigation, thus lowering water tables and reducing yields of municipal wells. Even in this region where most water use is for agriculture, the impacts of golf courses and housing developments on water demand are also a concern. For the residents of the Sonoran Desert Border Region HERO, flooding remains the greatest worry as upstream development and its impermeable surfaces increase downstream flows. The issue with upstream development is not only the quantity of floodwaters, but also the potential water pollution that can occur because enhanced flows often capture toxins from industry, wastewater from overflows at treatment plants, and open sewage from residential areas with inadequate sanitation systems.

The second theme to emerge from the student interviews addresses the role that public institutions play in regulating resources, influencing sensitivities to drought, and affecting adaptive capacity. Stakeholders mentioned that the structure provided by institutional presence and regulation has both positive and negative effects on water resources. At most sites, interviewees believed that contingency plans help reduce water system sensitivities to drought (Theme 2: Result Statement 3). These plans are often state-mandated and require water systems to have specific strategies in place for acquiring additional water resources in case of emergency. They also include plans generated at the state level for overall water planning and local farm-level crop and irrigation planning. Such water planning provides adjustments to temporary changes in water availability that may result from climate variations such as droughts. Although there is uncertainty as to whether specific contingency plans are sufficient to handle all impacts of climate variation on water resources, the results indicate a consensus among respondents that such plans are important for dealing with climate.

The results also note that not defining water rights clearly among various user groups increases local sensitivities to drought (Theme 2: Result Statement 4). Here, the lack of structure provided by institutional presence or regulation influences sensitivity. A wide range of contexts exists in the four HERO sites that contribute to ambiguous water rights. The situation is most dramatic in the drier western HEROs. In the High Plains Ogallala HERO, water rights are clearly defined and water use is strictly monitored. Most water rights, however, are set to a maximum annual use during a five-year period, with a "use it or lose it" provision. As a consequence, most users tend to pump more water than their needs. Although rare, cuts in water rights can occur when one farmer's use of groundwater inhibits a neighbor's access to sufficient water for irrigation. Negotiation is needed to minimize friction when multiple users stress the resource. Presently, the state government is considering plans to retire water rights in order to lengthen the usable life of the overall groundwater resource. Kansas is also considering plans to allow community acquisition of water rights in advance of projected population growth. In the Sonoran Desert Border Region HERO, groundwater is unambiguously regulated in the western portion of the Arizona study area but not over the whole HERO, while surface water rights are not regulated at all. The imbalance in water regulation causes great confusion where the two water sources are linked within watersheds, which typically occurs in floodplains. There is also a growing user population that relies on personal wells that are not regulated because each one pumps at relatively low rates. Many of these wells are situated in floodplains, thereby collectively compromising the capacity of the aquifer and the integrity of the regulatory regimes.

The interview results further indicate that the structure provided by institutional presence and regulation can be perceived as a double-edged sword, pitting resource sustainability concerns against local autonomy and individually perceived adaptive capacity (Theme 2: Result Statement 5). To illustrate, towns in the Central Massachusetts HERO feel this double edge. Although the Massachusetts Water Management Act was initiated to ensure safe yields and resource sustainability, the permitting process for an individual town to expand its supplies can last several years. This delay sometimes ends up forcing towns to consider hook-ups to another town's water systems instead of expanding their own supplies. The costs associated with such short-term coping strategies typically result in a net handicap of local adaptive capacity. Small to medium-sized CWSs in the Central Pennsylvania HERO also feel the internal conflict of institutional presence and regulation. State enforcement of federal regulation compels these systems to install expensive water filtration and monitoring equipment for a resource that they see as already safe. Farmers in the High Plains-Ogallala HERO experience similar uneasiness. They see regulation as necessary to prolong access to the groundwater resource, but also regard it as burdensome and not necessarily linked to local needs (see also Harrington 2001). The Sonoran

Desert Border Region HERO is keenly aware of this problem, too, because the border operates under two sets of federal, state, and local governments. At times, state or federal agencies stymie efforts of local resource managers who agree across the border. In sum, regulation is in place to control resource use and ensure the sustainable use of precious resources for the greater public. Regulation also poses outright limits to use, while requiring constant attention to frequently amended and refined specifics. Although the intent of regulation is at least in part to reduce regional sensitivities by managing resources sustainably, stakeholders also perceive regulation as presenting challenges for individual stakeholder adaptation because it can limit local autonomy.

Still, institutional structure can be important in enhancing adaptive capacity because it has the ability to affect the distribution of financial and technical resources, facilitate communication and interagency coordination, develop and enforce regulation, and disseminate information (Theme 2: Result Statement 6). Institutions also provide resource managers with services, including help with proposals for grants and other forms of aid in times of stress or even emergency. The student interviews found that in the Central Massachusetts HERO study site, a state law designed to address affordable housing has likely inadvertently slowed the growth rate of household water demand. The enactment in the late 1960s of Chapter 40b of the Massachusetts General Law restricts further low- or medium-density development if their affordable housing stock falls below 10% of total housing stock. In some towns where the ratio is under this threshold, new development appears to be slowing down, as is the water demand that would come with the population in these new developments. In another example, in the Central Pennsylvania HERO, higher-level institutions are responsible for enforcing the Safe Drinking Water Act, which has caused many CWSs to switch from surface water sources to groundwater, which is less sensitive to drought and therefore a beneficial adaptation (O'Connor *et al.* 1999). In farming areas of the High Plains–Ogallala HERO, institutions provide information through extension programs, administer federal farm financial and management aid, promote conservation through crop subsidies and the Conservation Reserve Program, and supply information on weather and climate outlooks, all of which aid agricultural mitigation of or adaptation to hydroclimatic stresses.

The circumstances under which institutions are most constructive in facilitating local adaptation often reveal the importance of individual agency – the third theme uncovered through the student interviews – as an important catalyst (Theme 3: Result Statement 7). Of equal weight to the influences of institutional structure on adaptive capacity is the personal initiative of individuals and collective actions of groups of individuals with common interests. The role of individual-level agency in adaptive capacity is most pronounced when stakeholders work to increase public awareness and proactively participate in planning and land-management decisions within institutions at all governmental scales. Thus, in general, the more that local political processes are democratic, transparent, and allow for such participation, the greater the adaptive capacity. In the Central Pennsylvania HERO, for instance, four officers of a very small CWS refused to submit to the Safe Drinking Water Act regulations and risked jail time to maintain local autonomy. The "Madisonburg Four" ultimately did not go to jail because the villagers rallied around their leadership and signed on as officers of the water system, daring the state to put the entire village in jail. The result of their action caused the state to work with the water system to bend the "one size fits all" regulation and to reach a compromise that maintained a degree of autonomy, saved the village thousands of dollars, and made the village water supply safer, more secure, and more adaptable in the face of drought. In another example from the Sonoran Desert Border Region HERO, one group of common property owners has worked collectively with local government to counter spontaneous land invasion by adopting a land development process that is proactive and planned. Agency is also seen in the willingness of individuals to take risks and be proactive and in the amount of political influence that individuals carry within their stakeholder groups.

The fourth theme uncovered by the student interviews acknowledges the physical environment as a factor in water system sensitivities. At each site, respondents noted specific influences on sensitivity associated with the local physical geographies (Theme 4: Result Statement 8). The Central Massachusetts HERO relies on surface water for its population, yet many rivers are dammed, leaving little physical room for more reservoirs. The karst topography of the Central Pennsylvania HERO makes it difficult to keep groundwater resources pure due to the high porosity of the landscape. The agricultural activity of the High Plains–Ogallala HERO is limited by the region's low rainfall and virtual absence of surface water; the health of the regional economy is inextricably linked to the finite groundwater resource. The obvious challenges for the Sonoran Desert Border Region HERO are the regional aridity and the transborder flows of the river basins.

Physical landscape challenges shape the ways in which adaptation has unfolded thus far and highlight the particulars of human–environment relations in each site's landscape. A result of the student interviews is the understanding of stakeholders that adaptations have and continue to occur, and that although these adaptations are the result of human decision-making, they often are responses to the physical characteristics of the environment (Theme 4: Result Statement 9). Their understanding of adaptive capacity centers largely on the physical manifestations of adaptations, such as the plethora of dams in the Central Massachusetts HERO, the shifts to more water-efficient irrigation techniques and to dryland agriculture in the High Plains–Ogallala HERO, and the utilization of micro-basin water management in parts of the Sonoran Desert Border Region HERO. Stakeholders have a much smaller appreciation for the role that human structure and agency has in shaping vulnerability and adaptation.

Vulnerability Assessment Evaluation

As discussed in Chapter 8, the HERO project established two criteria for evaluating whether our research has achieved its goals: saturation and credibility. In this section, we apply these criteria to evaluate the cross-site findings discussed in the preceding section.

Criterion 1: Saturation

Did we collect enough data and analyze it sufficiently to answer our research question? Saturation curves for all four HEROs and for each summer's student interviews are shown in Figure 10.1a–d. For the most part, the curves for 2003 and

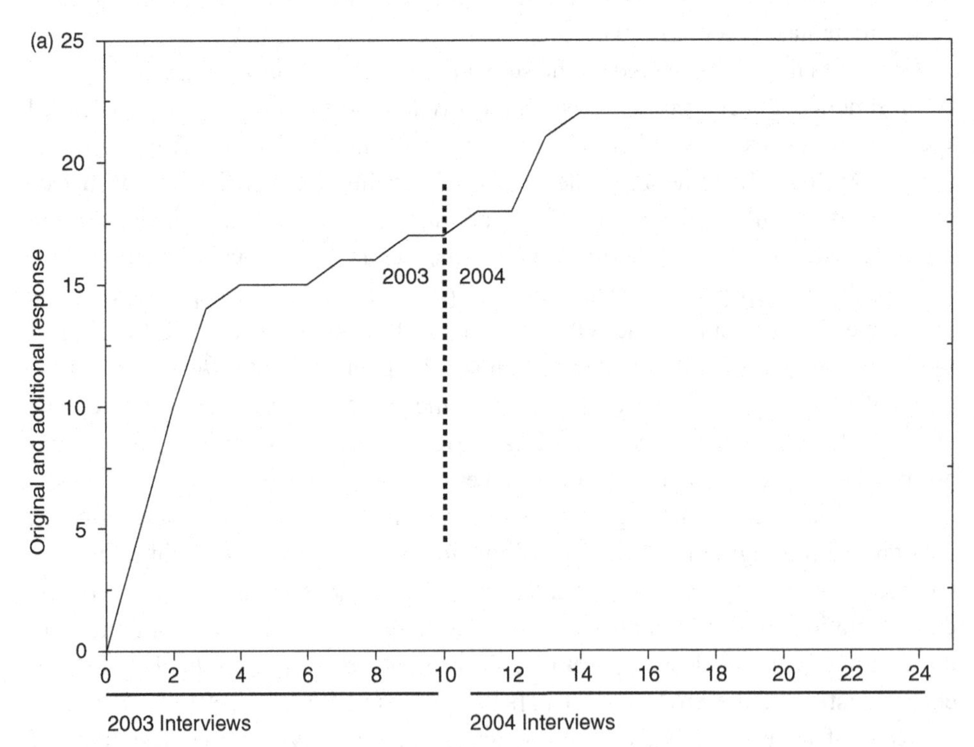

Figure 10.1. Interview saturation curve for (a) Central Massachusetts HERO; (b) Central Pennsylvania HERO; (c) High Plains–Ogallala HERO; (d) Sonoran Desert Border Region HERO.

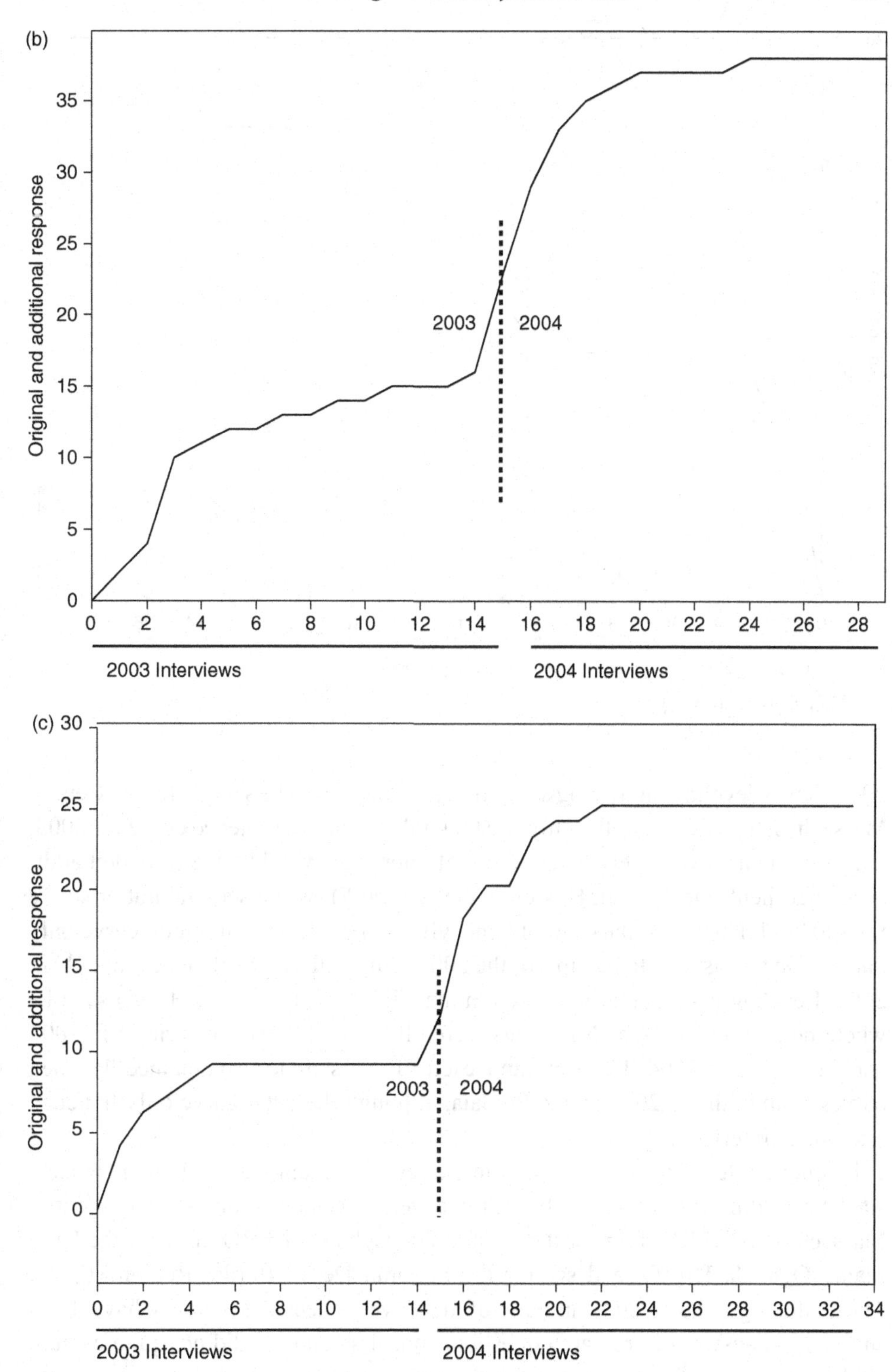

Figure 10.1. (cont.)

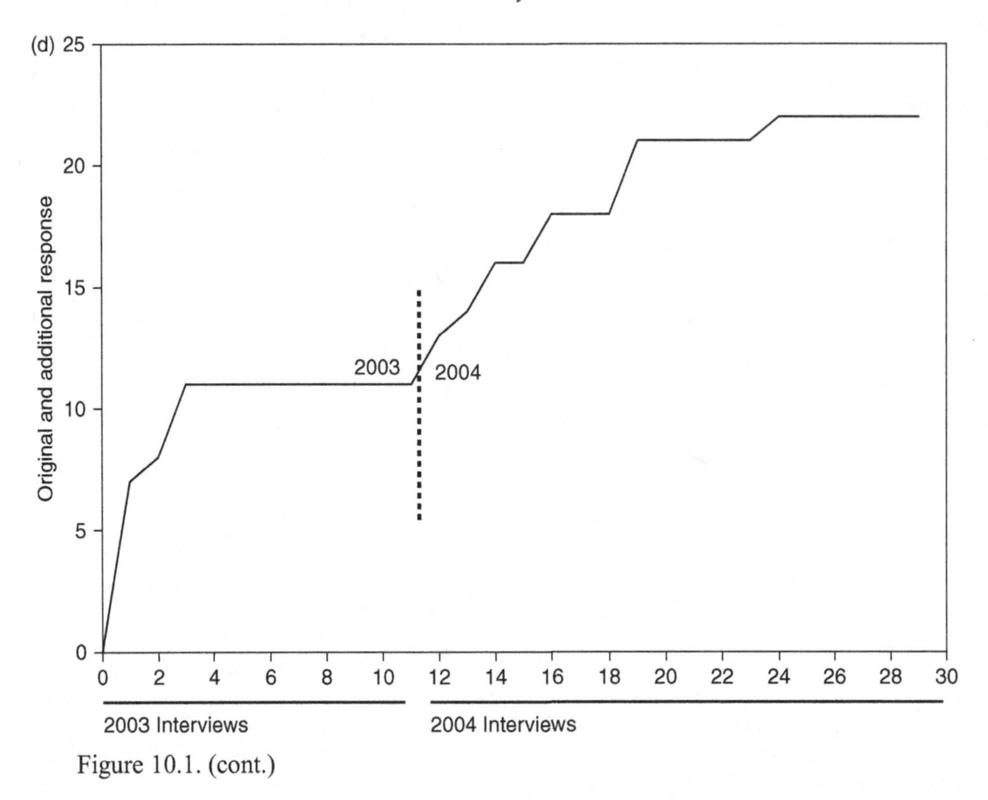

Figure 10.1. (cont.)

2004 show a leveling, indicating saturation in both years. The sample for the Central Massachusetts HERO was the only site that did not attain a flattened curve for 2003, indicating that it was likely that additional interviews would have provided additional pertinent information. However, the Central Massachusetts saturation curve for 2004 did flatten. For the Central Pennsylvania HERO, the saturation curve only flattened in the last two transcripts of the 2003 sample; the curve flattened quickly in 2004. Leveling is particularly marked in the High Plains–Ogallala HERO sample, where no additional research findings were obtained after five interviews in 2003 and after seven in 2004. The Sonoran Desert Border sample also obtained flattened curves from both the 2003 and 2004 data, although the latter curve only flattened after many interviews.

Despite the leveling of the saturation curves, we obtained new findings at each site by rereading the transcripts. In all, there were three new findings for the Central Massachusetts HERO, four for the Central Pennsylvania HERO, nine for the High Plains–Ogallala HERO, and six for the Sonoran Desert Border Region HERO. Thus, although the overall amount of data we collected seems to have been sufficient to answer our research questions, our first analysis did not quite capture all the information that was present in the data.

Table 10.3 *Description of the sample for Criterion 2*

	Worcester County	Centre County	High Plains–Ogallala	Southeast Arizona
City/regional planner or manager	4	2	3	1
Developer	2			
Rancher/farmer			1	2
Agricultural Extension			1	
Federal Conservation Programs			2	
State environmental scientists/agencies	1		2	
Conservation advocate		1		
Community Water System manager	3	4	2	6
Soliciting method	Individual interviews	Individual interviews, focus groups	Individual interviews	Individual interviews
Reference sample size	27	29	33	30
Validation sample size (as % of Reference sample size)	10 (37%)	7 (24%)	11 (33%)	9 (30%)

Criterion 2: Credibility

How credible are our results to stakeholders? A description of the stakeholder sample used to apply Criterion 2 is depicted in Table 10.3. Sample sizes varied among the sites, but each represented at least one-quarter of their respective original interview sample sizes. The stakeholders who participated in this second round of interviews included ranchers, conservation advocates, planners, federal and state program officials, independent developers, and community water managers. In general, stakeholders responded enthusiastically to the process, with the exception of one individual in the Kansas sample, who made little effort to engage in the results statements or respond to them (either negatively or positively). As a result this participant was dropped from the sample used in analysis.

Overall, 67% of responses agreed with assessment result statements to some degree, with only 10% disagreeing and the remaining 23% giving a response that was neither affirmative nor negative (Table 10.4). It is tempting to conclude that these numbers suggest the assessment results from the four sites are generalizable to other places we have not examined. However, the number of stakeholders at each HERO who responded with "agree" or "disagree" varied significantly. The Central

Table 10.4 *Cumulative results of Criterion 2*

			Response	
Stakeholder response to assessment results	Total responses	Total possible responses	Proportion of total sample	Proportion of sample that did not respond in affirmative or negative (left blank or made other comments)
High Plains–Ogallala, KS: Agree	115	209	55%	
High Plains–Ogallala, KS: Disagree	8	209	4%	41%
Worcester, MA: Agree	124	190	65%	
Worcester, MA: Disagree	24	190	13%	22%
Santa Cruz County, AZ: Agree	120	171	70%	
Santa Cruz County, AZ: Disagree	25	171	15%	15%
Centre County, PA: Agree	113	133	85%	
Centre County, PA: Disagree	15	133	11%	4%
Total Agree	472	703	67%	
Total Disagree	72	703	10%	23%

Pennsylvania HERO sample had the largest proportion of "agree" responses at 85%, followed by the Sonoran Desert Border Region HERO sample at 70%. The High Plains–Ogallala HERO and Central Massachusetts HERO samples reported lower agreement rates. Stakeholders who indicated neither agreement nor disagreement either chose to make no comment, or added a comment more applicable to their context, or noted the result statements were not applicable to their context. Nearly half of the respondents in the High Plains–Ogallala HERO responded in these ways, perhaps suggesting that site-specific context in this region is more important than in the other sites.

Using averages of the percentage agreement with the result statements, the greatest agreement across sites was, from highest to lowest, numbers 9, 3, and 7 (Table 10.5). These result statements each recorded on average 80% or greater agreement. Result Statement 9 is an acknowledgement that regional adaptations continue to occur over time and that physically observable manifestations of these adaptations are more salient than invisible human drivers. Result Statement 3 relates specifically to the role of institutional structure in reducing sensitivity to drought via mandated emergency water supply plans. Although there was some speculation

Table 10.5 *Site-specific results of Criterion 2 (in percentage agreement with statements in Table 10.2)*

	Statement number	MA	PA	KS	AZ
Population issues and sensitivity to drought	1a	70	86	55	89
	1b	70	100	45	11
	1c	60	100	45	44
	2	50	71	91	89
Institutional presence and regulation	3	90	100	73	78
	4	50	86	55	78
	5	80	57	82	89
	6a	80	100	45	89
	6b	70	100	36	89
	6c	50	100	27	78
	6d	50	100	36	78
	6e	90	100	36	78
Agency	7	80	57	82	100
Physical geography	8	50	86	82	56
	9	100	86	91	78

among the respondents about the actual effectiveness of current contingency plans (not shown), there was substantial agreement about their general role as important in buffering the effects of drought and other climate variations. In contrast, Result Statement 7 refers directly to the role of agency as a potential driver of adaptation – specifically the capacity of individuals to be proactive, to work in stakeholder groups, and to participate in planning and land management decision-making.

The remaining result statements garnered much less agreement (Table 10.5). In Result Statements 1 and 2, the Central Massachusetts HERO respondents thought the relationships among population, water demand, and drought to be of lesser consequence than the respondents from the Central Pennsylvania HERO. In stark contrast, the respondents from the High Plains–Ogallala HERO and Sonoran Desert Border Region HERO largely disagreed with these results statements. In southwestern Kansas, water system sensitivities are principally related to agricultural use, water rights, and demand. Revenue based on population was not seen as critical; population growth was thought to increase demand for a limited resource, and the income of residents was mentioned as more important than population-driven revenue increases. In southern Arizona, overall water demand in relation to scarce water resources remains such a prominent concern that even the prospect of increased revenues from growing populations is not perceived as sufficient to ameliorate vulnerability.

Respondents in the Central Massachusetts HERO saw little relationship between ambiguous water rights and sensitivities to drought (Result Statement 4). Although respondents from the Central Pennsylvania HERO expressed strong agreement with this statement and think that ambiguous water rights do create sensitivities to drought, respondents from the High Plains–Ogallala HERO sided with the central Massachusetts counterparts and do not consider ambiguous water rights a significant problem. Sonoran Desert Border Region HERO respondents' views were closer to those of central Pennsylvania than to those of central Massachusetts or southwestern Kansas.

Result Statement 5, which addresses relationships among regulation, sensitivity, and adaptive capacity, scored high in the Central Massachusetts, High Plains–Ogallala, and Sonoran Desert Border Region HEROs, but surprisingly low in the Central Pennsylvania HERO.

Respondents displayed mixed levels of agreement concerning the role of institutions in facilitating adaptive capacity, be it through access to resources, communication, regulatory enforcement, emergency resource, or dissemination of information (Result Statement 6a–6e). Respondents in the High Plains–Ogallala HERO expressed doubt as to the positive influence of institutions on adaptive capacity. In contrast, the Central Pennsylvania and Sonoran Desert Border Region HERO respondents found the role of government to be helpful in providing or facilitating adaptive capacity. Overall, the Central Massachusetts HERO respondents tended to be somewhat less enthusiastic about the positive role of government than those respondents from central Pennsylvania or the Arizona border region.

On the face of it, the lack of agreement among Result Statements 4, 5, and 6 suggest conflicting sentiments concerning the role of government in promoting adaptive capacity. This conclusion may be driven by local conservative political ideals favoring a limited role of government in people's daily lives, even though government is providing important functions such as delivering ample and clean water. This conclusion could also be the result of poor wording of the consensus result statements from 2003 and 2004, which could have confused respondents. When Vulnerability Assessment Evaluation uncovers such disagreement, follow-up interviews are required in order to make sense of these findings; we did not conduct such interviews because the project ended about this time.

The Central Pennsylvania HERO was the only site largely to disagree with the result statement concerning agency and adaptive capacity (Result Statement 7). The other three HEROs all agreed that agency is important to adaptive capacity. Rural Central Pennsylvania is deeply conservative, but because of its Anabaptist roots, it is also deeply communal and highly suspicious of individual agency. Although there were no Anabaptists in the sample, these results may reflect their influence on that element of the local character.

Interestingly, Result Statement 8, which states that physical geography has considerable influence on water resource sensitivity to weather and climate, did not achieve consensus in the Central Massachusetts and Sonoran Desert Border Region HEROs. At first, this finding seems ironic because these two sites represent the extremes in climate among the four HERO sites. A closer look, however, suggests that respondents in both locations see the roles of institutional regulation and human agency as critical to sensitivity and adaptive capacity, thus perceiving that human action is capable of compensating for more extreme climatic and physical geography characteristics. Not surprisingly, water resources in both sites are heavily engineered.

Conclusions

There were two principal objectives of this chapter. The first objective was to illustrate the HERO Vulnerability Assessment Evaluation technique presented in Chapter 8. By developing this technique, HERO aimed to fill a problematic gap in the literature on human–environment interactions in general and vulnerability in particular: validation. At least where the dominant focus of vulnerability research projects is social science, there appear to be very few, if any, vulnerability assessments that have engaged in a systematic and transparent post-hoc criticism of their own results. If vulnerability researchers are to promote and defend their results, the results must be able to withstand reasonable challenges to their validity. Yet, the multidimensional and dynamic nature of vulnerability means that measuring vulnerability is difficult (at best), which means that establishing the validity of selected measurements will also be difficult.

The HERO Vulnerability Assessment Evaluation technique is designed to gauge the strength of the results using two criteria: saturation and credibility. Saturation is designed to determine if we had collected enough data and analyzed it sufficiently to answer our research question. This criterion therefore roughly corresponds to what social scientists term *internal validity*, a measure of the extent to which the results are internally consistent. Our operationalization of this evaluation concept involved rereading the original transcripts (from interviews conducted in 2003 and 2004), and plotting the findings derived from each transcript against the number of transcripts reread. In this way, we could then compare the findings from the rereading exercise against the findings derived from the original 2003 and 2004 analyses to see if our original analyses had failed to report important results.

Credibility is designed to determine how well our findings (derived from the 2003 and 2004 interviews) corresponded to what the interview participants in fact told us in the original interviews. This criterion therefore roughly corresponds to what social scientists term *external validity*, a measure of the extent to which the results correspond to the concept we purport to be measuring. To operationalize this concept, we presented our findings to a sample of the original interview participants in each site.

Taken together, putting these two criteria into operation represents a helpful analytical foundation for future vulnerability assessments, whether or not the assessments are conducted in the context of a networked, multi-site collection of study sites. The results of the evaluation provide both producers and consumers of research with the basis for promoting and defending a given set of results.

The second objective of this chapter was to summarize the insights on vulnerability provided by the various research initiatives conducted over the duration of the HERO project. The primary aim of the HERO project was not to produce a definitive statement on the substance of vulnerability, but instead to produce an analytical infrastructure and methodological basis for future researchers to make such substantive statements. Accordingly, for example, the Rapid Vulnerability Assessment technique was introduced in Chapter 8 and illustrated in Chapter 9 in terms of a single vulnerability dimension – adaptive capacity – rather than in terms of all three dimensions. Nonetheless, the HERO team has assessed each of the three vulnerability dimensions in each of the four study sites. Consequently, we are in a position to draw some conclusions about land-use-induced vulnerability to hydro-climatic variation and change in each of the sites, as well as across the sites.

For both the Central Massachusetts and Central Pennsylvania HEROs, overall regional vulnerabilities should be relatively low due to, first, the relative abundance of moisture and the lack of intra- and inter-annual variation in precipitation and, second, these two regions' economic diversities and low levels of dependence on agriculture (which can be a water-intensive sector of the economy) relative to the overall local economies. The Central Pennsylvania HERO, however, includes significant numbers of Anabaptist farmers that rely on agriculture for their livelihoods. The vulnerability of this subpopulation relative to non-Anabaptist farmers is not immediately apparent from the data collected by the HERO project. Non-Anabaptist farms use more modern technologies and greater energy and material inputs, and the primary family income often comes from off-farm activities. In contrast, Anabaptists have far stronger social networks, are accustomed to frugality, and cope with climate variations using traditional farming strategies involving less technology – but they typically have much less off-farm income. Thus, although non-Anabaptist farmers may be quicker and better able to adopt irrigation, Anabaptists may have less need for this adaptation strategy.

For the High Plains–Ogallala HERO, groundwater use will decrease in the future as the aquifer depletes, likely leading to greater sensitivity to drought as the local economy evolves to dryland farming. Dryland agriculture is much more sensitive to climate than irrigated agriculture because supplemented water eliminates climatically induced variations in water supply. Making the situation worse is the dependency the regional socioeconomic system has developed on irrigated agriculture, including not only the economic system that supports farming, but also the massive feedlot and dairy industries that count on crops to feed the animals. If aquifer depletion results in mass conversion to dryland farming, then these industries must import feed or leave the area, with the latter option potentially causing significant socioeconomic dislocation. Under a scenario of significantly higher fuel costs, such as occurred in mid 2008, importing feed may not be a viable option.

In the Arizona portion of the Sonoran Desert Border Region HERO, the farm population appears to be sensitive to drought effects. Overall, however, this portion of the study area should have a relatively small negative response to drought because most of the population live in urbanized areas and are not dependent on agriculture for their livelihoods. If populations continue to expand and other economic activities grow, rapid increases in water consumption in urbanized areas could create significant vulnerabilities among all sectors of the region. Conditions are worse in the Mexican portion of the study area. Both rural and urban inhabitants on this side of the border are likely more vulnerable than their US counterparts, because while they are equally exposed to drought, they appear to be more sensitive and to possess lower adaptive capacity as a result of lower average incomes. Moreover, the urban population is particularly sensitive to drought because water delivery systems are poor and cannot keep up with the exploding population.

In the end, were exposure (to drought) to increase in frequency or severity, a better adaptation for agriculture in most areas would be to change crop types rather than to increase their dependency on irrigation. Even with that adaptation, climate history suggests that both High Plains–Ogallala and Sonoran Desert Border Region HEROs are due for major droughts that could last for decades. Coupled with climate change projections for greater aridity in the Great Plains and with population projections for greater growth in the Desert Southwest, agriculture in these two HERO regions appears to be especially vulnerable. With respect to public water supplies, it is true that few people are suffering health effects from a lack of water or are unable to take showers or wash their dishes. Nonetheless, public water supply systems in each of these four areas appear to be, on balance, experiencing a decline in adaptive capacity with respect to drought. The days of large-scale supply augmentation (i.e., the building of large dams and reservoirs) appears to have come to an end; this development affects the Massachusetts, Pennsylvania, and Arizona HERO sites. For the Kansas HERO site, adapting to water supply challenges by augmenting supply has historically taken a different approach – mining the Ogallala Aquifer. This option also appears to be increasingly less viable with each passing year. As a result, all four sites appear to be increasing in vulnerability to the effects of hydroclimatic variability not so much because of a change in environmental conditions (i.e., exposure), but in response to changing social adaptive capacities. Of course, society could implement a novel approach to coping with water supply stress through an innovation (be it technological, regulatory, or even behavioral), at which point this overall assessment of vulnerability may need to be modified. But until such an event occurs, these four US locations appear to be increasingly vulnerable. As such, although these vulnerabilities may appear modest in the larger scheme of problems that municipalities have to manage on a daily basis, this situation suggests that these populations may find it difficult to cope with an unexpected and independent exogenous shock – such as a dramatic rise in energy prices – and associated ripple effects throughout the local economies and societies.

References

Harrington, L. M. B., 2001. Attitudes towards climate change: major emitters in southwestern Kansas. *Climate Reseach* **16**(2): 113–122.

O'Connor, R. E., B. Yarnal, R. Neff, R. Bord, N. Wiefek, C. Reenock, R. Shudak, C. L. Jocoy, P. Pascale, and C. G. Knight, 1999. Weather and climate extremes, climate change, and planning: views of community water system managers in Pennsylvania's Susquehanna River Basin. *Journal of the American Water Resources Association* **35**(6): 1411–1419.

Sorrensen, C., C. Polsky, and R. Neff, 2005. The Human–Environment Regional Observatory (HERO) Project: undergraduate research findings from four study sites. *Geographical Bulletin* **47**(2): 65–72.

Part V

11

The mounting risk of drought in a humid landscape: structure and agency in suburbanizing Massachusetts

COLIN POLSKY, SARAH ASSEFA, KATE DEL VECCHIO, TROY HILL, LAURA MERNER, ISAAC TERCERO, AND ROBERT GILMORE PONTIUS, JR.

Introduction

This chapter explores the vulnerability of two areas, located in central and eastern Massachusetts (Figures 11.1 and 11.4), to the effects of drought. Consistent with the dominant trend in the climate change and global environmental change literatures, we define vulnerability in terms of three principal dimensions: exposure, sensitivity, and adaptive capacity (Turner *et al*., 2003; Parry *et al*. 2007). This chapter explores the exposure and sensitivity of the region by referencing the local climate, social and biophysical landscapes, and human drivers of landscape change. Adaptive capacity is discussed in terms of the factors associated with, on the one hand, groups of people and elements of the social power structure (e.g., government), and, on the other hand, individual people and small groups of individuals. These two sets of factors are termed, respectively, structure and agency. Understanding structure and agency is important for understanding the vulnerability of different places, or of a given place over a period of time.

This chapter consists of a vulnerability assessment of the Central Massachusetts study site, completed in 2004, and of the Eastern Massachusetts study site, completed in 2005. The later research builds on the earlier research. Each case study starts with a description of local changes in land- and water-use patterns, and ends with a description of exposure, sensitivity, and adaptive capacity (i.e., vulnerability), with a special focus on the relative roles of structure and agency.

Defining vulnerability

In colloquial terms, vulnerability refers to the potential for harm associated with a hazard (Kates 1985; Cutter 1996). The recent scholarly literature on climate

Sustainable Communities on a Sustainable Planet: The Human–Environment Regional Observatory Project, eds. Brent Yarnal, Colin Polsky, and James O'Brien. Published by Cambridge University Press.

change and global environmental change has defined vulnerability as a function of exposure, sensitivity, and adaptive capacity (McCarthy *et al.*, 2001; Turner *et al.*, 2003). *Exposure* refers to the extent, duration, frequency, and severity of an environmental hazard within a specific geographic region, and to the human and environmental systems exposed. *Sensitivity* refers to the amount of cost or damage incurred or potentially incurred following exposure to the hazard. *Adaptive capacity* refers to the ability to modify the degree of loss or damage incurred following exposure to the hazard.

These dimensions of exposure, sensitivity, and adaptive capacity jointly describe necessary and sufficient conditions for a place to be vulnerable. Thus, a place may be vulnerable if it is exposed to a stress, but only if it is also sensitive. Similarly, a place that is exposed and sensitive is not vulnerable if it can effectively adapt either in anticipatory or reactive modes to avoid undesirable consequences. Clearly, these concepts are mutually constituted and therefore difficult to untangle. For example, exposure can itself be a function of sensitivity, sensitivity a function of adaptive capacity, and adaptive capacity a function of exposure and sensitivity. Yet, the attempt to parse the events and processes of human–environment interactions in a given place is nonetheless valuable. For the purpose of this chapter, exposure and adaptive capacity are analyzed separately (as much as possible), and sensitivity is treated as overlapping each of those two dimensions. (For more in-depth discussions of vulnerability, see Polsky *et al.*, Chapter 5 and Chapter 8.)

Central Massachusetts

Central Massachusetts landscape

The Central Massachusetts study area comprises ten towns, located amid the rolling hills that characterize the region and partition it into five major watersheds (Figure 11.1). The region has a predominately continental climate with oceanic influences due to its close proximity (~45 miles) to the Atlantic Ocean. The region sits atop a plateau contributing to the significantly higher annual snowfall (~67 inches) and precipitation (~46 inches) rates than in the surrounding, lower-altitude areas to the east and south. The Blackstone River, the major river in the area, has its headwaters in central Massachusetts and drains into Narragansett Bay in Providence, Rhode Island (~45 miles to the south). The Blackstone River contributed to the development of Worcester into a national industrial leader in the early nineteenth century, because of its significant hydropower resources: the river drops some 450 feet in elevation over a short distance. Moreover, the Narragansett Bay served at the time as one of the country's most active shipping ports. Together, these

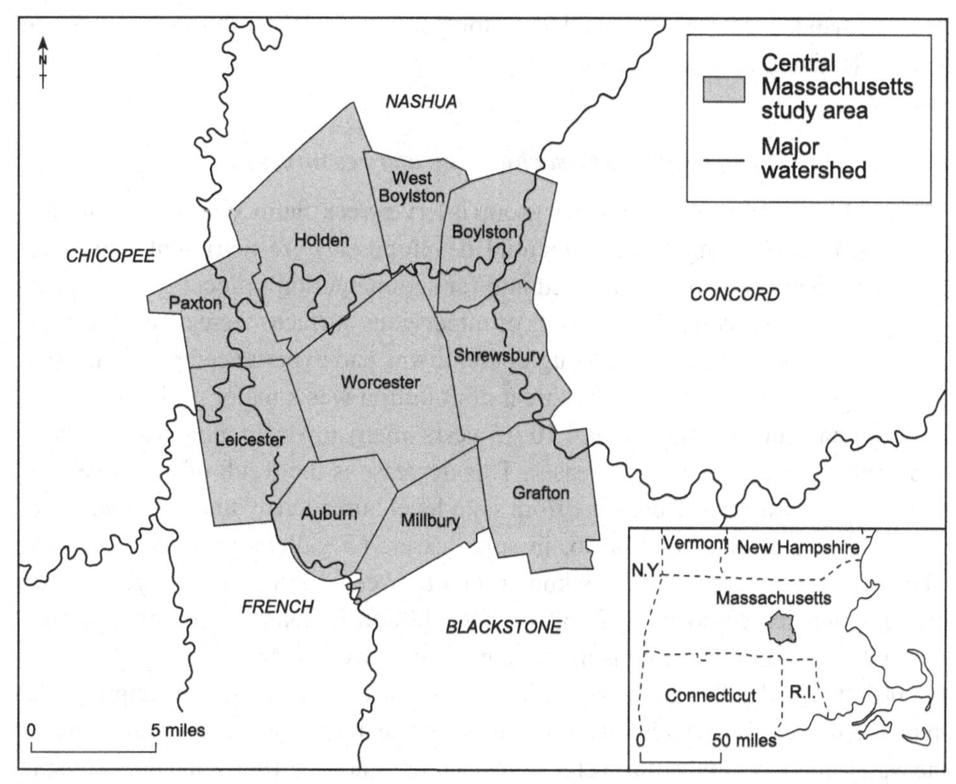

Figure 11.1. Central Massachusetts study area.

factors enabled Worcester to develop into the home of a significant industrial base, defined largely but not exclusively by textile mills.

Central Massachusetts land-use change

The ten towns comprising the study area averaged a population increase of 22% over the period 1970–2000 (United States Census Bureau 1973, 2000). This growth has been propelled by the development of significant transportation infrastructure joining Worcester with Boston. A wave of westward suburbanization from Boston has resulted, with people moving to central Massachusetts while maintaining jobs in the Greater Boston/Providence metropolitan area. However, during largely the same period (1971–1999) the ten towns averaged a 38% increase in built area, with one town, Boylston, registering a 60% increase (Massachusetts EOEA 1971, 1999). Land consumption outpacing population growth hints at a phenomenon that has been pervasive in the broader New England region: low-density residential development. Such residential development patterns, and their associated increases in water demand, pose potential challenges for Community Water Systems (CWSs) to adapt effectively to climatic variation.

Central Massachusetts water resources

In central Massachusetts, advocacy group interviewees claimed increasing impervious surfaces (arising from residential development) are detrimental to water quality in the Blackstone watershed and for sedimentation in local ponds (specifically, Coes Reservoir). The impacts of impervious surfaces were discussed most in connection with Worcester, though other towns had experienced problems; five out of eight respondents said the threat of pollution was a major problem.

Based on data covering the past 10–15 years, many towns in the study area have experienced water demand decreases. This decrease is the result of intensive conservation campaigns and pressure from state-level bureaucracy to lower town-level residential water consumption to, in some cases, 65 gallons per capita per day. Though the data are sparse, it is known that between 1958 and 1966 yearly total consumption increased from 7.762 to 8.403 billion gallons in Worcester (Hardy 1966). In 1991 total water consumption in Worcester was 8.485 billion gallons, and in 2002 that had fallen to 8.128 billion. Thus, in terms of total consumption, the trend has been stable which is a bit worrisome since the population in Worcester and the water-intensive industrial sector have declined since the 1960s. In contrast to the City of Worcester, populations are growing quickly in the surrounding towns. Six of eight CWS managers responded to a question about population growth impacts on demand saying it was a primary concern of theirs.

CWS managers in the area typically would respond to water demand increases associated with population growth by promoting infrastructure (supply) expansions. However, the challenges associated with state permitting processes for adding new water sources have generally discouraged CWS managers from supply augmentation in the past two decades (Platt 1995); instead the managers have focused on demand-side management (typically through periodic summer water-use restrictions). Further complicating the options available to CWS managers is the financial structure under which all but one of the CWSs in our study area operate: revenues are directly linked to water consumption. As a result, CWS managers find themselves in a "catch-22" situation: they have to tell consumers to reduce their consumption in order to supply water to new residents, but in doing so they reduce the funding available for needed infrastructure maintenance. The outcome is, in the words of one local town official, a "fine balancing act," in which towns are forced to implement watering bans and restrictions even in times of average or above average rainfall.

The restriction on supply augmentation has reduced the capacities of water managers in the smaller towns to respond to population growth. Worcester by contrast has always had greater economic and political resources to draw on compared to the outlying towns, and less need to expand infrastructure since it already enjoys the use of a system of reservoirs and, since the late 1960s, a supply relationship with the vaunted Boston water supply system, the Metropolitan Water Resources Authority (MWRA) (previously the Metropolitan District Commission, MDC).

Land development in the one metropolitan area in the study region, Worcester, has been associated with the presence of brownfields (polluted parcels of land). These parcels are generally avoided by developers because of the potential legal liability if someone becomes ill from inhabiting the newly developed but still polluted site. Consequently, development assumes an "infill" pattern, whereby seemingly every available piece of city land is transformed into a residential or commercial property. The problem with this pattern, for Worcester at least, is that the vast majority of such parcels are on steep slopes. Developing such areas modifies the local hydrologic cycle by contributing to high-velocity runoff and soil erosion. Such first-order consequences can affect local drought vulnerability.

For example, in the 1960s a prolonged rainfall deficit in Worcester coincided with the construction of the Worcester Airport, which was to be located atop a steep hill. The construction caused erosion that compromised one of Worcester's main water sources of the time, Lynde Brook Reservoir, located below the construction site. As a result, the city of Worcester immediately lost a substantial proportion of its local water supply precisely when it needed it most – during a period of meteorological drought. The city felt forced to respond by turning to the MWRA system for emergency supply augmentation, a relationship which has since emerged into a permanent feature of the local water planning landscape: Worcester can count on the vast water reserves of the MWRA in times of need. In short, local development patterns have directly influenced local drought sensitivities.

Central Massachusetts: two recent climate events and their impacts

The Central Massachusetts study region has experienced several moments of climatic stress in the twentieth century (Hill and Polsky 2005). Two droughts are discussed in this chapter, the landmark drought of 1963–66 and a more recent drought in 1999, which was of significantly lesser duration and intensity, yet produced non-trivial impacts.

Beginning in the mid-1960s, rainfall in central Massachusetts fell to unprecedented lows (Hardy 1965), with reservoir levels corresponding closely (Figure 11.2). The state was entering what the United States Geological Survey deemed "the severest drought on record in the Northeastern United States" (Paulson *et al.* 1991). In addition

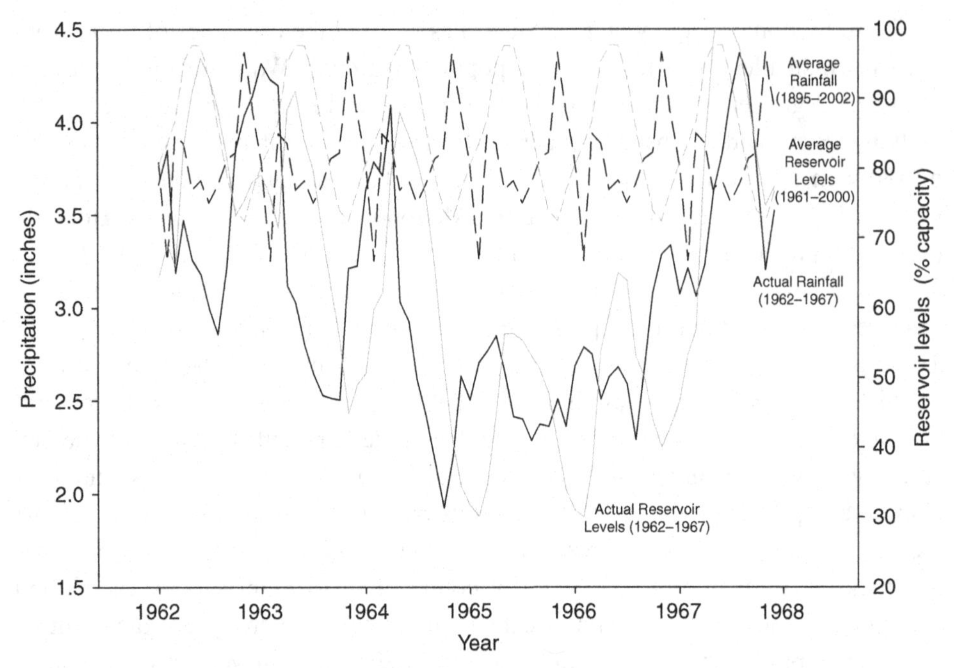

Figure 11.2. Actual and average monthly rainfall and reservoir levels, Worcester, Massachusetts (1962–67).

to low rainfall, Worcester was further hindered by the previously mentioned closure and contamination of Lynde Brook Reservoir due to clay runoff from nearby development at the Worcester Regional Airport. The contamination decreased available water by as much as 1 billion gallons in 1963, and by several millions of gallons more in 1964 and 1965 (Hardy 1964, 1965). Unusually low rainfall coinciding with the contamination forced Worcester to reduce demand and augment supply simultaneously.

To reduce demand Worcester implemented conservation measures, primarily outdoor watering and car-washing bans (Russell *et al.* 1970). From 1964 to 1965, overall consumption was reduced by about 500 million gallons (5.6%) (Hardy 1966). However, before reducing demand, the city sought out emergency water sources. In October of 1963, the Worcester Bureau of Water was authorized to make arrangements for tapping into Wachusett Reservoir, one of the principal sources of metropolitan Boston's water supply. During that year alone, the city spent $60 000 (approximately $372 000 in 2004 dollars) for MDC water, and also spent an additional $15 000 (nearly $93 000 in 2004 dollars) pumping an extra 723 million gallons from its own Quinapoxet Reservoir. In 1966, Worcester signed into an agreement with the MDC to build a pumping station at Shaft 3 of Quabbin Aqueduct, so it could tap into the reservoir whenever necessary in the future.

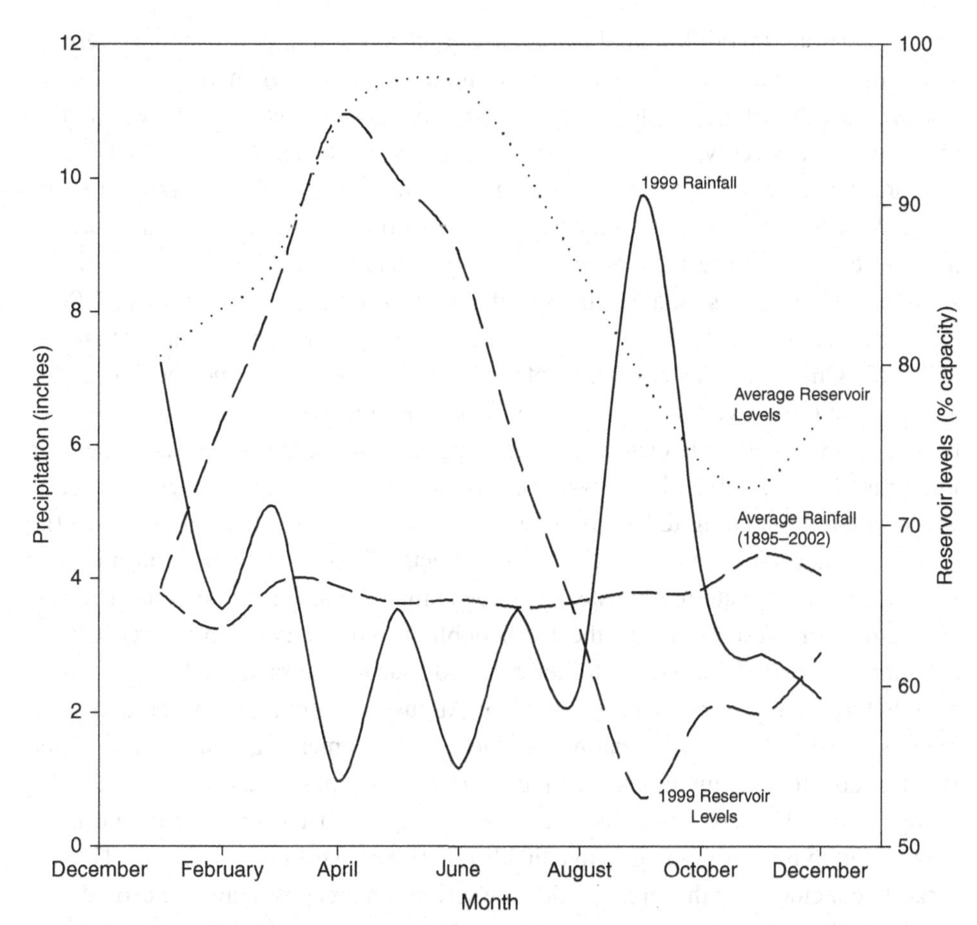

Figure 11.3. Precipitation and reservoir levels: average and 1999 values for Worcester, Massachusetts. (Source: Worcester Department of Public Works.)

While the meteorological drought of 1963–66 resulted in reservoir levels as low as 31.5% of capacity (Figure 11.2), the drought was exacerbated by the loss of a main water source (Hardy 1964). Due to limited availability of information on surrounding towns, our discussion of the 1960s drought focused on Worcester. It is important to recognize that the nine towns surrounding Worcester in all likelihood experienced more far-reaching impacts than Worcester proper due to unequal access to emergency water supplies.

The drought of 1999 was of a shorter duration with a less dramatic drop in rainfall and reservoir levels than the 1960s drought, but it significantly affected central Massachusetts all the same. April 1999 was the second driest April ever recorded to date in Worcester; rainfall in June and August was also below average (Figure 11.3). However, the area received average precipitation during May and July, and for the year as a whole the rainfall total essentially equaled the long-term annual average. Thus, three months of below average rainfall – coming when they did, in the summer months when people are trying to maintain their lawns – sufficed to drop reservoir levels steeply, to close to 50% of capacity in September (Shaw 1999).

Worcester responded to this situation with increasingly serious measures. In June, Worcester requested voluntary restrictions on residential water use, which had the effect of lowering consumption significantly (~25% by the end of the summer) but also resulted in the significant loss of revenue (some $350 000) (Kotsopoulos 1999). In July, the city began buying emergency water from the MWRA's Quabbin Reservoir (at a total cost of $1.2 million), and the Worcester City Council approved a voluntary outdoor watering ban (McDonald 1999). As the drought was ending in October, the city began repairing the Wachusett Reservoir Pumping Station, although the water could only be used as a last resort since it was unfiltered, violating federal regulations on water quality (Kotsopoulos 1999).

In surrounding towns there were similar effects. Shrewsbury and Auburn implemented voluntary watering bans early in the summer, and these bans soon became mandatory. In West Boylston, the town publicly requested voluntary reductions (Magiera 1999a). The town of Holden rejected four formal requests for mandatory bans before finally implementing them for August (Magiera 1999b). Holden also spent some $583 000 investigating prospects for augmenting supply from Poor Farm Brook to prevent future shortages. Thus, the impacts associated with the drought of 1999 were not unlike those associated with the drought of the mid-1960s. Yet, in meteorological terms, the droughts were strikingly different. Thus, it is fair to conclude that this area would face serious challenges if another drought on the scale of the 1960s event were to occur. We also conclude therefore that local sensitivity to the effects of drought has increased in recent decades. In terms of how society is responding to this increased sensitivity, there is a discernible shift from a focus decades ago on supply-side responses to the more recent emphasis on addressing demand first (developing what one CWS manager dubbed a "culture of conservation").

Central Massachusetts: structural underpinnings

"Structure" in this context refers to the rules and regulations, both formal and informal, which are designed to guide how people individually and collectively use their water. State legislation on water resources appears to have negatively affected adaptive capacity in the Central Massachusetts study site. In 1983 Massachusetts passed the Interbasin Transfer Act (ITA), to manage the transfer of water between watersheds and reduce human alterations of stream flows. The ITA encourages demand management by making water supply augmentation a time-consuming process that can be undertaken only after conservation measures and other alternatives have been explored (DCR 2005). Two years later, the Water Management Act (WMA) was implemented with the intention of managing demand by "protect[ing] water resources by limiting withdrawals to a 'safe yield,'" as judged by environmental impacts (CZM 2005).

These policies are intended to increase local water management adaptive capacities under the assumption that protecting the local surface- and groundwater supply bases would enhance water system management options in times of low precipitation. However, these state-level structural forces have not necessarily had the intended effect. For example, the withdrawal permits given to each CWS are based in part on water consumption levels between 1981 and 1985. This "baseline withdrawal" standard has generated criticism for (unintentionally) rewarding towns with excessive consumption during the 1981–85 period (Glennon 2002). In addition, the permitting process (which can take as long as a decade) can represent a burden on towns lacking substantial administrative and legal resources. As a result, towns are (again, unintentionally) given the incentive to respond to drought situations by establishing a connection either with the MWRA or with another town that has connections to the MWRA. Towns on the MWRA system are not necessarily held to the same consumption cap as towns that supply their own water, so the incentive to reduce consumption – i.e., the original intent of these structural forces – is lost.

Central Massachusetts: the role of agency

"Agency," i.e., the ability of individuals to take actions to modify a water system and/or the physical landscape, also merits consideration when trying to understand a given water system's vulnerability to the effects of drought. Developers are the first of three principal classes of agents found to be significant players in shaping local vulnerabilities in central Massachusetts. Developers have discretion as to the prominence of conservation devices and the use of landscaping techniques that reduce runoff and impervious surfaces. These actions can reduce the strain development puts on CWSs, allowing the CWSs flexibility and enhancing their ability to respond to the effects of climatic variation. Interviews revealed that developers engaged in these measures only to the degree required by local by-laws or state legislation. Half of the developers interviewed derided current environmental regulations (such as protections for wetlands) as hindrances to "efficiency" in development, and 75% thought that their clientele, though aware of conservation measures, would not prefer water- and energy-efficient homes if such features increased the cost of the home. This is likely due to the perceptions held by the developers themselves; they expressed a general lack of concern about water availability, often citing recent abundant rainfall and efficient management by water departments. It is also true that home-buyers are sensitive to price.

Town planners and advocacy groups are other agents found to impact adaptive capacity in this study area. These "agents" work as part of the social "structure," but in central Massachusetts they reflect agency more than structure. Planners review and influence by-laws regarding development projects in their town and influence open space preservation. Many towns in Massachusetts do not have sustained planning efforts, as most towns are so small that they cannot afford to pay a staff of professional planners on a fulltime basis. As such, towns may not be able to avail themselves of the adaptive options theoretically open to them through the planning mechanism. The state steps into this vacuum to assist with those matters, by promoting planning through region-level (i.e., multi-town) planning commissions; the planning commission in this study area is the Central Massachusetts Regional Planning Commission (CMRPC). The dilemma is that these planning commissions possess little if any legal authority. Their primary role is to serve a convening function, to catalyze people in various towns interested in growth management to meet and brainstorm possible solutions. The CMRPC stays away from advocating water-related policies seeing it as "primarily a domain of state government." Even without an explicit focus on water availability, helping to protect open space and supplementing towns' resources will advance the towns' adaptive capacity. Thus, town planners contribute in theory to improving local adaptive capacity – indeed, that is their very mission – but their ability to bring about significant change is limited by their limited budgets and authority.

Advocacy groups serve a similar role to the planners. They often work with individuals to gather political support for enacting conservation easements, monitoring water quality, and providing a voice for the watershed when (often volunteer-staffed) town planning boards and conservation commissions are considering applications for new residential developments. However, advocacy groups in central Massachusetts were found to be more focused on issues of water quality rather than quantity.

Eastern Massachusetts

Eastern Massachusetts landscape

The Eastern Massachusetts study region exhibits a similar but slightly more moderate climate and topography compared to Central Massachusetts, moderated by close proximity to the Atlantic Ocean (approximately 5 miles) and less physical relief of the coastal plain. The selected study area towns – Danvers, Middleton, and Topsfield – lie completely or partially within the Ipswich River Watershed (Figure 11.4). The three towns collectively cover less than 40 square miles

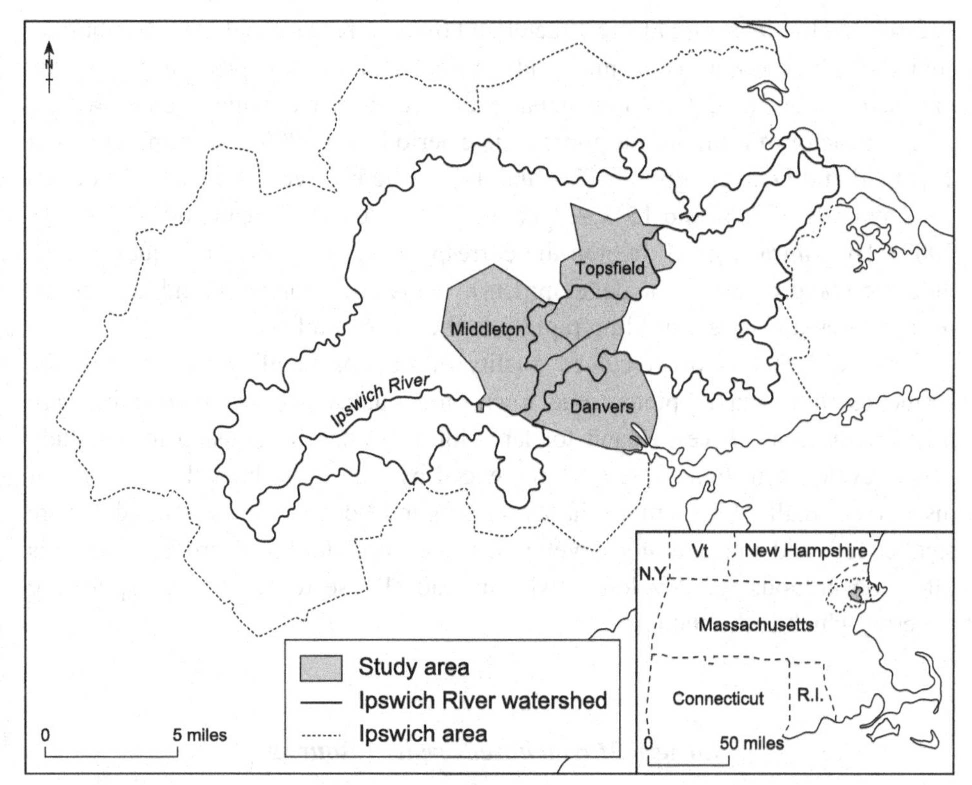

Figure 11.4. Ipswich River Watershed area.

(US Census Bureau 2000), with a joint population of about 39 000 residents (US Census Bureau 1999, 2000). The source of the Ipswich River is in the wetlands of the town of Burlington, and the river drains into Plum Island Sound to the northeast. The watershed is dominated by forests of white pine and mixed hardwoods, mushrooms, ferns, wildflowers, rushes, and mosses, and diverse terrestrial and riparian mammals, birds, reptiles, fish, amphibians, and invertebrates (IPSWATCH 2005).

Eastern Massachusetts land-use change: from agriculture to impervious surface

In the Eastern Massachusetts study area, all three towns have transitioned from agriculture to residential land use in the last century. Danvers, Middleton, and Topsfield are all "bedroom communities" – towns where people live and raise families, but commute into larger towns or cities for employment opportunities – as they are located only ~20 miles from Boston. The dominant development style today is large-lot, single-family housing, with average lot sizes equal to

1–2 acres. The towns are also characterized by features associated with suburban sprawl, such as single-use zoning and car-dependent landscapes. For two of the towns, the rate of land development has exceeded the rate of population growth by at least a factor of four. For the approximate period 1990–2000, the populations of Topsfield and Danvers grew by 7% and 4%, while undeveloped land was developed at a rate of 32% and 15%, respectively (US Census Bureau 2000; MassGIS 2002). By contrast, in Middleton the corresponding rates were of equal magnitudes (57% growth, 57% development). Thus, development and associated resource demands are unfolding rapidly in the entire study area.

The social factors of perceived quality of life for families and retirees, the aesthetics of a quaint, picturesque town, and quality schools funded by high property taxes all drive demand for large-lot single-family housing in the study area. Development trends are also influenced by zoning by-laws that prohibit or discourage smaller lots or multi-family housing in Middleton and Topsfield. Where smaller lots and multi-family development are possible, the approval process is often too arduous for developers who instead choose to build more sprawling patterns, which is the status quo.

Eastern Massachusetts water resources

An important characteristic of these three towns is their complete dependency on the Ipswich River Watershed for their water supply. This dependency places these towns at a disadvantage relative to some of their neighbors in the watershed; several watershed towns have established (or are strongly considering establishing) connections to the MWRA, Boston's water supply, and therefore are not affected as heavily by variations in streamflow in the watershed. The recent growth of these three towns has led to increasing rates of water withdrawal and decreasing rates of groundwater recharge. In turn, falling groundwater and surface water levels have been observed. This development has also been associated with increasingly degraded water quality, as anthropogenic pollutants are more concentrated in an environment of falling water levels, and are more prevalent to begin with in an environment of growing human presence (IPSWATCH 2005).

The three area towns examined in this study pump groundwater and surface water from the watershed for municipal use. Topsfield operates two groundwater wells and Danvers and Middleton receive their water from three joint surface water sources and two joint groundwater sources (Table 11.1). The Danvers–Middleton relationship is unusual for the region, as one town (Danvers) owns all of the other town's (Middleton) water sources, but each town operates its own

Table 11.1 *Population and water system information by town*

Town	Water source	Number of days permitted	Consumption permitted (millions of gallons per day)	Population (2000)	Population growth (1990–2000)	Percent increase in residential land cover (1971–99)
Danvers	3 Surface water, 2 groundwater	365	3.83	25 212	4%	15%
Middleton	Water supply owned by Danvers	See Danvers	See Danvers	7 744	57%	57%
Topsfield	2 Groundwater	365	0.66	6 141	7%	32%

Source: US Census Bureau 2000; IRWA 2005; MassGIS 2005.

water system as required under state law. However, the municipal water supply does not serve all residents in any of our study towns. Approximately half of Middleton's residents and 20% of Topsfield residents rely on private wells for their water.

Much of the development in the study area limits water recharge to the Ipswich River Watershed through impervious surfaces and water-exporting infrastructure (namely, the wastewater export system). As development progresses, areas of impervious surface such as roads and rooftops increase, preventing precipitation from infiltrating the ground and recharging the groundwater. In developed areas, rainfall moves quickly over the surface to streams that export it rapidly from the watershed. Wastewater export also occurs from areas that have sewers by a process of inflow and infiltration, in which clean water enters the sewer system from intentionally channeled gutter systems (inflow) or seeps into cracks in pipes (infiltration) and is transported to sewage treatment facilities and recharge destinations outside the watershed.

As groundwater recharge within the watershed diminishes, water withdrawal from the watershed increases with development, further lowering water levels. The low-density residential development of the study area is associated with greater per capita water demand than medium- and high-density development. In our area, the most significant water use is outdoor watering, particularly in the irrigation of the large lawns that typically accompany houses built on large lots. Water use in the summer months typically doubles or triples winter water use in the study towns (EOEA 2002; Mackin and Wagner 2002). Consequently, water levels tend to be lowest in summer months when precipitation-driven recharge is lowest and irrigation-related withdrawal is highest.

Another development-related water quality problem in the Ipswich River Watershed is polluted runoff. With greater runoff across impervious surfaces, more contaminants are carried to rivers and streams because pollutants are not filtered out by the ground percolation process. The impact of contaminants is enhanced by lower water levels because less water is available to dilute contaminants allowing them to persist at higher concentrations. Contaminants including salt and petroleum enter waterways from runoff from roads and fecal bacteria from septic systems can make the water unfit for human use and unsuitable for various natural organisms. Runoff from fertilized lawns and leaky septic systems may cause eutrophication in the Ipswich River and its tributaries rendering sections unsuitable for recreational use. The decomposition of dead plant matter lower in the water column also causes dissolved oxygen levels to fall, a problem exacerbated by low flow, reduced water levels, higher water temperatures, and streams being reduced to stagnant ponds. Many species of aquatic organisms are unable to survive in water with low dissolved oxygen concentrations. This results in generalist species of "pond fish" eliminating the "river fish" such as trout in the Ipswich River (E.1 2005; IPSWATCH 2005). Interestingly, even though the same runoff and pollution processes presented here apply to both the Eastern and Central sites, the relatively advanced water cleaning infrastructure in Worcester diminishes this concern there.

Eastern Massachusetts: climate events and their impacts

Towns within the Ipswich River Watershed experience frequent drought watches and warnings and some towns, such as the three examined here, experience voluntary and mandatory restrictions on water use almost every summer. The most serious drought on record, in meteorological terms, for the Ipswich River Watershed occurred between May of 1965 and October of 1966 (Northeast Regional Climate Center 2003). This drought spurred the development of many organizations, rules, and regulations to limit the effects of future droughts by increasing and protecting water supplies within the Ipswich River Watershed. Yet, despite these efforts, drought remains a problem for the Ipswich River basin, even though the 1960s meteorological conditions have not recurred.

From the late 1990s to the early 2000s, many towns in the Ipswich River Watershed have again been faced with drought. There were periods of low- and no-flow on the Ipswich River in the summers of 1995 (Kearns 2004), 1997 (Horsley 2003), and 1998–99 (Horsley 2003). According to the non-point source action strategy of the Massachusetts Department of Environmental Protection, several brooks used by some

Ipswich towns also experienced low flow problems. The 1998–99 drought caused water bans to be implemented, but the bans were too late to protect the river's aquatic life (Horsley 2003).

As in central Massachusetts, problems of water shortage can be managed by some combination of controlling the amount of water demanded and increasing water supply. Water management for some towns in the Ipswich River Watershed is more flexible than the three towns examined here because the former are connected to water sources outside the watershed. In particular, a number of towns depend on the MWRA as their main water source. Towns unable to augment readily their water supply have to be more creative about ways of managing demand and supply in times of drought. Municipal water system managers in the study area have some tools to influence water demand, but their ability to expand water supply is highly limited because their water comes exclusively from within the Ipswich River Watershed. Danvers and Middleton are actively seeking new water sources in order to continue with their desired development plans.

Eastern Massachusetts: structural underpinnings

Many structural forces shape the response of agents to the current condition of the Ipswich River. These forces are largely the same as those enumerated for the case of Central Massachusetts above. However, the ways in which these structural forces affect water management differ somewhat between the two regions. For example, even though the Federal Safe Drinking Water Act (SDWA), which sets legal limits on the levels of certain contaminants in drinking water (EPA 2005), applies to the entire state, this structural feature was not mentioned by respondents in central Massachusetts as important. By contrast, in the Eastern Massachusetts study area, water system managers voiced frustration with the lack of funding they receive for the contaminant monitoring and treatment they are required to conduct. Compliance with any costly regulation reduces the resources a local water system has to devote to other measures, such as managing the effects of drought. The EPA (2005) estimated in 1993 that the annual costs of complying with the SDWA would cost an average of $52 000 per year for testing and upgrading equipment for each water system. Management of all three towns' water systems reports being short-staffed and under-funded, and therefore struggling to keep up with water supply maintenance and drinking water standards. Thus, the structural force of federal water supply policy (in this case, the SDWA) has an indirect effect on local adaptive capacities insofar as it represents an unfunded mandate for an institution with limited resources to begin with.

Respondents in the Eastern Massachusetts study area agreed with their counterparts in the Central Massachusetts study area by claiming that the state ITA and

WMA laws (see section on central Massachusetts structural forces) deter towns from looking for new sources. Again, such an outcome is consistent with the intent of those laws. Yet, this program has (unintentionally) reduced the adaptive capacity of local water systems with respect to managing the effects of drought: water system managers reported that in technical terms expanding supply is often a relatively straightforward and low-cost endeavor. Thus the ITA and WMA laws represent barriers to relatively low-cost water management options. Thus if it were not for these laws, local water managers would have an easier time enacting anticipatory adaptations, i.e., preparing for future periods of low rainfall.

These laws also appear to have affected local adaptive capacity in eastern Massachusetts by discouraging population growth from what it may have been in the absence of the laws. This outcome is significant because population growth typically adds to the revenue base of the local water system, thereby providing additional resources for preparing for times of stress, such as droughts. On the other hand, however, population growth increases the aggregate demand on the water resource base, which would suggest a heightened sensitivity to periods of low rainfall in the first place. Further research is needed to determine the net effect of diminished population growth on water system management adaptive capacity.

Study participants in eastern Massachusetts also mentioned Comprehensive Permits Law (Chapter 40B) as a source of stress on the water supply. This state law of course also applies to the Central Massachusetts study site, but respondents there did not report this law as having more than a theoretical impact on their adaptive capacities. Chapter 40B was enacted in 1969 in response to the shortage of low- and moderate-income housing in the state. It allows the state to override a town's decision to deny a development proposal – if the proposal contains at least a small number of low-income units – for those towns that do not currently have at least 10% of their housing stock as "affordable" (Commonwealth of Massachusetts 2003a). The significant feature of this law from the water management perspective is that Chapter 40B developments – by satisfying an affordable housing need – are exempted, as an incentive, from many, if not all, zoning restrictions, including some that relate to environmental protection or water-use efficiency. In these cases state-level structural incentives to produce water-efficient residential developments are lost. Consequently, per capita water use will probably not decline in towns with a significant presence of Chapter 40B developments, with the result that the town is left with a residential landscape that is less water efficient than it could be, at the same time that the town has to abide by state-level water withdrawal limits.

Another state-related reform in housing and zoning policy is Chapter 40R, the Smart Growth Zoning and Housing Production Act of 1994, which "encourage[s] smart growth and increased housing production in Massachusetts" (Commonwealth of Massachusetts 2003b). This law provides financial incentives to encourage adoption in the Commonwealth of zoning overlay districts. The goal is to allow mixed-use development and to encourage the use of existing building structures (Commonwealth of Massachusetts 2003b). Chapter 40R could help the area's adaptive capacity to drought by promoting low-impact development and water efficient residences. However, to date there have been few Chapter 40R developments and thus the associated effects on sensitivity and adaptive capacity remain to be seen.

Eastern Massachusetts: the role of agency

In this study region, the principal agents were planners, developers, advocacy group members, and members of the Metropolitan Area Planning Council (MAPC). Advocacy groups such as the Ipswich River Watershed Association (IRWA) are considered agents, as the group exhibits a small membership compared to the various levels of government agencies involved indirectly or directly in water management. Members in the IRWA exercise their agency by independently deciding to test water and monitor new developments to ensure developers are adhering to the appropriate rules and regulations. The group has also sued the state, claiming that the state has not fulfilled its obligation to protect local water resources. Members working within advocacy groups also offer suggestions and guidelines to local planners, such as through the creation of water conservation programs and methods.

The MAPC members (like the CMRPC) are responsible for creating suggested models to show people how to build in accordance with zoning laws and in ways that might achieve specific objectives, such as reducing "sprawl" or conserving water. The MAPC is also involved in drafting a "regional plan" every ten years, which is a policy instrument designed to increase cooperation and communication among towns. The most common topics of concern to the MAPC are water, housing, transportation, and economic development in general; the MAPC actively reaches out to communicate their findings to planners, selectmen, and the general public.

Planners are local officials responsible for recommending guidelines and regulations to planning boards for new developments and zoning changes to protect the landscape and to ensure developments conform to local regulations. Of interest in this study area is the fact that some of the town planners are not professionally trained as such. These towns are small and are not likely to be able to fund a full-time professional staff. Instead, the planning process is discharged by a group of rotating volunteers, some of whom are quite knowledgeable about the process but others of whom are not. Thus, the impact – positive or negative – of "planners" in managing water system vulnerabilities in a given town may be overstated. The planners are of course aware of their lack of formal training in those cases where it applies, and take care to "proceed carefully" to avoid "unintended consequences". However, an understaffed planning board that wishes to guide local development in a way contrary to the wishes of a developer who enjoys deep resources (such as legal aid) is likely to lose the battle with the developer, with the result that the town develops in, for example, a water-intensive, rather than a water-efficient, manner.

The role played by developers is identical in eastern and central Massachusetts. Developers in theory have the ability to develop using smart growth methods or conventional methods (low-cost, environmentally inefficient technologies and methods). Developers in the region predominantly choose conventional methods leading to 1- and 2-acre parcel developments dominating the landscape. The reason is simple: the costs, in time and money, of building in resource-efficient ways are often not the lowest-cost option available to the developer.

Central and Eastern Massachusetts considered together

The imposition of regulations on local actors has not been very effective at promoting drought-related adaptive capacity. The developers we interviewed had to comply with various environmental regulations and go through time- and resource-intensive processes to show compliance with laws or get the necessary permits. As a result, they felt overly constrained. One developer wished to experiment with mixed-use and alternative forms of development such as smart growth but felt that those goals were complicated by the need to focus on means rather than on the product itself. The consensus was that developers would only do the minimum required to satisfy the regulations instead of proactively addressing environmental issues. It is unclear whether local adaptive capacity would benefit more from continuing to rely largely on structurally created regulations or by relying more on the local agency of individuals (not only advocacy groups but also developers).

In other cases, structure and agency have worked together and produced a positive impact. For example, the state Community Preservation Act (structure), which if passed by individual communities (largely a reflection of agency), allows them to receive state funds to assist communities with open space protection, historic landmark preservation, and affordable housing production. A cabinet-level state organization, the Executive Office of Environmental Affairs

(EOEA), has conducted 'buildout analyses' (a product of structure), showing what each community will look like (under certain assumptions) when all its land is developed according to current zoning (EOEA 2000). This presents towns with the levels of water consumption, population, and developed land they will face if maximum buildout were to occur under the present zoning configuration. This information is provided free of charge on the Internet, and is exhaustively described so that individuals (a reflection of agency) can use it if they wish. This information is too costly to produce for individuals. By working together in this way, towns are encouraged through major incentives to preserve open space, which reduces disturbance to the hydrologic cycle, and to assess development. The state benefits by being able to focus funding on communities that are already engaging these ideas and have developed a local support base.

Institutional elements have the ability to supplement local agents' resources and knowledge bases such as in the case with the buildout analyses and the creation of regional planning commissions while also unifying the actions of local planners and CWS managers around common principles. The EOEA's recently issued Water Policy is an example of such balanced interactions: extensive efforts were made to solicit and incorporate input from a diverse range of stakeholders. Some interviewees were (and continue to be) involved in this process, which gathered input from stakeholders and the public to produce a set of recommendations to guide communities in integrating land-use and water planning for sustainable management of water resources. Reliance on voluntary cooperation could potentially leave some towns behind (specifically those without the resources to dedicate to following voluntary principles); it also has the potential to solidify a new, more unified relationship between municipalities and the state. Advances achieved through this cooperation have been stronger and provided more comprehensive and coordinated support for the region's adaptive capacity than advances resulting from structure or agency acting alone.

Thus, we conclude that in the rapidly suburbanizing the study areas, a synergy of structure and agency will prove to be crucial for reducing drought-related vulnerabilities, as top–down and bottom–up influences on adaptive capacity, taken separately, each have limitations. Policies and principles articulated at the state level can offer a framework in which towns can manage development in a way that does not sacrifice the ability of local water systems to adjust to droughts and floods. The formulation of the EOEA's Water Policy could be the most promising example of this synergy due to its efforts to decentralize input and incorporate a diverse range of stakeholders in its creation and execution. Yet the EOEA is the very agency responsible for imposing some of the regulatory constraints (particularly with respect to water consumption caps for towns in "stressed" basins), so there is a natural tension between competing objectives. As such, improvements in adaptive capacity achieved through approaches that leverage both structure and agency have been stronger and more comprehensive than cases where structure-only (top–down) or agency-only (bottom–up) approaches have been dominant.

References

Commonwealth of Massachusetts (The), 2003a. *General Laws of Massachusetts: Chapter 40B. Regional Planning.* Accessed at www.mass.gov/legis/laws/mgl/gl-40b-toc.htm.

Commonwealth of Massachusetts (The), 2003b. *Chapter 40R, the Smart Growth Zoning and Housing Production Act.* Accessed at www.mass.gov/dhcd/40R/default.htm.

Cutter, S., 1996. Vulnerability to environmental hazards. *Progress in Human Geography* **20**(4): 529–539.

Department of Conservation and Recreation (DCR), 2005. *Interbasin Transfer Act.* Accessed at www.mass.gov/dcr/waterSupply/intbasin/lawsregs.htm.

Glennon, R., 2002. *Water Follies: Groundwater Pumping and the Fate of America's Fresh Waters*. Washington, D. C.: Island Press.

Hardy, C. B., 1964. *Report of the Bureau of Water of the City of Worcester for the Year Ending December 31, 1963.* Worcester, MA: Bureau of Water.

Hardy, C. B., 1965. *Report of the Bureau of Water of the City of Worcester for the Year Ending December 31, 1964.* Worcester, MA: Bureau of Water.

Hardy, C. B., 1966. *Report of the Bureau of Water of the City of Worcester for the Year Ending December 31, 1965.* Worcester, MA: Bureau of Water.

Hill, T., and C. Polsky, 2005. Suburbanization and adaptation to the effects of suburban drought in rainy central Massachusetts. *Geographical Bulletin* **47**(2): 85–100.

Horsley, S. W., 2003. *Ipswich River Watershed Management Plan.* Accessed at www.horsleywitten.com/ipswich/REPORT.pdf.

IPSWATCH, 2005. *Ipswich–Parker Suburban WATershed Channel.* Accessed at www.ipswatch.sr.unh.edu/.

Ipswich River Watershed Association (IRWA), 2005. Accessed at www.ipswichriver.org/river.html.

Kates, R. W., 1985. The interaction of climate and society. In *Climate Impact Assessment: Studies of the Interaction of Climate and Society*, eds. R. W. Kates, J. H. Ausubel, and M. Berberian, pp. 3–36. Chichister: John Wiley.

Kearns, M., 2004. Low Flow Inventory, Sept. Accessed at www.mass.gov/dfwele/river/rivlow_flow_inventory/ipswich.html.

Kotsopoulos, N., 1999. Dry weather this year cost Worcester $1.2 million. *Worcester Telegram and Gazette*, October 23, p. A.4.

Mackin, K., and L. Wagner, 2002. *Ipswich River Basin Water Conservation Report Card: Ipswich River Basin on Water Conservation and Water Efficiency.* Accessed at www.massaudubon.org/PDF/NEWS/IPSWICH_RIV_REP.pdf.

Magiera, M. A., 1999a. Auburn water use restricted. *Worcester Telegram and Gazette*, July 21, p. B.2.

Magiera, M. A., 1999b. Water bans appear to be working. *Worcester Telegram and Gazette*, August 28, p. A.2.

Massachusetts Executive Office of Environmental Affairs (EOEA), 1971. *Land Use Datalayer.* Accessed at www.mass.gov/mgis/lus.htm.

Massachusetts Executive Office of Environmental Affairs (EOEA), 1999. *Land Use Datalayer.* Accessed at www.mass.gov/mgis/lus.htm.

Massachusetts Executive Office of Environmental Affairs (EOEA), 2000. *Buildout Maps and Analysis*. Accessed at commpres.env.state.ma.us/community/cmty_list.asp.
Massachusetts Executive Office of Environmental Affairs (EOEA), 2002. *Securing Our Water Future: Ensuring a Water Rich Massachusetts*. Accessed at www.mass.gov/envir/mwrc/pdf/waterpolicyinstitute.pdf.
MassGIS, 2002. *Datalayers/GIS Database: Land Use*. Accessed at www.mass.gov/mgis/lus.htm.
MassGIS, 2005. *Converted Land*. Accessed at www.mass.gov/mgis/lus.htm.
Masssachusetts Office of Coastal Zone Management (CZM), 2005. *Water Management Act*. Accessed at www.mass.gov/czm/envpermitwatermanagementact.htm.
McCarthy, J. J., O. F. Canziani, N. A. Leary, D. J. Dokken, and K. S. White (eds.), 2001. *Climate Change 2001: Impacts, Adaptation and Vulnerability. Contribution of Working Group II to the Third Assessment Report of the Intergovernmental Panel on Climate Change (IPCC)*. Cambridge: Cambridge University Press.
McDonald, C., 1999. Worcester taps Quabbin water: emergency supply likely to cost city nearly $10,000 a day. *Worcester Telegram and Gazette*, July 10, p. A.3.
Northeast Regional Climate Center, 2003. Palmer Drought Severity Index. Period of record: January 1895 – July 2003. Accessed at www.nrcc.cornell.edu/drought/MA_drought_periods.html.
Parry, M. L., O. F. Canziani, J. P. Palutikof, P. J. van der Linden, and C. E. Hanson (eds.), 2007. *Climate Change 2007: Impacts, Adaptation and Valnerability*. Cambridge: Cambridge University Press.
Paulson, R. W., E. B. Chase, R. S. Roberts, and D. W. Moody, 1991. *National Water Summary 1988–89: Hydrologic Events and Floods and Drought*, US Geological Survey Water-Supply Paper No. **2375**: 327–334.
Platt, R. H., 1995. The 2020 Water Supply Study for Metropolitan Boston: the demise of diversion. *Journal of the American Planning Association* **61**: 185–200.
Russell, C. S., D. G. Arey, and R. W. Kates, 1970. *Drought and Water Supply: Implications of the Massachusetts Experience for Municipal Planning*. Baltimore, MD: Johns Hopkins University Press.
Shaw, K. A., 1999. City may ask for water use limits. *Worcester Telegram and Gazette*, June 22, p. A.1.
Turner, B. L. II, R. E. Kasperson, P. Matson, J. J. McCarthy, R. W. Corell, L. Christensen, N. Eckley, J. X. Kasperson, A. Luers, M. L. Martello, C. Polsky, A. Pulsipher, and A. Schiller, 2003. A framework for vulnerability analysis in sustainability science. *Proceedings of the National Academy of Sciences of the USA* **100**: 8074–8079.
United States Census Bureau, 1973. *The U.S. Census 1973, Volume 1, Part 23*. Accessed at www2.census.gov/prod2/decennial/documents/1970a_ma-01.pdf.
United States Census Bureau, 1999. *The U. S. Census* 1999. Accessed at http://factfinder.census.gov.
United States Census Bureau, 2000. *The U. S. Census* 2000. Accessed at http://factfinder.census.gov.
United States Environmental Protection Agency (EPA), 2005. *Safe Drinking Water Act*. Accessed at www.epa.gov/safewater/sdwa/index.html.

12

A diverse human–environment system: traditional agriculture, industry, and the service economy in central Pennsylvania

BRENT YARNAL

Introduction

The central Pennsylvania study region is a land of natural and human contrasts that add important dimensions to the HERO project. It has rugged hollows and hills in the western part of the region, but broad valleys and low ridges in the eastern part. It has a humid climate with warm summers and cold winters. It has rich, thick soils in the valleys, but poor, thin soils on the forested hills and ridges. It is prone to flooding, yet is also surprisingly prone to drought. It is stunningly diverse socio-economically, with coexisting agrarian,[1] industrial, and post-industrial economies. The Central Pennsylvania HERO investigators focused their research on the heart of central Pennsylvania, Centre County, because it possesses all of these characteristics within a relatively small area.

This chapter describes Centre County's physical and human landscapes and its vulnerability to hydroclimatic extremes, specifically floods and droughts. It focuses on the exposure, sensitivity, and adaptive capacity of two important venues for human–environment interaction: emergency management and water supply management.[2] The chapter starts by painting a picture of the physical and human landscapes on which these interactions between people and their environment take place.

The physical landscape

Centre County lies in the geographical center of Pennsylvania (Figure 12.1; see also Figure 8.2), covering 1108 square miles (2870 square kilometers). The county sits astride parts of two major physiographic provinces: the Appalachian Plateau to the west and the Ridge and Valley region to the east (Figure 12.1). The Appalachian Plateau covers roughly the western and northern half of the county. The plateau has been highly dissected by stream erosion, resulting in numerous hollows separated

Sustainable Communities on a Sustainable Planet: The Human–Environment Regional Observatory Project, eds. Brent Yarnal, Colin Polsky, and James O'Brien. Published by Cambridge University Press.

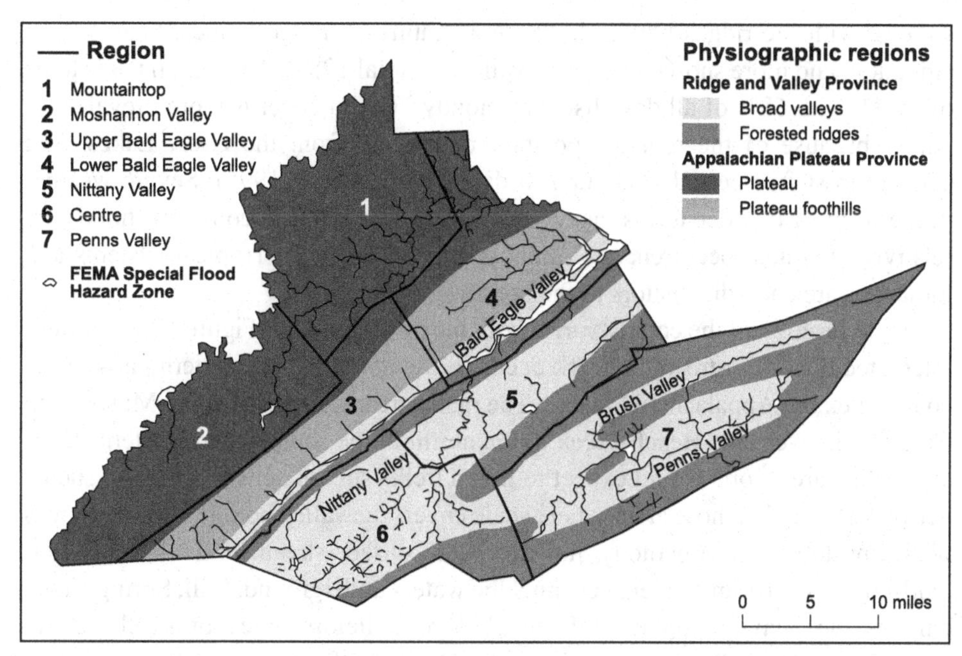

Figure 12.1. Central Pennsylvania Human–Environment Regional Observatory (SRB-HERO) study area showing physiographic regions and stream network. (Source: Centre County Government 2003.)

by plateau remnants, which produces a rugged topography that gives the impression of a mountainous landscape. In reality, the plateau consists of gently warped, but highly eroded sedimentary strata, including large fields of bituminous coal. Less than 50% of the coal in Centre County has been mined, leaving large reserves (Marsh and Lewis 1995). In contrast to the undisturbed Appalachian Plateau, the Ridge and Valley Province formed from the intense folding of flat-lying sedimentary beds as a result of the collision of North America with Africa during the Paleozoic Era (Faill and Nickelsen 1999). Consequently, the Ridge and Valley is a region of elongated, low ridges with intervening broad, rolling valleys; both ridges and valleys stretch for many tens of miles (kilometers) from southwest to northeast across the southern and eastern half of the county. The ridges generally consist of very hard, slowly eroding sandstones, whereas the valleys are made up primarily of softer, more easily weathered limestone, with occasional large pockets of shale and smaller deposits of iron-bearing sandstone (Faill and Nickelsen 1999).

Centre County has a humid continental climate, with the 39 inches (1000 mm) of annual precipitation distributed somewhat evenly throughout the year. Average annual temperature is approximately 50 °F (10 °C), with summer mean temperatures of 71 °F (21.5 °C) and winter mean temperatures of 26 °F (−3.5 °C) (Yarnal 1995). There are striking variations in weather between the ridges and valleys, with the ridges having lower temperatures, stronger winds, heavier precipitation, and more snowfall than the valleys (Yarnal 1989). The region tends to be overcast, with 68% of all days listed as cloudy.[3] Cloud cover is more prevalent in winter because of the region's position downwind from the Great Lakes. The average frost-free period is about 140 days (Yarnal 1989, 1995). Severe weather in the form of thunderstorms, heavy snow, and ice storms is common; floods are relatively frequent occurrences (Yarnal *et al.* 1997, 1999). Tropical systems and tornadoes are sporadic visitors to the county.

The hydrology of the county varies with the physiography (Figure 12.1). A well-integrated dendritic stream network occupies the northern and western parts of the county, i.e., the Appalachian Plateau. The main streams in this area are Moshannon Creek to the west and Beach Creek to the north. In the southern and eastern Ridge and Valley areas, one would expect to find a trellis stream network. In fact, such a network is found in those areas where shale covers the valley bottoms. In large areas with limestone (karst) geology, however, few surface streams exist because networks of sinkholes and caverns channel the waters underground. Still, Spring Creek empties the heavily populated State College–Bellefonte area of south–central Centre County and Penns Creek drains the Penns Valley region of eastern Centre County, both of which are primarily karst landscapes. Bald Eagle Creek runs the length of the county from southwest to northeast, occupying the transitional zone between the Appalachian Plateau and the Ridge and Valley province.

The study area's natural vegetation consists mainly of tall, broadleaf deciduous trees known as Appalachian Oak Forest (Abrams and Nowacki 1992). Cultivated areas are geographically intermittent because of variable soil quality, which depends on the parent strata in which the soils formed (Miller 1995). In general, upland areas in both the Appalachian Plateau and Ridge and Valley have thin soils with low nutrient status. Narrow stream valleys in the Appalachian Plateau tend to have better soils, but little total area is available for agriculture. In contrast, the broad valleys of the Ridge and Valley tend to have thick, rich soils, with the best soils found in the dominant limestone areas and lesser soils found in shale pockets. Valley soils formed in iron-bearing sandstones are useless for agriculture.

The human landscape

Centre County is a complex, sparsely populated mixture of agrarian, industrial, and post-industrial activities. Its population was roughly 136 000 in 2000, when the HERO project started, and is estimated to have grown to approximately 141 000 by 2006 (US Census Bureau 2008a). Population growth was 10.7% for the entire decade of the 1980s and 8.8% for the 1990s, with most of that growth taking place in the Centre Region. The Centre Region – which is home to Penn State

University and consists of the Borough of State College and the surrounding townships of College, Ferguson, Halfmoon, Harris, and Patton – is the most urbanized part of the County, with 57% of its population formally classified as urban. The Centre Region is accordingly the most densely populated part of the county with a population of over 42 000 students and over 37 000 permanent residents. Most remaining areas in the county are lightly populated and classified as rural. The average population density of the county in 2000 was 123 people per square mile (48 people per square km) as compared to 274 people per square mile (106 people per square km) for the state of Pennsylvania (US Census Bureau 2008a). The university population's influence is evident in the average age of residents in the county. The mean age for Centre County is roughly 27 years, whereas the means for surrounding counties range from 32 to 36 years (Denny *et al.* 2003).

Despite the great diversity that marks the county's physical and human landscapes, the area is not diverse racially. In 2000, over 91% of county residents were white, compared to 81% for Pennsylvania and 75% for the United States. Of the less than 9% of the county population that was non-white, 4% was Asian, 2.6% was African–American, 1.1% was more than one race, and all other groups totaled less than 1% each (Centre County Planning Office 2004; US Census Bureau 2008b).

Outside the Centre Region, Centre County is in many respects an integral part of Appalachia. The population in the Moshannon Creek and Beach Creek watersheds of the Appalachian Plateau is particularly economically depressed, which may account for the slow economic growth and population decline occurring in these areas (Simpkins 1995). It is the large number of college students living in the county, however, that skews poverty levels and incomes downward. College students generally have very low incomes and high apparent poverty rates that distort the overall economic characteristics of the county (Centre County Government 2003). Nevertheless, county per capita incomes remain higher than those of surrounding counties, while unemployment rates are significantly lower, largely because of the vigor and stability of the university.

Eighteenth- and nineteenth-century economic development in Centre County was based on farming and extracting timber and metals. In the late nineteenth century, numerous coal mines and a host of industries and services supporting coal mining sprung up in the Appalachian Plateau portions of the county, continuing the extractive focus of the economy. Diverse types of manufacturing became important across the county in the mid twentieth century. In the last one third of that century, non-extractive industries and services became prominent components of Centre County's portfolio, with the booming growth and economic health of Penn State University driving the economy. During that time, manufacturing employment decreased steadily, only employing 10.6% of the county labor force in 2000; post-2000 plant closures further decreased that number significantly and nearly eliminated manufacturing from Centre County. Coal mining and ancillary sectors were important into the 1980s, but completely stopped by the 1990s despite the presence of exploitable coal reserves; only large-scale limestone quarrying continues the county's mining tradition today. Agriculture and forestry still occupy much of the land, but together with mining only employed 1.7% of the county's workers in 2000. In contrast to this gloomy picture, the relatively high-paying professional and educational service sectors employed 7.5% and 36.2% of workers, respectively, in 2000. When taken together with information services (2.4%), financial, insurance, and real estate services (4.0%), arts, entertainment, recreation, accommodation, and food services (10.8%), public administration (3.5%), and other services (3.7%), services accounted for 68.1% of all county employment in 2000.

There is a strong spatial bias in these employment data. The non-student population living in the Centre Region tends to have relatively high-paying professional, educational, and information-based employment, whereas a large proportion of the population living outside this core region works at the university in lower-paying, but secure service positions with excellent fringe benefits. Large numbers of low-paying service, construction, and retail workers commute daily from outlying areas in Centre County and adjacent counties into the Centre Region for their jobs, causing intense but short-lived traffic congestion. These commuters come from the depressed economies of the Appalachian Plateau to the west and north, where coal mining was important, from the declining industrialized urban centers surrounding Centre County, such as Altoona in Blair County, and from the declining agricultural economies to the east. The Centre Region is central Pennsylvania's most dynamic growth pole.

Centre County is conservative politically, religiously, and culturally. Republican votes have dominated for the last several decades (Williams 1995), except in the Democratic hotspot of State College Borough. Conservative Protestants form the largest religious group in the county. Enclaves of Amish and conservative Mennonites occupy the agricultural areas east of the Centre Region (Zelinsky 1989, 1995). These Anabaptist populations have a long history in the county, first coming into the area in the late eighteenth century.

It is possible to summarize the human geography of Centre County by dividing it into three socioeconomic/sociocultural populations following three different stages of development: agrarian, industrial, and post-industrial (Kates *et al.* 1990). The agrarian population lives mainly in the Penns Valley region of eastern Centre County. Farms are traditional, non-industrial, family-based units that are small by American standards. Amish and Mennonites (i.e., Anabaptists) occupy many of these farms, with the Amish using horses and mules to power farm vehicles and horse-drawn buggies for local transportation. "English" (i.e., non-Anabaptists) occupy the remainder of the farms and rely on gas-powered, but small-scale farm equipment and vehicles. The "English" find it difficult to make a living from on-farm activities, so most families have members working off-farm, commuting to the Centre Region daily. The Anabaptist and "English" populations have lived cheek-by-jowl for over two centuries and share many common cultural traits, especially a deep social, cultural, and religious conservatism and a distrust of government authority. Family and community ties tend to be strong in these populations.

The second population is industrial and occupies the Appalachian Plateau in the western and northern portions of the county. These people settled the area in the late nineteenth century/early twentieth century, working in the coal mines and the support industries for the coal mining and steelmaking that dominated western Pennsylvania until the late twentieth century. With the collapse of coal and steel, the region suffered economic depression, massive unemployment, and significant social problems. Many remnant towns and villages persist, but with no local work; population is declining with those still living in the area commuting long distances to work in low-paying service jobs, often in the Centre Region. In contrast to the agrarian people on the opposite end of the county, they are a much coarser people.

Occupying the central ground between the agrarian population to the east and the industrial population to the west is the third wave of development in Centre County – the post-industrial population. Penn State was founded as the Farmers High School in 1855, with the town of State College and the five townships growing around the institution over time. Growth exploded after The Pennsylvania State College became a university in 1953, roughly coinciding with the restructuring of Pennsylvania's industrial economy into one dominated by services. The post-industrial economy of the Centre Region has been fueled by the propinquity of Penn State University, which offers a highly educated labor pool. Indeed, educational attainment in the Centre Region as measured by proportion of the population with high school, baccalaureate, and postgraduate degrees far exceeds that of the remainder of the county, state, and nation (Centre County Planning Office 2004). The result is an upwardly mobile population forming an economic, social, and cultural island quite unlike anything else in central Pennsylvania. As the Centre Region continues to thrive, it grows outward, putting pressure on the remaining people and land of Centre County.

Vulnerability to hydroclimatic extremes

As in the other three HEROs, water is key to understanding vulnerability in the Central Pennsylvania HERO. Water is not only essential for human health and many of the region's economic activities, but also a threat to human health and the economic activities, with floods and droughts having significant impacts on the region's past, present, and future. The following section covers the three dimensions of vulnerability – exposure, sensitivity, and adaptive capacity – in Centre County, focusing on hydroclimatic extremes (floods and drought), especially in relation to community water systems (CWSs).[4]

Exposure

Centre County, like much of Pennsylvania, is exposed to severe floods from several flood-producing mechanisms (Yarnal *et al.* 1999). In the eastern United States, Pennsylvania ranks first in flood-related deaths and second in number of floods (LaPenta *et al.* 1995). The state is home of the infamous Johnstown floods of 1889, 1936, and 1977, which killed over 2300 people, and the 1972 Tropical Storm Agnes flood, which was one of the costliest disasters in United States history (Yarnal *et al.* 1999). In 1996 alone, Pennsylvania experienced five presidential disaster declarations for floods that resulted from five different mechanisms: a rain-on-snow event, a mesoscale convective complex, a summer squall line, a tropical storm, and an early winter storm (Yarnal 2004). In that same year, Centre County suffered from the January rain-on-snow event experienced by the rest of the state and from a June 17 event with localized, severe convection that dumped over 5 inches (125 mm) of rain near the Penn State campus in 40 minutes. The year 1996 was not alone in the flood record: for instance, the county also saw significant floods in 1936 from a major rain-on-snow event and in 1972 from Tropical Storm Agnes.

Despite this propensity for floods, Centre County is also exposed to various degrees of drought a few times per decade. These droughts can last from seasons to many years. An example of a seasonal drought was the short-lived, intense drought of summer and early fall 1995. Throughout these few months, the storm track diverted from its normal course and Centre County received essentially no rainfall; cloudless skies increased the evapotranspiration rates and further reduced available water. The drought broke suddenly in late October with a shifting of the storm track and the onset of the wettest period in Susquehanna River basin history (Yarnal 2004). An example of multi-year drought occurred in the 1960s when the entire mid-Atlantic region faced the driest decade in history (Yarnal and Leathers 1988). Similar to the 1995 drought, the record 1960s drought resulted from a large-scale diversion of the Polar Front and broke with a significant shift in the storm track in 1970 that brought the wettest decade in the regional record.

Sensitivity

Despite repeated exposure of the area to floods and droughts, many elements of the Central Pennsylvania HERO are sensitive to these hydroclimatic variations. The focus here is on two important elements of central Pennsylvania's human–environment interactions – emergency management and CWS – and reasons for their sensitivity.

Emergency management and floods

Centre County could be sensitive to natural hazards because its local emergency management system is relatively untested. Similar to other states, emergency management in Pennsylvania has a clear hierarchy. By federal and state law, each of Pennsylvania's 2567 municipalities – that is, each township, borough, or city – must have an emergency management services coordinator and plan. Many of these local coordinators are unpaid political appointees or volunteers with little to no training as an emergency management professional. Each of the state's 67 counties must have a professional County Emergency Management Coordinator to coordinate emergency management services and training of local coordinators and staff members. County coordinators work under the authority of the Pennsylvania Emergency Management Agency (PEMA). PEMA manages this system and coordinates local, county, state, and federal regulations. The Federal Emergency Management Agency (FEMA) sits atop the entirety of emergency management in the United States.

The responsiveness of this post-September 11, 2001 system is not clear because it has yet to be tested. Knuth (2004) found that local responders – fire services, emergency medical services, and police – are responsible for most day-to-day emergency response in Centre County, although they are not under the direction of FEMA. The municipal emergency managers mandated by FEMA and enforced by PEMA serve a political function but may or may not be active in coordinating response. In contrast, the Centre County Emergency Coordinator is very active in both coordinating emergency response and conducting hazard mitigation planning. That office was in place long before 2001.

Because there are many ways to assess sensitivity to a hazard such as flood or drought, it is helpful to focus on a specific case. Here, the focus will be on what may appear to be an unusual sensitivity – sensitivity to flooding outside of floodplains – which parts of Centre County experienced with a major, regional rain-on-snow flood in January 1996 and a heavy convective storm on June 17, 1996 (Yarnal *et al.* 1997, 1999). Blocked or under-designed sewer and retention pond systems, water channeled by roads or parking lots into low-lying areas, or simple overland flows in areas where infiltration rates were less than rainfall or snowmelt rates were common problems that resulted in flooding away from floodplains during these events.

One indicator of sensitivity to floods outside of floodplains is presence or absence of flood insurance. The first line of economic defense against exposure to floods is insurance. In the United States, however, traditional homeowners' policies cannot cover water damage from floods; only insurers working within the National Flood

Insurance Program (NFIP) can cover such claims (Insure.com 2008). For individuals to be eligible to purchase a flood insurance policy from NFIP, their community must participate in the Floodplain Protection Program (FEMA 2007). For communities to be eligible for this program, they must have some of their incorporated area designated as a Special Flood Hazard Area, in which case the purchase of flood insurance is mandatory for the protection of property located in that area. Special Flood Hazard Areas are universally located in floodplains.

As noted, many of the areas flooded by the January and June 1996 floods in Centre County were not in floodplains. Many of the homeowners, renters, and businesses affected by the floods were not located in communities eligible to participate in the Floodplain Protection Program, so they were not insured for flood damages. For those located in communities that did participate in this program, it was possible to purchase a relatively inexpensive flood insurance policy. Because there is such a strong perception that floods only occur in floodplains, however, this possibility was largely overlooked. An informal phone survey of emergency managers and insurance agents in Centre County revealed that none were aware that it was possible to purchase such a policy. In short, government flood insurance did not directly cover a significant proportion of the population affected by flood damage, thus revealing a critical sensitivity of the study area.

Beyond insurance, disaster aid and disaster loans provide much of the social safety net for flood victims. The amount and type of aid available primarily depends on the institutional level of the disaster declaration; in the United States, the most common designations parallel the county, state, and federal levels of government. Generally, to participate in disaster relief and recovery, state emergency management agencies require a request from a county-level agency, while FEMA needs to be asked for help by a state agency (FEMA 2007, 2008). Only then can the President make a disaster declaration, which makes funding available to victims.

County- and municipal-level disaster designations follow guidelines established by the disaster division of the Small Business Administration (US Government Printing Office 2008). For a county or municipality to be declared a disaster area and to be eligible for aid or loans, at least 25 of a combination of its businesses and houses must suffer damage equaling at least 40% of their replacement value. The number can be as low as three if three businesses employ at least 25% of the area's workforce. A problem develops when storms are highly localized because flood damage can be intense over a very small area, but few homes or businesses might be damaged. Unless the amount of damage meets the Small Business Administration's disaster designation threshold, disaster victims cannot benefit from institutionalized relief and recovery. Therefore, homeowners and businesses are sensitive to the size of the flood and where the flood strikes – especially if the flood is outside an NFIP floodplain.

Water supply is especially sensitive to the problem of localized flooding. For example, when the state of Pennsylvania does not declare a county to be a disaster area, individuals or businesses affected by flooding do not receive certain emergency planning benefits. Damages from the June 17, 1996 severe convective event were too localized and not sufficiently widespread to warrant the state to declare Centre County a disaster area. Subsequently, free well-sample testing – available to well owners after a state disaster declaration – was not available to the many private homeowners whose wells had been flooded and were likely to suffer contamination.

When a county fails to achieve disaster status, there is still some protection against the effects of out-of-floodplain floods. For example, the state requires CWSs to maintain emergency management plans. As part of these plans, some water authorities have developed cooperative agreements with neighboring systems to ensure safe water supply under disaster conditions. Also, most CWSs work closely with county emergency managers, especially during floods, so although money may not be forthcoming, advice and a helping hand are still available. Nonetheless, and in summary, flood relief and recovery efforts associated with the out-of-floodplain floods of 1996 demonstrate that institutional safety nets can fail to protect households, businesses, and even water authorities, thereby increasing the sensitivity (and therefore vulnerability) of all who fall through the gaps in those nets.

Community Water Systems

There are many reasons that Central Pennsylvania's CWSs are sensitive to climate variation and change. Central Pennsylvania HERO investigators discovered at least six influences on CWS sensitivity: water source type, infrastructure age, regulatory compliance, emergency plans, source backup, and manager characteristics.[5]

CWSs that rely on surface water are more sensitive to weather and climate than are systems that rely on groundwater (see also O'Connor *et al.* 1999). Surface water systems are exposed directly to weather-related problems, such as sedimentation, and respond quickly to increased or diminished inputs, such as storm flow. In contrast, groundwater is shielded from the weather and responds more slowly and conservatively to inputs. Nevertheless, the karst geology of Centre County complicates this relationship. Most of the CWSs in the county are groundwater systems, but because of the many sinkholes, caverns, and disappearing streams found in the valley floors where people live and draw water, surface water reaches the subsurface much more quickly and directly than it does in less porous geological settings. Consequently, local groundwater is often legally under surface water influence so that state and federal regulations treat it as surface water, often requiring expensive filtration systems typically associated with surface systems. Thus, sensitivity to weather and climate in Centre County's CWSs depends not only on the surface-water–groundwater dichotomy, but also on the split among groundwater systems experiencing surface water influence and those groundwater systems shielded from surface water influence. Surface water is most sensitive, groundwater under surface water influence has moderate sensitivity, and groundwater with no surface water influence is least sensitive to weather and climate.

Infrastructure age has a major influence on CWS sensitivity to weather and climate. Nearly half of the CWSs whose managers were interviewed by the HERO team have pipes that are more than 100 years old – including some that are wooden – and therefore prone to failure and constant repair. Weak pipes are susceptible to extreme cold and the increased pressure generated by heavy rains or floods. Because of their age, systems often do not have the necessary shut-off valves to isolate leaks and breaks. Older systems often lack individual meters on homes and businesses, further making it difficult to identify leaks and points of failure in the systems. Older CWSs also tend to have smaller pipes, thus making it difficult to provide sufficient water to new hook-ups on the system.

The US Environmental Protection Agency formulates water quality regulations and the Pennsylvania Department of Environmental Protection enforces those regulations. An important goal for these agencies is to help make it possible for CWSs to provide plentiful and safe drinking water to its customers. Yet, the independent-minded, conservative ethos in Central Pennsylvania often leads local people to resist regulations because the regulations are perceived to be an undue threat to local autonomy. Those CWSs that are out of compliance with regulations are more sensitive because many regulations specifically target water-quality problems resulting from extreme weather and climate. For example, CWSs drawing groundwater influenced by surface water are required to filter their water because heavy rainfall or floods could easily wash farm wastes and other pollutants into the groundwater. The rural Madisonburg CWS has this type of system, but rather than comply with state regulations and reduce sensitivity to weather extremes, first the water board, then the entire village risked prison to maintain local sovereignty (Gibb 2000).

One particular problematic regulation is the requirement for each CWS to have a plan to provide clean water to customers during emergencies. Some systems are simply out of compliance, but more often systems download a boilerplate plan from the Internet, fill in the blanks, and thereby comply with the letter of the law. These systems consequently are much less able to cope with weather and climate extremes and are therefore much more sensitive than those CWSs that develop emergency plans tailored to their context.

For the last decade or more, Pennsylvania state and county water officials have been promoting the concept of *regionalization*: the process of combining system infrastructure and/or management, either temporarily during emergencies or permanently. Should a system fail because of weather or climate, infrastructure collapse, or other problems, the regionalization plan provides a backup. Regionalization also allows adjacent systems to share increasingly expensive infrastructure, which systems often need to install to be compliant with state and federal regulations. Benefits also include improved funding from the state and federal agencies that promote regionalization. In Centre County, some systems have readily embraced regionalization, while others have resisted all attempts to regionalize because they fear losing local autonomy. Regionalized systems are much less sensitive than most stand-alone systems.

Some of the most important indicators of CWS sensitivity to weather and climate are the characteristics of its manager (see also O'Connor *et al.* 1999, 2005; Dow *et al.* 2007). Better-educated managers are aware of a wider range of management options and tend to be willing to explore those options. Experienced water managers (who are not necessarily better educated) usually know what to do when faced with weather or climate stress. Part-time and volunteer managers and operators often have other jobs and sometimes cannot devote sufficient time towards the water system. Managers who have experienced weather and climate problems are much more likely to think that their system is sensitive, although those perceptions of vulnerability decrease as the events become farther away in time. Managers' attitudes and beliefs also affect sensitivity: managers who resist regulation and regionalization on ideological grounds, or who think that their systems are not vulnerable to weather and climate now or in the future, fail to take advantage of opportunities to decrease the sensitivity of their CWS.

Conclusions about sensitivity

The examples of emergency management and CWSs identified several important components of sensitivity to weather and climate in central Pennsylvania. One of the most important lessons to be drawn from these examples is that most of these sensitivities relate directly to human choice – decisions about how to organize an emergency management system; how to define eligibility for flood insurance, disaster aid, and disaster loans; how to build and maintain water system infrastructure; how to respond to legal requirements and policy recommendations, or how to manage a water system. Only a few of the sensitivities were determined by the physical nature of the system. Thus, the sensitivity dimension of vulnerability appears to be dominated by the "human side" of the human–environment interactions observed at this HERO.

Adaptive capacity

There are many ways to view the adaptive capacity of Central Pennsylvania's emergency management and community water systems. One approach is to reduce the components of adaptive capacity (Chapter 5) to two broad themes that emerged through the HERO research: access to resources, and perceptions and attitudes.

Access to resources

Having access to resources is essential to adapting to hydroclimatic extremes in weather and climate. The three most critical resources in central Pennsylvania are finances, social capital, and knowledge and experience. Emergency managers and public officials need funds to pay for hiring and training personnel involved in emergency planning and response, to purchase, maintain, and replace equipment and facilities, and to keep up and comply with the regulations issued by the US Department of Homeland Security and its sub-agency, FEMA. Similarly, finances are topmost on the agendas of water managers and public officials because they need money to maintain, repair, or upgrade infrastructure, to hire managers, operators, accountants, and lawyers, and to pay for testing, drilling, and conducting day-to-day operations. The ability of emergency management or community water systems to fund these items influences their capacity to reduce exposure or sensitivity. Well-financed emergency management systems have full-time, well-trained personnel with new, well-maintained equipment; they have up-to-date emergency plans and are in full compliance with FEMA requirements. CWSs with excellent access to funds have well-maintained, new water infrastructure in good repair; they have a full complement of professional staff making sure that the system complies with regulations and provides safe, plentiful water with minimal interruption. Poorly financed emergency management or CWSs often find it difficult to meet these standards, thus exposing more members of the community to hydroclimatic extremes and increasing their sensitivity to those extremes.

When financial capital is in limited supply, it is possible to substitute social capital for money. When there is limited money for professional staff, volunteers can give their time, providing service, training, and expertise. When equipment is not available, community members can donate physical labor, access to equipment, and even fuel. If the number of volunteers, the amount of time they give, the quality of their work and knowledge, and the nature of the equipment and fuel they provide are first-rate, then the systems will suffer little by having insufficient access to funds. Unfortunately, in most cases social capital is an inadequate proxy for financial capital. Some types of equipment are absolutely necessary to carry out emergency or water operations so it is impossible to substitute human labor for that equipment. Moreover, it is often true that the number of volunteers are inadequate, the amount of time they can give is insufficient, the quality of the work is lower than the quality of a trained professional, the level of expertise is low, the available equipment is inappropriate, and the quantity of fuel is too little to get the job done well or to bring the system to compliance.

Knowledge and experience are important parts of social capital, but these two components go beyond that narrow understanding. For example, one of the most important ways of reducing sensitivity in a CWS is to have a knowledgeable, experienced operator or manager who has first-hand familiarity with the weather or climate extreme in question. When experience is not available, textbook knowledge can be adequate to get the CWS through the extreme event. Many times, however, CWSs are managed and operated by part-time volunteers who are inexperienced and, because they donate their time above their regular employment, have little time for training and textbook learning, thereby putting the system at greater risk during an extreme event. Even on a day-to-day basis when weather and climate are normal, having CWS managers with training and experience trumps lack of education and inexperience because the system stays on a more even keel and has better adaptive capacity during times of stress. The same factors hold true for local emergency management.

Many examples from the Central Pennsylvania HERO demonstrate that access to resources is important to adaptive capacity. For instance, most CWSs in Centre County are public, but a few are private and the private systems typically do not have access to state- or county-level funds for infrastructure improvements. These private CWSs therefore must raise all money needed for capital improvements through their water rates, which is problematical in the more impoverished parts of the county where they operate.[6] Thus, infrastructure improvements reducing exposure or sensitivity to weather and climate extremes are difficult to establish in the private systems. In another example of the importance of access to resources, many of the dozens of very small CWSs in the rural portions of Centre County have no paid employees and rely solely on community volunteers to manage the systems. Often, the state requires expensive tests and equipment to bring the CWS into compliance with regulations, which necessitates grant applications to raise the funds for these small, relatively poor systems. The volunteers often do not have the time, education, or inclination needed to fill out this tedious paperwork, so the water systems that most need the grants are least likely to receive them (Jocoy 2000). In their study of volunteer firefighters, Yarnal and Dowler (2002/2003) found the same problem holds true in emergency management throughout rural Pennsylvania. In short, our work and the work of O'Connor *et al.* (1999), Jocoy (2000), Knuth (2004), Dow *et al.* (2007), and others find that small CWSs and emergency management systems in central Pennsylvania are resource-poor, which compounds with other problems to reduce their adaptive capacity and increase their vulnerability. Small systems have a small customer base upon which to draw funds, meaning that they have no way to pay for experienced managers and people skilled at paperwork. They rely on volunteerism to run the system, resulting in dated infrastructure, uneven system maintenance and performance, difficulty in procuring grants, and trouble staying in compliance with regulations.

Perceptions and attitudes

Much of the vulnerability – especially the adaptive capacity – of central Pennsylvania's emergency management and community water systems stems from the perceptions and attitudes of decision-makers, including individual homeowners, business owners, and managers. Considerable effort has gone into understanding the perceptions of central Pennsylvania's CWS managers regarding vulnerability to weather and climate extremes and to climate change (O'Connor *et al.* 1999, 2005; Dow *et al.* 2007). Emerging from that body of work is the finding that water managers who feel the most vulnerable are those who have experienced system damage or disruptions from extreme weather or climate in the recent past: the more recent the experience, the stronger the perceptions of vulnerability and the greater the willingness to include management practices that reduce vulnerability. Events that are more distant in time result in weaker feelings of vulnerability and less likelihood of taking action to reduce vulnerability. Managers who have never experienced adversity from weather and climate tend not to feel vulnerable. Some scientists attribute such perceptions to *optimism bias* – the tendency for people to overestimate the likelihood of positive events (in this case, not experiencing negative consequences from weather and climate extremes) and to underestimate the likelihood of negative events. In central Pennsylvania, managers of groundwater systems tend to think that their systems are not vulnerable to weather and climate extremes, while managers of surface systems in the same region – who have in fact suffered negative impacts from weather and climate – tend to feel vulnerable. As noted elsewhere in this book, surface water systems are more vulnerable to weather and climate extremes, so there are good reasons for the existing perceptions and biases. Nevertheless, groundwater systems are not invulnerable to weather and climate, so there is reason to be concerned about these perceptions.

The public's attitude concerning the true value of water was an issue that persistently surfaced during the HERO research. Perhaps because water tends to be abundant in the region, the opinion of many Centre County residents is that water should be a free, unlimited good. These people resist the notion of paying for water and, accordingly, for supporting the human and physical infrastructure needed to deliver that water. In response to this common attitude, CWS rates are very low with two predictable consequences: the CWSs receive insufficient income to run the systems and there is little monetary incentive to conserve the resource. The need for financial capital to increase adaptive capacity was discussed above. Conserving the resource is theoretically not important in a well-watered place, but because all of Centre County's CWSs are exposed to droughts and most are sensitive to small downturns in water availability due to the age of their infrastructure, conservation is often necessary. Therefore, there is dissonance between the attitudes of much of the public and the mission of the CWS to deliver water.

Public attitudes about local autonomy also cause discrepancies between local government and higher levels of government. With over 2500 independent municipalities in the state, Pennsylvania has a tradition of local rule, which each township and borough defends jealously; nowhere is local rule more warily guarded than the small municipalities of Centre County. Many of the conflicts over water are really disputes over local autonomy (e.g., Pascale 1997). One flashpoint has been regionalization. As noted earlier in this chapter, the state and county have promoted regionalization as a way to ensure a more secure water supply and to help cash-poor, smaller CWSs share expensive infrastructure. By definition, however, regionalization reduces local autonomy. Moreover, small municipalities see regionalization as an unwanted intrusion imposed by county and state government on local government. Ironically, our research suggests that the CWSs that would benefit the most from regionalization are often the most likely to resist it.

It is important to note that the Anabaptist segment of the population, which is perhaps the most independent, conservative group in the United States, may benefit from their attitudes and beliefs about autonomy. The Anabaptists are not only inward-looking, but also extremely self-supportive: when adversity sets in, this community pulls together to solve the problem, finding ways to adjust and adapt within the framework of their religious dogma (Kraybill 2001). The Anabaptists have little money, but tremendous social capital and strong, decisive, and experienced decision-makers, and thus have much greater adaptive capacity than the neighboring "English" population.

Conclusions

The Central Pennsylvania HERO is a diverse place physically and socioeconomically. Ideologically conservative rural areas rooted in the eighteenth, nineteenth, and early twentieth centuries systems surround the dynamic, progressive twenty-first-century growth pole in the center of the county. Although the entire county population is exposed to extremes in weather and climate, the most sensitive parts of the county are the rural areas outside the Centre Region. Sensitivity results from both physical and human factors, but the majority of factors affecting sensitivity relate to human choice, ranging from decisions on eligibility for flood insurance or disaster aid to compliance with regulations and policy. Characteristics of decision-makers in key roles (e.g., managers or policy-makers) have an especially significant impact on sensitivity. In central Pennsylvania, adaptive capacity is largely determined by access to essential resources: funds, social capital, and knowledge and experience of decision-makers. The perceptions and attitudes of the county's residents, including decision-makers, also have a fundamental influence on adaptive capacity, with the conservative, locally focused ideology reducing adaptive capacity in general – with a notable exception involving the county's Anabaptist population. In the end, as theory would suggest (see Chapter 5), vulnerability in central Pennsylvania results from a combination of natural and environmental components, but the human determinants override the physical ones in all but the most extreme cases.

Notes

1. The term *agrarian* here refers to pre-industrial agriculture that relies primarily on animal and human power instead of power from industrialized sources, such as fossil fuels. Industrialized agriculture, which does depend on these more advanced sources of power, comes under the heading *industrial*. The Amish farmers of the Central Pennsylvania HERO primarily use horse- and mule-driven farm equipment, whereas their "English" neighbors use gasoline-powered tractors and other farm equipment. Hence, agrarian refers to Amish and other Anabaptist populations in the study area.
2. The work in this chapter includes not only all the elements of the vulnerability assessments conducted in the other HERO study regions, but also assessment of emergency planning and flood types, which is crucial to understanding hydroclimatic variation in Pennsylvania because of the high incidence of floods.
3. The meteorological convention is that a day is "cloudy" if it has 6/10 or more cloud cover.
4. The vulnerability data and the interpretations of those data presented in this chapter were gathered over many years and two research projects. The severe storm and flood data and the associated emergency management information were part of the Susquehanna River Basin Integrated Assessment (Yarnal *et al.* 1997, 1999). The rest of the data and information came from the HERO project, with the emergency management analysis resulting from Knuth (2004). The physical hydroclimatic analyses used many traditional sources of such data (see Yarnal *et al.* 1997, 1999; Knuth 2004), whereas the human–environment data and analyses were based primarily on newspaper archival retrieval and interviews (Sorrensen *et al.* 2005). See also endnote 5.
5. These six sensitivities were uncovered by undergraduate students who interviewed managers at 14 Centre County CWSs in summer 2003. These students – Dominic DeFazio, Allyson Gatski, Tania Metz, and Morgan Windram –were part of the HERO Research Experiences for Undergraduates (REU) Site (Sorrensen *et al.* 2005; Yarnal and Neff 2007).
6. Companies purchased the private community water systems to realize a profit. The infrastructure improvements discussed here come out of the companies' overall profits, not out of the funds associated directly with the local CWS where the improvements take place. Although the companies reduce their taxes (and thereby improve their profit margins) through depreciation of capital stock, operating losses, etc., they are slow to make improvements unless complying with regulations or responding to physical necessity because these improvements reduce company profits.

References

Abrams, M. D., and G. J. Nowacki, 1992. Historical variation in fire, oak recruitment, and post-logging accelarated succession in central Pennsylvania. *Bulletin of the Torrey Botanical Club* **119**: 19–28.

Centre County Government, 2003. *Centre County Comprehensive Plan*: Comprehensive Plan Phase I, Adopted December 2003. Accessed at www.co.centre.pa.us/planning/compplan/default.asp.

Centre County Planning Office, 2004. *Centre County Fact Sheet*. Accessed at www.co.centre.pa.us/planning/data.asp.

Denny, A., B. Yarnal, C. Polsky, and S. Lachman, 2003. Global change and central Pennsylvania: local resources and impacts of mitigation. In *Global Change in Local Places: Estimating, Understanding, and Reducing Greenhouse Gases*, Association of American Geographers Global Change in Local Places Research Team, eds., pp. 122–140. Cambridge: Cambridge University Press.

Dow, K., R. E. O'Connor, B. Yarnal, G. J. Carbone, and C. L. Jocoy, 2007. Why worry? Community water system managers' perceptions of climate vulnerability. *Global Environmental Change* **17**: 228–237.

Faill, R. T., and R. P. Nickelsen, 1999. Appalachian Mountain section of the Ridge and Valley Province. In *The Geology of Pennsylvania*, ed. C. H. Shultz, pp. 269–285. Harrisburg, PA: Pennsylvania Geological Survey.

FEMA, 2007. *The Robert T. Stafford Disaster Assistance and Emergency Relief Act, as amended, and Related Authorities, Federal Emergency Management Agency*, June 2007. Accessed at www.fema.gov/about/stafact.shtm.

FEMA, 2008. *A Guide to the Disaster Declaration Process and Federal Disaster Assistance, Federal Emergency Management Agency*, March 2008. Accessed at www.fema.gov/library/viewRecord.do?id=2127.

Gibb, T., 2000. Townsfolk dare Pennsylvania's DEP to jail them over local water works. *Pittsburgh Post–Gazette*, January 16.

Insure.com, 2008. *Who Needs Flood Insurance?* Accessed at www.insure.com/articles/floodinsurance/who-needs-flood-insurance.html.

Jocoy, C. L., 2000. Who gets clean water? Aid allocation to small water systems in Pennsylvania. *Journal of the American Water Resources Association* **36**: 811–821.

Kates, R. W., W. C. Clark, and B. L. Turner II, 1990. The Great Transformation. In *The Earth as Transformed by Human Action: Global and Regional Changes in the Biosphere over the Past 300 Years* eds. B. L. Turner II, W. C. Clark, R. W. Kates, J. F. Richards, J. T. Mathews, and W. B. Meyer, pp. 1–17. Cambridge: Cambridge University Press.

Knuth, S. E., 2004. A vulnerability assessment of the Moshannon Creek Watershed, Pennsylvania. Unpublished B. S. Honors thesis. University Park, PA: The Pennsylvania State University.

Kraybill, D., 2001. *The Riddle of Amish Culture.* Baltimore, MD: Johns Hopkins University Press.

LaPenta, K. D., B. J. McNaught, S. J. Capriola, L. A. Giordano, C. D. Little, S. D. Hrebenach, G. M. Carter, M. D. Valverde, and D. S. Frey, 1995. The challenge of forecasting heavy rain and flooding throughout the eastern region of the National Weather Service. 1: Characteristics and events. *Weather and Forecasting* **10**: 78–90.

Marsh, B. and P. F. Lewis, 1995. Landforms and human habitat. In *A Geography of Pennsylvania*, ed. E. W. Miller, pp. 17–43. University Park, PA: The Pennsylvania State University Press.

Miller, E. W., 1995. Agriculture. In *A Geography of Pennsylvania*, ed. E. W. Miller, pp. 183–202. University Park, PA: The Pennsylvania State University Press.

O'Connor, R. E., B. Yarnal, R. Neff, R. Bord, N. Wiefek, C. Reenock, R. Shudak, C. L. Jocoy, P. Pascale, and C. G. Knight, 1999. Community water systems and climate variation and change: current sensitivity and planning in Pennsylvania's Susquehanna River Basin. *Journal of the American Water Resources Association* **35**: 1411–1419.

O'Connor, R. E., B. Yarnal, K. Dow, C. L. Jocoy, and G. J. Carbone, 2005. Feeling at-risk matters: water managers and the decision to use forecasts. *Risk Analysis* **25**: 1265–1275.

Pascale, P. H., 1997. The impacts of the Safe Drinking Water Act on community water systems and changes in drought vulnerability. Unpublished M. S. thesis. University Park, PA: The Pennsylvania State University.

Simpkins, P. D., 1995. Growth and characteristics of Pennsylvania's population. In *A Geography of Pennsylvania*, ed. E. W. Miller, pp. 87–112. University Park, PA: The Pennsylvania State University Press.

Sorrensen, C., C. Polsky, and R. Neff, 2005. The Human–Environment Regional Observatory (HERO) Project: undergraduate research findings from four study sites. *Geographical Bulletin* **47**(2): 65–72.

United States Census Bureau, 2008a. *State and County QuickFacts: Centre County, Pennsylvania*. Accessed at http://quickfacts.census.gov/qfd/states/42/42027.html.

United States Census Bureau, 2008b. *American FactFinder: DP-1. Profile of General Demographic Characteristics: 2000, Centre County, Pennsylvania*. Accessed at http://factfinder.census.gov/servlet/QTTable?_bm=y&-qr_name=DEC_2000_SF1_U_DP1&-geo_id=05000US42027&-ds_name=DEC_2000_SF1_U&-_lang=en&-format=&-CONTEXT=qt.

United States Government Printing Office, 2008. *United States Small Business Administration Disaster Loan Program. Code of Federal Regulations, Title 13 – Business Credit and Assistance*, Part 123, March 2008. Accessed at www.access.gpo.gov/nara/cfr/waisidx_08/13cfr123_08.html.

Williams, A. V., 1995. Political geography. In *A Geography of Pennsylvania*, ed. E. W. Miller, pp. 154–164. University Park, PA: The Pennsylvania State University Press.

Yarnal, B., 1989. Climate. In *The Atlas of Pennsylvania*, eds. D. J. Cuff, W. J. Young, E. K. Muller, W. Zelinsky, and R. F. Abler, pp. 26–30. Philadelphia, PA: Temple University Press.

Yarnal, B., 1995. Climate. In *A Geography of Pennsylvania*, ed. E. W. Miller, pp. 44–55. University Park, PA: The Pennsylvania State University Press.

Yarnal, B., 2004. Informed scenarios of climate change in the Mid-Atlantic Region. *Penn State Environmental Law Review* **12**: 127–134.

Yarnal, C. M., and L. Dowler, 2002/2003. Who is answering the call? Volunteer firefighting as serious leisure. *Leisure/Loisir* **27**: 161–189.

Yarnal, B., and D. J. Leathers, 1988. Relationships between interdecadal and interannual climatic variations and their effect on Pennsylvania climate. *Annals of the Association of American Geographers* **78**: 624–641.

Yarnal, B., and R. Neff, 2007. Teaching global change in local places: the HERO Research Experiences for Undergraduates program. *Journal of Geography in Higher Education* **31**: 413–426.

Yarnal, B., D. L. Johnson, B. J. Frakes, G. I. Bowles, and P. Pascale, 1997. The flood of '96 and its socioeconomic impacts in the Susquehanna River Basin. *Journal of the American Resources Association* **33**: 1299–1312.

Yarnal, B., B. Frakes, I. Bowles, D. Johnson, and P. Pascale, 1999. Severe convective storms, flash floods, and global warming in Pennsylvania. *Weather* **54**: 19–25.

Zelinsky, W., 1989. Religion. In *The Atlas of Pennsylvania*, eds. D. J. Cuff, W. J. Young, E. K. Muller, W. Zelinsky, and R. F. Abler, p. 91. Philadelphia, PA: Temple University Press.

Zelinsky, W., 1995. Cultural geography. In *A Geography of Pennsylvania*, ed. E. W. Miller, pp. 132–153. University Park, PA: The Pennsylvania State University Press.

13

Fossil water and agriculture in southwestern Kansas

LISA M. BUTLER HARRINGTON, MAX LU, AND
JOHN A. HARRINGTON, JR.

Introduction

Among the four areas investigated as part of the HERO project, semi-arid southwestern Kansas is the most reliant on agriculture. This region faces far different issues with respect to land-use/land-cover change, vulnerability to environmental stress (including hydroclimatic variability and change), and sustainability than do densely settled areas and those locales with low economic reliance on agriculture. In both this region and other non-urban parts of the country, populations in many rural counties and small towns are declining, adjustments to economic globalization are taking place, and fluctuations in forcing by coupled human and natural systems are continuing to affect agricultural success. Changes faced by farming regions also vary among those places with generally sufficient rainfall to grow most important crops, those that receive little rain and lack supplemental sources of water, those reliant on renewable surface water sources, and those reliant on declining groundwater sources. Much of southwestern Kansas is reliant on declining groundwater resources, but some areas lack sufficient ground and/or surface water for use in farming.

The name High Plains–Ogallala (HPO-)HERO recognizes this agricultural region's physical identity and its reliance on the Ogallala and other aquifers. Over the last 30 years, the research site has developed a rich literature that connects the people and land of southwestern Kansas (e.g., Worster 1979; Warren *et al.* 1982; Reisner 1986; Sherow 1990; Kromm and White 1992, 2001; White 1994; White and Kromm 1995; Opie 2000; Bloomquist *et al.* 2002; Harrington *et al.* 2003; Broadway and Stull 2006). As such, HPO-HERO represents a continuation of the long-term human–environment research regarding resource use, communities, and change in the Great Plains generally and this portion of the Great Plains in particular.

Sustainable Communities on a Sustainable Planet: The Human–Environment Regional Observatory Project, eds. Brent Yarnal, Colin Polsky, and James O'Brien. Published by Cambridge University Press.

Study area

Physical attributes

HPO-HERO consists of 19 counties located in southwestern Kansas, encompassing about 15 880 miles^2 (41 120 km^2) (Figure 13.1). The primary study area lies at the center of the American High Plains and is in the heart of the former Dust Bowl. Elevation ranges from 2500 to 3500 feet (760 to 1070 m) above sea level, but there is little internal topographic relief for most of the area. Gas and oil fields underlie the area, with important but declining production.

Highly variable precipitation amounts generally average less than 21 inches (545 mm) annually through the region. Potential evapotranspiration exceeds precipitation (Sophocleus 1998), so only crops adapted to low moisture levels grow reliably without irrigation supplements. Extremity and inter-annual variability are the key characteristics of weather in the High Plains.

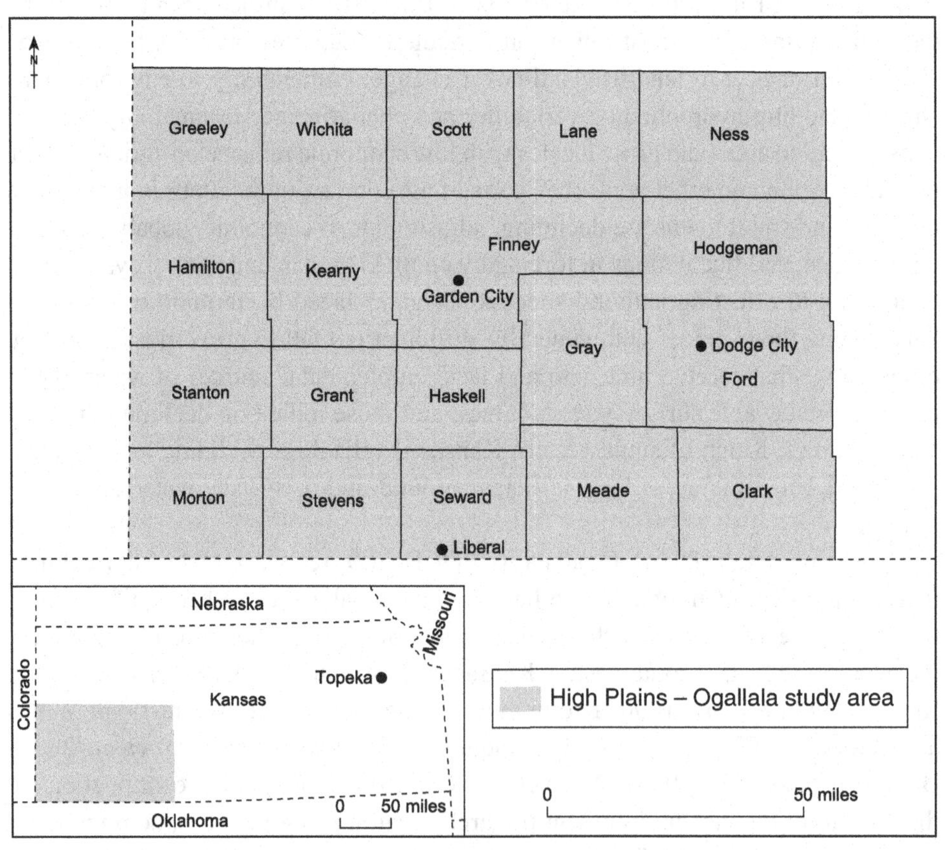

Figure 13.1. High Plains–Ogallala Human–Environment Regional Observatory (HPO-HERO) study area.

Typical natural vegetation was shortgrass prairie, sand-sage prairie in areas of Holocene sand dunes, and narrow bands of riparian woodland (Kuchler 1976). Soils are generally good for agricultural production, although some areas of sandier soils are unsuitable for crops without relatively high applications of fertilizer and water. Wind erosion is of greater concern than water erosion in this region (Leathers and Harrington 2000).

Relatively little surface water is available in the study area. Most water supplies, whether for domestic, agricultural, or industrial use, are based on the High Plains aquifer system, including the Ogallala formation and, to a lesser extent, alluvial groundwater. There are two rivers, the Cimarron and the Arkansas, but flow is highly variable to non-existent depending on the time of year and the climate of a given year. For many years, aboveground Arkansas River flow was rare due to upstream irrigation diversions in Colorado. With legal rulings and negotiations between Kansas and Colorado, there has been renewed flow, although salinity has increased (Whittemore 2000; Harrington and Harrington 2005).

Exposure to hydroclimatic hazards (see Chapter 10) largely takes the form of intermittent drought. In a region that receives marginal rainfall on average and is dependent on agricultural activities, drought represents a major concern. Localized flooding can occur in places, generally as flash or urban floods, but flood problems are infrequent. Other hydroclimatic problems include hailstorms, which have more localized effects than drought, and the timing of precipitation with respect to planting and harvest needs.

Social and economic characteristics

Euro-American settlers moved into the High Plains in the late nineteenth century, with the 1862 Homestead Act (Riebsame 1990) and the construction of the transcontinental railroads serving as major impetuses. Activities were mainly oriented around crop farming when and where sufficient water was available, but in some areas cattle ranching has been dominant from the outset. Variable precipitation and repeated droughts resulted in periodic land abandonment, particularly near the end of the nineteenth century and during the 1930s Depression and Dust Bowl.

Since the 1860s, total population increased over each census period, except the 1890s and 1930s when the region was severely affected by droughts (Figure 13.2). Overall population of the region grew at a somewhat stable, moderate rate since 1940. According to the 2000 Census, slightly more than 156 000 people lived in the 19-county area, yielding a population density of only 3.89 persons km^{-2}; the US average is 29.2 persons km^{-2}. While the region as a whole is sparsely populated and rural, three regional urban centers, or 'Ogallala Oases' (White 1994) account for just below half of the region's total population. Garden City had a population of 28 451

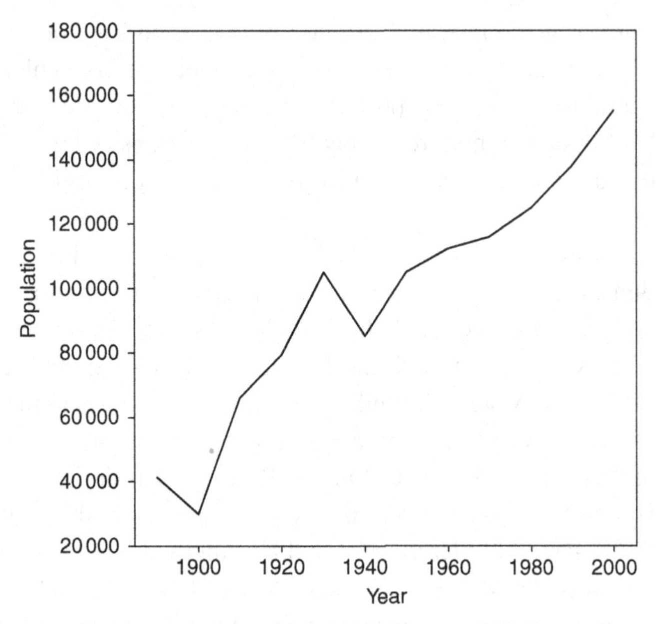

Figure 13.2. Regional population, 1890–2000. (Source: US Census Bureau 2000.)

in 2000, while Dodge City had 25 176 and Liberal had 19 666, for a combined population of 73 293.

Census figures for the largest three counties, Finney, Ford, and Seward, show 2000 Census populations of 40 523, 30 548, and 22 510, respectively (Figure 13.3). Finney, with the Garden City urban area, has grown especially rapidly for the High Plains. It more than doubled in population between 1960 and 1990. Dodge City (Ford County) and Liberal (Seward County), similarly important local centers, have also seen major growth (White 1994). In contrast, some counties have continuously lost residents since the 1930s Dust Bowl era: Ness declined from a high of 8358 for the 1930 Census to 3454 in 2000, and Clark declined from 4796 to 2390 in the same period. Greeley had the smallest county population in the most recent census, at 1534. These and other small counties in the region have estimated populations at even lower levels for 2004 (US Census Bureau 2006).

Areas without significant groundwater resources continued to rely on dryland agriculture and cattle grazing through the twentieth century. Locations lacking groundwater resources also are the places that have lost population to communities with good access to water (see White 1994). In a sense, dryland agricultural areas have been vulnerable to competition from cities that have grown with irrigation-reliant industries.

The region has seen not only major changes in population size and intra-regional distribution, but also changes in ethnic composition. In 1980, with a total population

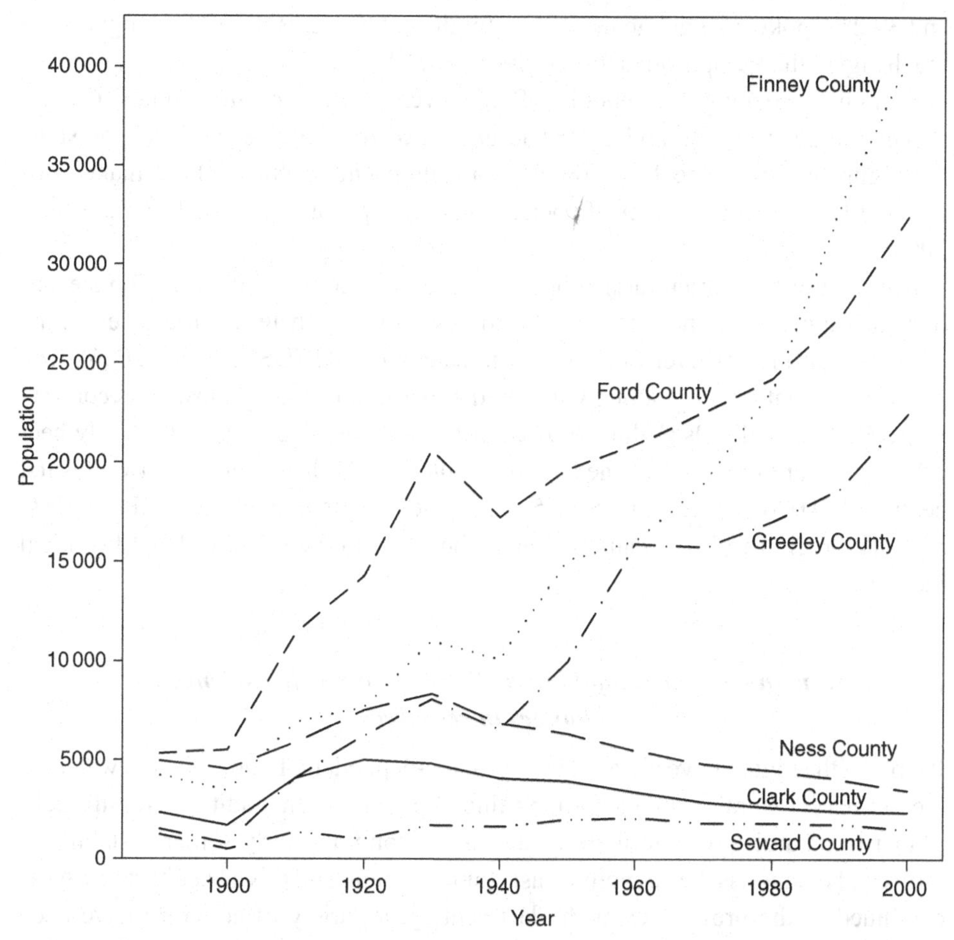

Figure 13.3. Population trends in six selected HPO-HERO counties. (Source: US Census Bureau 2000.)

of 124 912, the populace was 93% white, 0.3% Asian, and 8.3% Hispanic. With passage of the federal Indochinese Migration and Refugee Act of 1975, an influx of immigrants from Laos and Vietnam increased the Asian population of southwestern Kansas, particularly in the Garden City, Liberal, and Dodge City areas. By 1990, the proportion of the white population (138 064) had decreased to 85.8% of the total, although greater in absolute number than in 1980; Asians made up 1.8% of the population; and Hispanics reached 15.5% of the total regional population. Hispanics, in particular, arrived to take jobs with expanding meat-packing operations (Broadway and Stull 2006). By 2000, the white population had dropped to 77% of the total, the Asian population appeared to have stabilized at about 1.7%, and the Hispanic population had risen to 31.6%. For Finney County in 2000, nearly one-quarter of the population was foreign-born (23.7%)

and 39.2% spoke a language other than English at home. Both of these figures are much higher than proportions for Kansas as a whole (i.e., 5% foreign-born and 8.7% non-English-speaking households). Sixty-seven percent of the Finney County population 25 years old and older had graduated from high school; for the state, this figure is 86%. The fast-growing population and shifting ethnic makeup of the area has led to a number of social adjustment problems (Broadway and Stull 2006).

Agriculture and agricultural support services are the basis of the region's economy. Southwestern Kansas contains the top five counties in agricultural sales for the state, accounting for over $8.7 billion in sales in 2002 (USDA ERS 2006). The region's population grew along with rapid expansion of the agribusiness economy in the 1970s and 1980s (White 1994), but for decades, livestock, particularly beef cattle, have far outnumbered the human population. At the dawn of the twenty-first century, 154 000 people and over 2.5 million beef cattle inhabited the HPO-HERO area. The region truly is 'cattle country,' both in numbers and in local historical identity.

Human–environmental interactions: resource dependence and changes in resources

People of the region have always been resource-dependent. Land, soil, and water are the primary natural resources supporting the region. In addition, fossil fuels, primarily natural gas, contribute to the local economy. Early settlers established farms and ranches as the economic basis for the region, and this agricultural base has continued to the present. Over the last century, a variety of activities have been added to growing crops and raising livestock, but many of the manufacturing, sales, and service activities are in place because of their relationship to agricultural production (Kromm and White 2001; Harrington *et al.* 2003). A vertically integrated agribusiness sector now characterizes the region's economy.

Human action has transformed the HPO-HERO study region considerably over the last 150 years (Riebsame 1990). Groundwater-based irrigation, which started playing a major role in the region's agriculture in the 1950s, is largely responsible for the extent of transformations observed today.

The region faces many challenges for the future, including declining groundwater resources, changing ethnic composition, and unpredictable energy prices, as well as continued climate variability and potential climate change. How to mitigate adverse effects of these factors to reduce the region's vulnerability and maintain the vitality of the local economy is a serious concern. There is much to be learned from studying historical human–environment interactions in this region (Glantz and Ausubel 1984; Kromm and White 2001; Knight *et al.* 2003).

Water

Water is the resource most critical to the economic and societal health of the region. Early settlers settled along river valleys, diverting surface water where it was available for irrigation and relying on rainfall further from streams (Sherow 1990). Farmers tapped alluvial aquifers for irrigation water soon after settling the area (Figure 13.4), but the entire region was vulnerable to inconsistent rainfall and periodic multiyear droughts for decades.

Borchert (1971) identified important Great Plains drought 'midpoints' in 1892, 1912, 1934, and 1953, with lesser droughts also occurring. Periods of extreme dryness also occurred in the 1970s and the early 2000s. The drought and Dust Bowl of the 1930s is by far the best-known hardship faced by the region (Worster 1979). Other important droughts took place in the late 1880s/early 1890s (Tannehill 1946) and the 1950s. The 1950s drought was meteorologically severe, but it did not have the same effects – or receive as much attention – as the 1930s drought. Borchert (1971) suggested that differing climatic conditions in the 1950s led to much less wind and, therefore, an absence of the extensive soil erosion and environmental damage that occurred during the Dust Bowl years. Soil conservation practices also improved greatly after the Dust Bowl, indicating significant progress in

Figure 13.4. Farmer irrigating sugar beets, Syracuse, Kansas, 1939 (Farm Security Administration photo, Lee Russell).

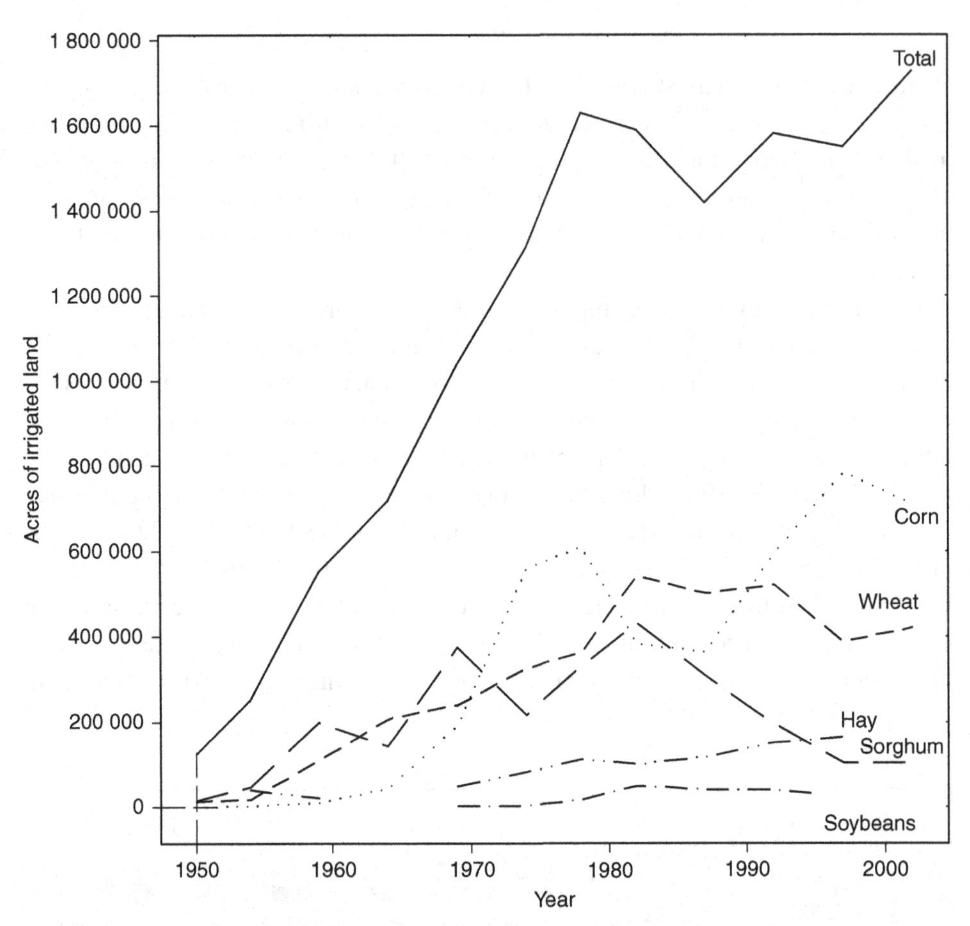

Figure 13.5. Growth of irrigation and related crop changes in HPO-HERO. (Source: USDA NASS, 2002.)

understanding of the environment and soils. These important adjustments resulted in an increase in adaptive capacity.

Use of deeper groundwater from the High Plains aquifer system, including the Ogallala formation, became widespread in the 1950s–1970s as new technology and cheap energy made it possible to pump deeper water (Kromm and White 1992). Irrigation increases production, as well as the stability and predictability of production, by reducing dependence on the variable precipitation regime. Irrigated cropland in the study area expanded from about 123 700 acres (50 060 ha) in 1950 to about 1037 800 acres (420 000 ha) in 1969 (Figure 13.5). In 2002, irrigated acreage amounted to 1416 800 acres (573 340 ha), or 42% of harvested cropland in the HPO-HERO (USDA NASS 2002). Ninety-six percent of freshwater use in the study area is for irrigated agriculture. Corn (for livestock feed), wheat, grain sorghum, alfalfa, and soybeans are the five most commonly irrigated crops in southwest Kansas today. Although irrigated agriculture is often the focus of attention, it is important to note that much of the region's 9 508 870 acres (3 848 100 ha) of farmland is devoted to dryland crops and range. Moreover, much of the region has never had sufficient groundwater to support irrigated agriculture and, in some areas that had tapped into groundwater resources for irrigation, farming has had to revert to dryland crops in recent years due to declines in water availability (see, e.g., Kettle 2003; Kettle *et al.* 2007).

Although locations not overlying the High Plains aquifer system typically have not had sufficient groundwater to support farming, their community water systems – as well as those water systems overlying rich aquifers – have relied exclusively on groundwater for drinking supplies (Harrington 2005). Droughts affect drinking water supply indirectly through their effects on other types of water use, mainly irrigation.

Soil

For southwestern Kansas, the status of the soil resource connects intimately to the status of water resources. Clearly, the extensive soil loss and drifting of the Dust Bowl resulted from severe drought. Just as clearly, the Dust Bowl also resulted from overuse of soil resources – spurred by the high price of wheat and the consequent expansion of wheat farming in the 1920s – during the relatively moist period just preceding the drought. Plowing and cultivation of large swaths of prairie left the soil open to extensive wind erosion and, when rain came again, water erosion. As a response to the Dust Bowl, the federal government began establishing soil and land conservation programs, including the Soil Conservation Service (now Natural Resources Conservation Service) in 1938. Government purchase of abandoned lands commenced in the 1930s, and became formal with the Land Utilization Project in 1937. Cimarron National Grassland in the southwestern corner of the HPO-HERO is a legacy of this program. In the 1950s, the federal government also established the Land Bank program, a precursor to today's Conservation Reserve Program (CRP) and other environmentally oriented set-aside programs.

Modern soil conservation practices and federal cropland reserve programs can serve to mitigate the effects of drought on both soils and farmer income (Nellis *et al.* 1997). Through CRP, farmers can temporarily stop working erodible or marginal lands in exchange for federal payments. Although there is evidence that CRP is less effective regarding soil protection than conservationists would hope (Leathers and Harrington 2000), there is also anecdotal evidence that CRP helps to stabilize farm income.

Livestock

The region has long been associated with major cattle drives to Dodge City, "the Queen of the Cow Towns." These drives actually only occurred for a few years following the decline of buffalo hunting. With a major blizzard in 1886 that resulted in significant losses of cattle and with the invention of barbed wire, the large drives ended and cattlemen closed the open range. Although several large ranches were established in southwestern Kansas, mixed farming (crops and livestock) was more important for supporting families during the first half of the twentieth century.

In contrast to ranches where ranchers raised cattle on range forage, farmers often purchased cattle in the fall, kept them in pens, and fed them over the winter to sell in the spring. Such cattle-feeding operations were small scale (Bussing and Self 1981). These operations began to expand and commercialize in the late 1950s (Bussing and Self 1981; Barnaby 1996). Today, large concentrated animal feeding operations (CAFOs) are economically very significant, although they take up a small proportion of the study area. Feedlots for beef cattle have been important for decades, reaching capacities of over 100 000 animals.

Four large packing plants process beef cattle in the area, shipping boxed beef across the United States and around the world. More recently, there has been dramatic growth in other animal-feeding operations, especially dairy cattle (growing to at least 55 800 head in 2002, about 19 times the 1985 population) and hogs (628 000 head in 2000, 14 times the 1995 population).

Natural gas

The sedimentary strata of the Hugoton Gas Area (HGA) underlie a large portion of the HPO-HERO study area (Figure 13.6) and are an important source of natural gas and, to a much lesser extent, petroleum. Fossil fuel extraction from the HGA, one of the world's largest gas fields, began in the 1920s and peaked in the early 1970s (Carr and Sawin 1996). Production has declined as the resources have depleted, with perhaps 75% of the available gas already extracted, although rule changes facilitated a minor resurgence in gas production in the mid-1990s (Figure 13.7a); oil wells also increased at the beginning of the twenty-first century (Figure 13.7b). Once estimated at over 435 psi, the average well-field pressure has dropped to an average of less than 60 psi (SERCC 2003). The cost of gas extraction, and therefore the cost of natural gas, is apt to continue rising as additional compression is needed.

The decline in natural gas availability and reservoir pressure is a major concern to local agricultural interests. In the past, farmers in the Hugoton region were able to run their irrigation pumps with natural gas. Although out-of-state energy companies own most of the gas and oil rights, landowners and extraction companies made deals allowing low-cost purchase at the wellhead. Reduction in the ability to use natural gas because of decreasing availability and reservoir pressure (SERCC 2003), as well as the rising costs of the fuel and the need for more energy to lift increasingly deeper water, combine to compromise farmers' abilities to utilize groundwater.

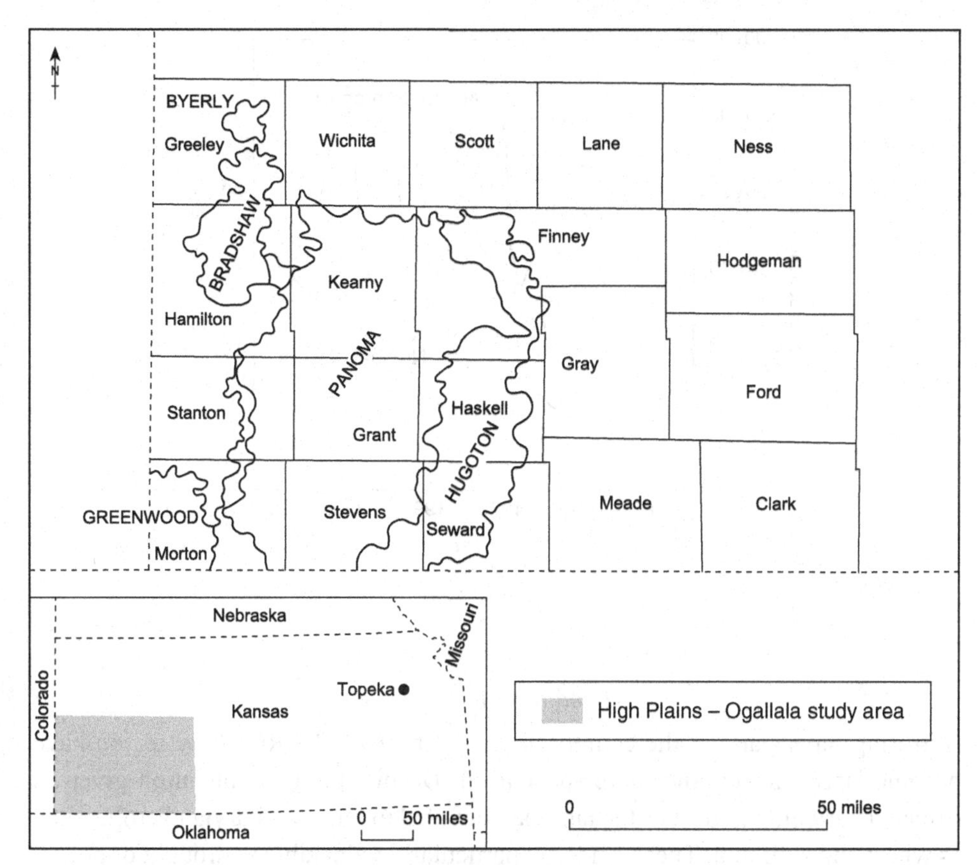

Figure 13.6. Hugoton Gas Area. (Source: Carr and Sawin 1996.)

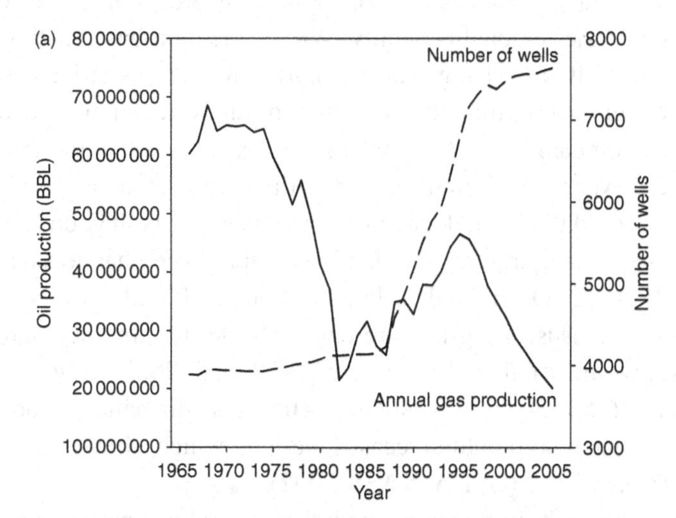

Figure 13.7. (a) Natural gas production from the Hugoton Gas Area, and (b) petroleum production from the HGA. (Source: KGS 2006.)

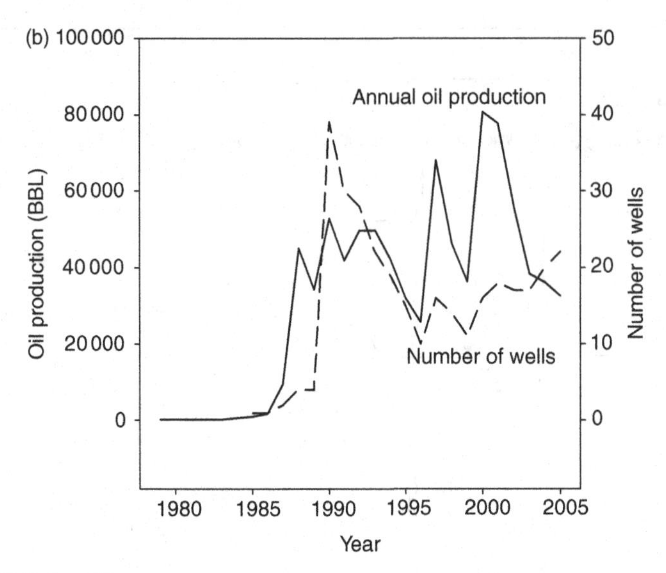

Figure 13.7. (cont.)

Land use/land cover

Mirroring other parts of the country, farms in the HPO-HERO study region have become larger, fewer, and more specialized. Despite the great attention given to irrigated agriculture, most of the lands are devoted to range and dryland crops, such as wheat and sorghum. These lands are particularly vulnerable to drought or poorly timed rainfall. Although they occupy less than 15% of the land-surface area, vertically integrated irrigated agriculture and confined animal feedlots are the basis for much of the regional economy. While irrigation-based activities are not vulnerable to precipitation changes in the short term, periods of increased dryness can become a significant problem if regional groundwater supplies decline, especially when accompanied by high energy prices (for water pumping) or low producer prices. Many local producers have diversified their investments, so this diversification would likely moderate financial impacts in both good and bad times.

Although an agricultural region, land-use/land-cover change is surprisingly dynamic in HPO-HERO. Probably related to both CRP withdrawals and groundwater depletion, Census of Agriculture data indicate that irrigated acreage in the 19-county study area fell 13% between 1978 and 2002, from 1 629 700 to 1 416 800 acres (659 530 to 573 340 ha). The most dramatic period of decline was the most recent agricultural census period, which showed a 9% decline between 1997 and 2002 (USDA NASS 2002).

Crops grown through time changed with the ability to irrigate (see Figure 13.5). Goodin *et al.* (2002) analyzed land-cover change for a six-county subset of the

HPO-HERO study area and for the period 1972 to 1992. The region and period under study captured much of the rapid expansion of groundwater irrigation in the part of the HPO-HERO with greatest groundwater availability. Utilizing Landsat MSS data, they classified land cover as pasture/prairie, warm-season crop (e.g., corn, sorghum), cool-season crop (wheat, alfalfa), or other. The investigators found that the pasture/prairie class declined by 20%, warm-season crop area was up 6%, and cool-season crop area increased 14%. Even crops that are more drought tolerant (e.g., wheat) were sometimes irrigated, along with more water-demanding crops like corn, since the additional moisture greatly increased yield.

Crop rotation is an important aspect of the study area. Satellite image analysis shows, for example, that crop rotation dominated land-cover change in Gray County for the period 1985 to 2001.

The processes of land-use/land-cover change have played out differently in subregions of western Kansas, and water resource availability has played a very important role in determining within-region variations. In areas where the remaining groundwater resource is too limited or too expensive to extract, production has switched from irrigated crops to dryland crops (Kettle 2003; Kettle *et al.* 2007). For example, agricultural land-use/land-cover change in Wichita County, where reduction in groundwater availability has been relatively dramatic, corresponds with changes to the groundwater resource. Land enrollments in the CRP have also become important in the study area. In Gray County, CRP acreage occupied about 9% of the landscape in 2001, with early CRP enrollments (1985–92) occurring in areas with more limited groundwater supplies. In contrast, Hamilton County, which lacks access to the High Plains aquifer system, had approximately 22% of its land in CRP in 2001 (Reker 2004). Southwestern Kansas has dominated state enrollments in the CRP (Leathers and Harrington 2000).

In a remote sensing change-detection study of two of the HPO-HERO counties, Reker (2004) found that, although overall land in CRP was stable, the actual geographic distribution of enrolled land shifted over time. Between 1985 and 2001, over 30 000 acres (12 000 ha) of early-enrolled CRP land in each of the two counties (Grant and Hamilton) converted back to agricultural use. From Grant County, which has better groundwater access, the proportionate change was much higher than for off-aquifer Hamilton County. This finding is consistent with the greater need for places lacking supplemental water access to adapt to climate conditions through strategies like land program enrollment; those with access to groundwater have more freedom to respond to market forces by bringing land back into production when water access is not a strong concern. Reker's findings are consistent with those of Leathers and Harrington (2000), who found that farmers often enroll particular parcels in CRP and similar programs, only to open other lands to cropping at the same time.

High Plains adaptation/vulnerability research

Many researchers have addressed adaptation – including the related topic of vulnerability – to climate conditions and associated stresses in the High Plains or (more frequently) the Great Plains (e.g., Rosenberg 1986; Easterling *et al.* 1993; Polsky and Easterling 2001; Bloomquist *et al.* 2002; Polsky 2004). Webb (1931) emphasized "the water problem" and the search for sources of moisture in the Great Plains. Of the southern plains, Worster (1979, 3) wrote that "Nothing that lives finds life easy under their severe skies; the weather has a nasty habit of turning harsh and violent just when things are getting comfortable. Failure to adapt to these rigors has been a common experience for Americans..."

Farmers have adapted to and overcome annually variable rainfall through irrigation technologies, new crop varieties, and other technological improvements (Rosenberg 1986; Sherow 1990; Kromm and White 1992; Easterling *et al.* 1993; Harrington *et al.* 2003). At the end of the Depression, research by the US Department of Agriculture concluded that Sublette, in Haskell County, was vulnerable to climate variation, economic downturns, and other natural and socioeconomic problems (e.g., Edwards 1939; Bell 1942). In follow-up research, Bloomquist *et al.* (2002) found Sublette to be robust in the short to medium term, given current availability and use of groundwater. In short, the community experienced a complete turnaround due largely to improved water availability resulting from changing technology. In a modeling study that investigated the response of the central United States MINK (Missouri–Iowa–Nebraska–Kansas) region to a 1930s-like drought, Easterling *et al.* (1993) considered the impact that agricultural technology would have on the region's vulnerability. The MINK study found that modern agricultural technologies help make current farming less vulnerable to drought than agriculture of the earlier twentieth century. Similarly, Polsky and Easterling (2001) and Polsky (2004) found that irrigated agriculture helps buffer the Ogallala region from impacts of and vulnerabilities to future climate variation and change.

Interviews with key informants in the HPO-HERO study area finds them concerned with reliance on limited water resources, depletion of non-renewable fossil fuels, social conflict, and demand and prices for agricultural products (Harrington 2005). Some resource specialists have long-term concerns with continued enlargement and vertical integration of agricultural enterprises because future economic restructuring would result in changes in farm income and further depopulation of rural areas and small towns. Nonetheless, because southwestern Kansans have survived hard times in the past, those interviewed believe they will be able to adapt to hard times if they happen again. Perceptions of environmental stresses and adaptability may vary across socioeconomic groups, including native-born residents and recent immigrants, so more research is needed to determine if such optimism is pervasive throughout the population.

Status and future change

The HPO-HERO study area is highly dependent on agriculture, and – because agriculture is one of the oldest forms of human–environment interaction – understanding the region's agricultural vulnerability is crucial to understanding its human–environment interactions. Agriculture is especially vulnerable to climatic hazards (Cross 2001; Cutter *et al.* 2003), so the following section focuses on the region's vulnerability to climate variation and change. The section breaks this exploration into the three dimensions of vulnerability: exposure, sensitivity, and adaptive capacity.

Exposure

The Great Plains is a region where precipitation is especially variable (Harrington and Wood 2002; Wood 2002; see also Chapter 10). Deep within that region, southwestern Kansas suffers periodic exposure to severe drought. As severe as they were, the droughts of the late 1800s, 1930s, and 1950s were short-lived iterations of dry patterns that lasted much longer and covered much larger areas during presettlement periods (Woodhouse and Overpeck 1998; Woodhouse *et al.* 2002).

The variable climate of the region is predicted to change. Some climate models suggest increased minimum and maximum temperatures over the Great Plains, with decreased nighttime cooling and warmer winters (Ojima *et al.* 2002). Perhaps more important, models indicate higher evapotranspiration and possibly decreased precipitation over the High Plains, which is cause for concern in a region with marginal rainfall (see also McCarthy *et al.* 2001, 738). Changes in exposure to hydroclimatological phenomena could take the form of increases in drought duration, frequency, and intensity, as well as of changes in frequency and intensity of storm events in summer or winter. Climate models also show the likelihood of a greater frequency of intense rainfall events for southwestern Kansas.

Since the economy is so strongly linked to agriculture, any of these changes would expose the area to increased economic disruption over background variability. Crops and livestock would be exposed to storm-related hazards, including blizzards, hail, lightning, wind, and tornadoes, as well as to drought and heat waves. Rosenzweig *et al.* (2002) also suggested that more intense rainfall events could lead to reduced crop production due to soil moisture effects.

In recent history, hailstorms have been the most significant, frequent, and widespread cause of storm damage to the region. A single hailstorm can cause hundreds of thousands of dollars in losses. In Ford County alone, 371 hail events with hailstones >2 cm diameter occurred from 1955 through 2003 (Harrington 2005). Hailstones at least 14 cm in diameter have been recorded in the study area.

Floods generally have not been a major hazard in HPO-HERO, although flooding and flash flooding can occur in some locations (Harrington 2005). Small stream and urban flash floods have been the greater hazard in recent decades, but hydroclimatic change and changes in water management might increase flood risks. Because of litigation by Kansas, Colorado now releases more water into the Arkansas River than in the 1940s to mid-1990s (Harrington and Harrington 2005). These flows, as well as any increases in precipitation intensity, may increase flood problems in the future. Because the rivers have been dry for many years, there may be a degree of complacency and misunderstanding of potential risk.

Sensitivity

Socioeconomic sensitivity to climate extremes is heightened because of the heavy economic dependence of southwestern Kansas on agriculture – when strong climatic variations cause the agricultural economy to suffer, the people suffer, too. Climate is only responsible for part of the region's sensitivity, however. If variations in the global, national, or regional economy weaken the local economy of southwestern Kansas, then relatively small climatic variations can have large impacts on the HPO-HERO socioeconomic system. Conversely, if the large-scale economy is strong, then a powerful climatic excursion might have limited effects on the study area. The following discussion of sensitivity focuses on sensitivity to climate, but the crucial importance of the large-scale socioeconomic must also be acknowledged.

Despite the overall sensitivity to climate, precipitation does not currently have a strong correlation with total agricultural productivity in southwestern Kansas because the expansion of irrigated agriculture disconnected production from rainfall. Before extensive irrigation, agricultural productivity was closely tied to precipitation; production in off-aquifer portions of the study region is still sensitive to variations in precipitation. Irrigated areas will become more sensitive than they are today as groundwater depletion continues.

Climatic variation beyond drought can mean trouble for the study area's crops. For example, wet spells can promote plant disease and can delay planting or harvest if they occur at the wrong times. Wind erosion associated with extended or extreme dry spells can be a problem for croplands and rangelands. Sensitivity to heavy rainfall includes not only the potential for urban flooding (currently a minor concern), but also problems with feedlot waste lagoons (possible surface water contamination) and filling of playas (crop loss).

Weather and climate conditions have potential health effects on human populations, as well as on crops. During droughts, blowing dust can be a health problem for people with respiratory sensitivities. As a group, the elderly may be most sensitive, but others, including children, may also have an elevated risk of respiratory health problems.

Livestock, like humans, can respond negatively to weather and climate. With drought and blowing dust, cattle can develop respiratory problems, and heat stress affects weight gain and milk production. Different livestock types and breeds feel heat and cold stress effects in various temperature ranges (see, e.g., Johnson 1987); thus, the sensitivity of livestock raised in southwestern Kansas depends on species and breed adaptability. Sensitivity to climatic conditions and disease may be especially high for cattle in CAFOs. Disease spreads more rapidly with concentration of animals, and certain diseases require slaughter of all exposed animals even if only one animal is directly affected. Dairy cattle operations may be more sensitive to drought than beef cattle feedlots because dairy animals require higher-moisture alfalfa and production of alfalfa declines with decreased water availability. Nevertheless, dairy cattle tend to have somewhat better shelter from weather extremes than beef cattle, and hogs have even more environmentally protective facilities.

The human population feels the impacts of climate variation and change through economic repercussions. Enrollment in various agricultural insurance programs, including crop insurance (Rosenzweig *et al.* 2002) and livestock insurance, reduces sensitivity to climatic stress. Income stabilization through government programs like CRP also can be of great help in reducing sensitivity to climate change. Additionally, CRP and other programs that leave soil undisturbed support efforts to mitigate global climate change through carbon sequestration (see Conant *et al.* 2001).

Adaptive capacity and resiliency

Years with below-average rainfall have negatively affected the economic well-being of the region, but farmers in southwestern Kansas and other parts of the Great Plains have mitigated this limitation through irrigation (Rosenberg 1986; Sherow 1990; Kromm and White 1992; Easterling *et al.* 1993; Harrington *et al.* 2003). Throughout much of southwestern Kansas, exploitation of plentiful groundwater resources has supported a healthy agricultural economy centered on production of feed, CAFO-based livestock, and meat processing. There is no question of the future status of the High Plains aquifer system, however: current pumping levels are not sustainable. Withdrawals in the last four decades sometimes averaged 38 cm per year or more, but recharge rates over the High Plains are less than

5 cm per year – and generally less than 2.5 cm per year over the HPO-HERO (KGS 2000). Although withdrawal rates are decreasing, the regional groundwater resource is essentially non-renewable over human time-frames. Thus, policies aimed at "planned depletion" rather than maintenance and long-term availability of the resource to agriculture will cause loss of the resource (see Sophocleus and Sawin 1998; Gilson *et al.* 2001). Adding to this problem, rising energy costs are reducing the practical availability of groundwater: more energy is required to pump from deeper levels, and energy costs have been increasing even more sharply than the aquifer levels have been declining.

Some actions have prolonged the life of the groundwater resource in southwestern Kansas. In interviews conducted by the HPO-HERO team, respondents frequently mentioned improvements in irrigation technology and efficiency, changes of crops based on water availability, shifts to dryland farming, diversification of activities, and enrolling in federal agricultural programs (e.g., CRP) as actions taken to adapt to the growing difficulty in pumping groundwater (Harrington and Harrington 2005). Changes in irrigation technology, including subsurface drip irrigation, and actions to improve efficiency of water use, including metering and crop changes, have reduced recent rates of decline to less than 15 cm per year (KGS 2000). For dryland farmers (either historically, or in those areas where groundwater has already become unavailable in practical terms), mitigation of drought conditions is difficult and must take the form of crop choice and enrollment in federal programs (e.g., the CRP and crop insurance). In those places reverting from irrigated agriculture, shifts to more dryland crops (e.g., wheat, sunflowers, and "dryland corn") are taking place. To help make such transitions, the state is establishing aquifer subunits within Groundwater Management Districts (GMDs) based on hydrologic and water use parameters, thereby targeting the most vulnerable areas for more intense management. A possible result of these changes will be greater state, as opposed to local, control.

Despite these improvements, it could be argued that, in an area where rainfall is marginal for most crops in nearly every year, drought in combination with temporary or permanent loss of groundwater could push producers over the brink from survival to failure. Failure of the farms could cause a domino effect through the agricultural system, affecting the CAFOs and then the meat-packing plants, as well as ancillary businesses servicing them. During past droughts, an important form of adaptation was migration from the region; it is likely that collapse of the local agricultural system would lead to future emigration. The largely ethnic immigrant populations working in meat-packing have less economic, institutional, and social support to fall back on than longer-term residents of the study area, so it is likely that they would be the first to emigrate. Adaptive capacity through relocation could indicate the low level of regional adaptive capacity to drought.

To counter this gloomy picture, it is likely that the agricultural socioeconomic system of southwestern Kansas will survive because the capacity to adapt to increasing drought and decreasing availability of groundwater should be high. Technological improvements to improve efficiency of water delivery continue, and work on developing of more drought tolerant crop varieties (e.g., "dryland sorghum" and "dryland corn") moves forward. Livestock populations have exceeded the capacity of the region to provide sufficient types and quantities of feed; livestock growers have adapted by importing feed from other regions by rail. In case of local drought-related crop failure or reversion to dryland agriculture, the industry could easily import 100% of its feed to southwestern Kansas. Consequently, reversion of irrigated land to dryland farming or to rangeland would not create a crisis in the socioeconomic system. Indeed, in some areas, range use has been continuous since settlement in the nineteenth century; Kincer (1923) found rangeland agriculture to be appropriate for the region, and this land use may still be one of the most suitable (Popper and Popper 1999). Dryland farming and range may not support the same population as more intensive forms of agriculture, but they do represent reasonable adjustments to regional climate and resource conditions. Similar to recent growth in non-beef livestock operations, other land uses that bring income to the study area are also likely to expand.

Thus, the confidence of locals is their adaptive capacity may be well placed. Mirroring this local confidence, and based on past responses to resource and climate concerns, some researchers indicate relative confidence in the ability of residents to adapt in the face of environmental change (Kromm and White 2001; Knight *et al.* 2003).

Recapitulation

Although the focus of this section has been on climate- and water-related vulnerability, it is important to emphasize that many things shape the vulnerability of any particular place. Water is the key environmental concern for southwestern Kansas, but other factors also influence local impacts and responses. These factors include several trends in North American agriculture identified by Easterling (1996): slow growth in domestic demand for agricultural products, uncertain rates of productivity increases, a weakening of North American comparative advantage, structural changes in the number and size of agricultural operations, decline of rural communities, protection of environmental values, and increasing scarcity of water supplies.

Conclusions

Society often considers vulnerability in terms of total numbers of casualties and total damages in dollars, so there can be a tendency to consider rural regions as having low vulnerability simply because any stress would affect fewer people and less of the built environment (Harrington 2005). As demonstrated in this chapter, rural regions can be as vulnerable to climatic variability and change as more populous areas, if not more so. One mitigating factor in the regional vulnerability of the HPO-HERO study region, however, is a "can-do" attitude that promotes adaptation and resiliency. Although drought recurs often in the region, agriculturalists have been able to adapt and overcome this obstacle, creating thriving agricultural economies through irrigation and diversification that includes dryland crops and livestock feeding. Ironically, one of the most important factors in reducing decadal-scale vulnerability – groundwater-based irrigation – will ultimately increase vulnerability of the current system as the resource depletes (see Dregne 1980). In addition to the ever-declining availability of groundwater, increasingly erratic weather from global climate change and continued economic globalization will probably cause further stresses to the local socioeconomic system. Although it is likely that the region will evolve and adapt to the new conditions, the evolutionary process could be difficult and painful.

References

Barnaby, M. F., 1996. The distribution of feedlots in southwestern Kansas: 1962–1995. Unpublished Masters thesis. Manhattan, KS: Kansas State University.

Bell, E. H., 1942. *Culture of a Contemporary Rural Community: Sublette, Kansas*, Rural Life Studies No. 2. Washington, D.C.: US Department of Agriculture Bureau of Agricultural Economics.

Bloomquist, L., D. Williams, and J. C. Bridger, 2002. Sublette, Kansas: persistence and change in Haskell County. In *Persistence and Change in Rural Communities*, eds. A. E. Luloff and R. S. Krannich, p. 23–43. Wallingford, UK: CAB International.

Borchert, J. R., 1971. The Dust Bowl in the 1970s. *Annals of the Association of American Geographers* **61**(1): 1–22.

Broadway, M. J., and D. D. Stull, 2006. Meat processing and Garden City, KS: boom and bust. *Journal of Rural Studies* **22**: 55–66.

Bussing, C. E., and H. Self, 1981. Changing structure of the beef industry in Kansas. *Transactions of the Kansas Academy of Sciences* **84**(4): 173–186.

Carr, T., and R. S. Sawin, 1996. *Hugoton Natural Gas Area of Kansas*, Public Information Circular No. 5. Lawrence, KS: Kansas Geological Survey. Accessed at www.kgs.ku.edu/Publications/pic5/pic5_1.html.

Conant, R. T., K. Paustian, and E. T. Elliott, 2001. Grassland management and conversion into grassland: Effects on soil carbon. *Ecological Applications* **11**(2): 343–355.

Cross, J. A., 2001. Megacities and small towns: different perspectives on hazard vulnerability. *Environmental Hazards* **3**: 63–80.

Cutter, S. L., B. J. Boruff, and W. L. Shirley, 2003. Indicators of social vulnerability to hazards. *Social Sciences Quarterly* **84**(2): 242–261.

Dregne, H. E., 1980. Task group on technology. In *Drought in the Great Plains: Research on Impacts and Strategies*, ed. N. J. Rosenberg, pp. 19–42. Littleton, CO: Water Resources Publications.

Easterling, W. E., 1996. Adapting North American agriculture to climate change in review. *Agricultural and Forest Meteorology* **80**: 1–53.

Easterling W. E., P. R. Crosson, N. I. Rosenberg, M. S. McKenney, L. A. Katz, and K. M. Lemon, 1993. Agricultural impacts of and responses to climate change in the Missouri–Iowa–Nebraska–Kansas (MINK) region. *Climatic Change* **24**: 23–61.

Edwards, A. D., 1939. *Influence of Drought and Depression on a Rural Community: A Case Study in Haskell County, Kansas*. Social Research Report No. VII. Washington, D. C.: US Department of Agriculture Farm Security Administration and Bureau of Agricultural Economics.

Gilson, P., J. A. Aistrup, J. Heinrichs, and B. Zollinger, 2001. *The Value of Ogallala Aquifer Water in Southwest Kansas*. Fort Hays, KS: The Docking Institute of Public Affairs, Fort Hays State University.

Glantz, M. H., and J. H. Ausubel, 1984. The Ogallala Aquifer and carbon dioxide: comparison and convergence. *Environmental Conservation* **11**: 123–131.

Goodin, D. G., J. A. Harrington, Jr., and B. C. Rundquist, 2002. Land cover change and associated trends in surface reflectivity and vegetation index in Southwest Kansas: 1972–1992. *Geocarto International* **17**(1): 43–50.

Harrington, L. M. B., 2005. Vulnerability and sustainability concerns for the U. S. High Plains. In *Rural Change and Sustainability: Agriculture, the Environment and Communities*, eds. S. J. Essex, A. W. Gilg, R. Yarwood, J. Smithers, and R. Wilson, pp. 169–184. Cambridge, MA: CAB International.

Harrington, L. M. B., and J. Harrington, Jr., 2005. When winning is losing: Arkansas river interstate water management issues. *Papers of the Applied Geography Conferences* **28**: 46–51.

Harrington, J. Jr., D. Kromm, L. Harrington, D. Goodin, and S. White, 2003. Global change and southwest Kansas: local emissions, non-local determinants. In *Global Change and Local Places: Estimating, Understanding, and Reducing Greenhouse Gases*, eds. Association of American Geographers Global Change in Local Places Research Team, pp. 57–78. Cambridge: Cambridge University Press.

Harrington, J. A., Jr., and M. J. Wood, 2002. Climate stability in the Great Plains. *Preprints of the 13th Conference on Applied Climatology*, 236–237. Boston, MA: American Meteorological Society.

Johnson, H. D. (ed.), 1987. *Bioclimatology and the Adaptation of Livestock*, World Animal Science Series No. B5. New York: Elsevier.

Kansas Geological Survey (KGS), 2000. *An Atlas of the Kansas High Plains Aquifer*. Accessed at www.kgs.ku.edu/HighPlains/atlas.

Kansas Geological Survey (KGS), 2006. *Digital Petroleum Atlas: Hugoton Gas Area Field General Information*. Accessed at http://abyss.kgs.ku.edu/pls/abyss/gemini.dpa_general_pkg.build_general_web_page?sFieldKID=1000148164.

Kettle, N., 2003. Groundwater depletion and agricultural land use change in Wichita County, Kansas. Unpublished Masters thesis. Manhattan, KS: Kansas State University.

Kettle, N., L. M. B. Harrington, and J. A. Harrington, Jr., 2007. Groundwater depletion and agricultural land use change in the High Plains: a case study from Wichita County, Kansas. *The Professional Geographer* **59**(2): 221–235.

Kincer, J. B., 1923. The climate of the Great Plains as a factor in their utilization. *Annals of the Association of American Geographers* **13**(2): 67–80.

Knight, C. G., S. L. Cutter, J. DeHart, A. S. Denny, D. G. Howard, S. Kaktins, D. E. Kromm, S. E. White, and B. Yarnal, 2003. Reducing greenhouse gas emissions: learning from local analogs. In *Global Change and Local Places: Estimating, Understanding, and Reducing Greenhouse Gases*, eds. Association of American Geographers Global

Change in Local Places Research Team, pp. 192–213. Cambridge: Cambridge University Press.

Kuchler, A. W., 1976. A new vegetation map of Kansas. *Ecology* **55**: 586–604.

Kromm, D. E., and S. E. White (eds.), 1992. *Groundwater Exploitation in the High Plains*. Lawrence, KS: University of Kansas Press.

Kromm, D. E., and S. E. White, 2001. Regional change in the American Ogallala High Plains. In *Developing Sustainable Rural Systems*, eds. K. Kim, I. Bowler, and C. Bryant, pp. 67–74. Pyongyang: PNU Press.

Leathers, N., and L. M. B. Harrington, 2000. Effectiveness of conservation reserve programs and land "slippage" in southwestern Kansas. *Professional Geographer* **52**(1): 83–93.

McCarthy, J. J., O. F. Canziani, N. A. Leary, D. J. Dokken, and K. S. White (eds.), 2001. *Climate Change 2001: Impacts, Adaptation and Vulnerability. Contribution of Working Group II to the Third Assessment Report of the Intergovernmental Panel on Climate Change*. Cambridge: Cambridge University Press.

Nellis, M. D., L. M. B. Harrington, and J. Sheeley, 1997. Policy, sustainability, and scale: the US Conservation Reserve Program. In *Agricultural Restructuring and Sustainability: A Geographical Perspective*, eds. B. Ilbery, T. Rickard, and Q. Chiotti, pp. 219–231. Wallingford, UK: CAB International.

Ojima, D. S., J. M. Lackett, and the Central Great Plains Steering Committee and Assessment Team, 2002. *Preparing for a Changing Climate: The Potential Consequences of Climate Variability and Change – Central Great Plains*. Report for the US Global Change Research Program. Fort Collins, CO: Colorado State University.

Opie, J., 2000. *Ogallala: Water for a Dry Land*, 2nd edn. Lincoln, NE: University of Nebraska Press.

Polsky, C., 2004. Putting space and time in Ricardian climate change impact studies: agriculture in the US Great Plains, 1969–1992. *Annals of the Association of American Geographers* **94**(3): 549–564.

Polsky, C., and W. E. Easterling III, 2001. Adaptation to climate variability and change in the US Great Plains: a multi-scale analysis of Ricardian climate sensitivities. *Agriculture, Ecosystems and Environment* **85**: 133–144.

Popper, D., and F. Popper, 1999. The Buffalo Commons: metaphor as method. *The Geographical Review* **89**: 491–510.

Reisner, M., 1986. *Cadillac Desert: The American West and its Disappearing Water*. New York: Viking.

Reker, R. R., 2004. Mapping Conservation Reserve Programs trends in southwest Kansas using geospatial techniques. Unpublished Masters thesis. Manhattan, KS: Kansas State University.

Riebsame, W. E., 1990. The United State Great Plains. In *The Earth as Transformed by Human Action: Global and Regional Changes in the Biosphere over the Past 300 Years*, eds. B. L. Turner II, W. C. Clark, R. W. Kates, J. F. Richards, J. T. Mathews, and W. B. Meyers, pp. 561–575. Cambridge: Cambridge University Press.

Rosenberg, N. J., 1986. Adaptations to adversity: agriculture, climate, and the Great Plains of North America. *Great Plains Quarterly* **6**: 202–217.

Rosenzweig, C., F. N. Tubiello, R. Goldberg, E. Mills, and J. Bloomfield, 2002. Increased crop damage in the US from excess precipitation under climate change. *Global Environmental Change* **12**: 197–202.

Sherow, J. E., 1990. *Watering the Valley: Development along the High Plains Arkansas River, 1870–1950*. Lawrence, KS: University of Kansas Press.

Sophocleus, M., 1998. Water resources of Kansas: a comprehensive outline. In *Perspectives on Sustainable Development of Water Resources in Kansas*, Kansas Geological Survey Bulletin No. 239, ed. M. Sophocleus, pp. 3–59. Lawrence, KS: Kansas Geological Survey.

Sophocleus, M., and R. S. Sawin, 1998. *Safe Yield and Sustainable Development of Water Resources in Kansas*, Public Information Circular (No.) 9. Lawrence, KS: Kansas Geological Survey. Accessed at www.kgs.ku.edu/Publications/pic9/pic9_1.html.

State Energy Resources Coordination Council (SERCC), 2003. *Kansas Energy Plan*, Open-File Report No. 2003–3. Lawrence, KS: Kansas Geological Survey.

Tannehill, I. R., 1946. *Drought: Its Causes and Effects*. Princeton, NJ: Princeton University Press.

United States Census Bureau, 2000. *The US Census Bureau*. Accessed at http://factfinder.census.gov.

United States Census Bureau, 2006. *State and County QuickFacts (Kansas)*. Accessed at quickfacts.census.gov/qfd/states/20000lk.html.

United States Department of Agriculture Economic Research Service (USDA ERS), 2006. *State Fact Sheets: Kansas*. Accessed at www.ers.usda.gov/statefacts/KS.HTM.

United States Department of Agriculture National Agricultural Statistics Service (USDA NASS), 2002. *2002 Census of Agriculture*. Accessed at www.nass.usda.gov/Census_of_Agriculture/index.asp.

Warren, J., H. Mapp, D. Ray, D. Klatke, and C. Wang, 1982 Economics of declining water supplies in the Ogallala Aquifer. *Ground Water* **20**(1): 73–79.

Webb, W. P., 1931. *The Great Plains*, 1981 reprint. Lincoln, NE: University of Nebraska Press.

White, S. E., 1994. Ogallala oases: water use, population redistribution, and policy implications in the High Plains of western Kansas, 1980–1990. *Annals of the Association of American Geographers* **84**(1): 29–45.

White, S. E., and D. E. Kromm, 1995. Local groundwater management effectiveness in the Colorado and Kansas Ogallala region. *Natural Resources Journal* **35**: 275–307.

Whittemore, D. O., 2000. *Water Quality of the Arkansas River in Southwest Kansas: A Report to the Kansas Water Office*, Contract No. 00–113, Upper Arkansas River Corridor Study, Kansas Geological Survey Open-File Report No. 2000–44. Lawrence, KS: Kansas Geological Survey. Accessed at www.kgs.ku.edu/Hydro/UARC/quality-report.html.

Wood, M. J., 2002. Spatial and statistical analysis of Palmer's Climatic Instability Index. Unpublished Masters thesis. Manhattan, KS: Kansas State University.

Woodhouse, C. A., and J. T. Overpeck, 1998. 2000 years of drought vulnerability in the central United States. *Bulletin of the American Meteorological Society* **79**(12): 2693–2714.

Woodhouse, C. A., J. J. Lukas, and P. M. Brown, 2002. Drought in the western Great Plains, 1845–56: impacts and implications. *Bulletin of the American Meteorological Society* **83**(10): 1485–1493.

Worster, D., 1979. *Dust Bowl: The Southern Plains in the 1930s*. New York: Oxford University Press.

14

Urbanization and hydroclimatic challenges in the Sonoran Desert Border Region

CYNTHIA SORRENSEN AND ANDREW COMRIE

Introduction

The Sonoran Desert Border Region HERO consists of two watersheds, the Santa Cruz River and the San Pedro River, as well as the counties and municipalities predominantly situated in these watersheds. Both watersheds straddle the United States–Mexico border with their rivers flowing north from Sonora, Mexico into Arizona, United States. On the Arizona side, Santa Cruz and Cochise Counties reside mainly in these basins and rely on the groundwater sources within the basins. On the Sonoran side, there are five municipalities: Nogales and Santa Cruz in the Santa Cruz Basin, and Cananea, Naco, and Agua Prieta in the San Pedro Basin. Most of the population in this border region lives in two urban transborder communities: Nogales, Arizona and Nogales, Sonora, situated on the western side of the study area and together referred to as Ambos Nogales; and Douglas, Arizona and Agua Prieta, Sonora situated on the eastern side. A third transborder community, Naco, Arizona and Naco, Sonora, located just west of Douglas/Agua Prieta, is very small. Other settlements of significant size dot the region, including Sierra Vista, Rio Rico, Douglas, and Benson on the Arizona side, and Santa Cruz and Cananea on the Sonoran side (Figure 14.1).

The Sonoran Desert Border Region is semi-arid to arid, with summer temperatures frequently reaching over 104 °F (40 °C). The region experiences bimodal winter/summer precipitation patterns resulting from midlatitude frontal systems in winter and from thunderstorms within the regional North American monsoon circulation in summer (Adams and Comrie 1997; Sheppard *et al.* 2002). Rainfall during the monsoon can cause localized flash floods, but monsoon precipitation generally evaporates quickly, thus limiting percolation into the water table. Winter rains and mountain snowfall are the major source of groundwater recharge and can also cause flash floods. Extended dry periods have caused droughts throughout the period for which climate records are available. Both precipitation and temperature show strong inter-annual and multi-decadal

Sustainable Communities on a Sustainable Planet: The Human–Environment Regional Observatory Project, eds. Brent Yarnal, Colin Polsky, and James O'Brien. Published by Cambridge University Press.

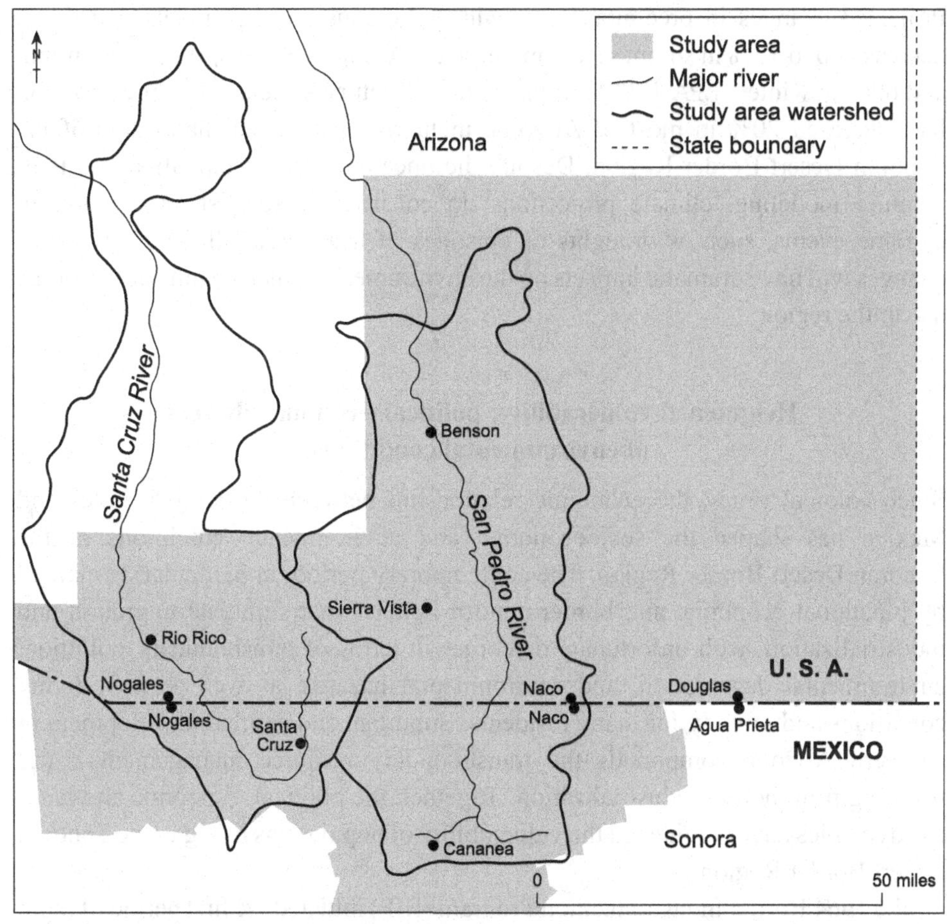

Figure 14.1. Location of the Sonoran Desert Border Region HERO, showing major watersheds and rivers, United States counties, and cities and towns mentioned in the text.

variability linked to Pacific Ocean sea surface temperatures (Sheppard *et al.* 2002; Brown and Comrie 2004).

Future climate projections for the Sonoran Desert Border Region are far from reaching consensus, but there is little doubt that the region, which has historically been affected by dramatic variations in weather and climate, will continue to be affected by climate events. Potential temperature changes exhibit the most consistency, with global climate models projecting average temperatures increases of 2–4 °F (1–2 °C) in spring and fall, and up to 5 °F (3 °C) in winter and summer (Watson *et al.* 1998). Century-scale mean temperature increases of ~4 °F (~2 °C) have already been observed just 60 miles (100 km) north of the region in Tucson, Arizona, principally due to higher overnight minimum temperatures (EPA 1998).

Projected changes in precipitation remain ambiguous among models, but could decrease up to 15% in summer, and increase on average 20% in spring, 30% in fall, and 60% in winter (EPA 1998). At present, the century-scale average precipitation has increased 20% in most of Arizona, including the United States side of the Sonoran Desert Border Region. Despite the uncertainty in precipitation and temperature modeling, climate projections do consistently suggest an increase in extreme events, such as droughts or episodes of heavy rainfall. These types of changes will have dramatic impacts on the environment, human health, and resource use in the region.

Heightened vulnerability: political–economic drivers of environmental conditions

Since colonial times, the economic relationship between the United States and Mexico has shaped the socioeconomic and environmental conditions of the Sonoran Desert Border Region. The contemporary period, in particular, is marked by binational economic and border control policies that stimulate migration and industrialization, with unfortunate outcomes in terms of transboundary pollution, environmental degradation, and environmental hazards, as well as poor living conditions and poverty for many residents. Suburban and exurban development in southern Arizona compounds the transboundary resource management issues resulting from border industrialization. Together, the political–economic and land-use dynamics have heightened the vulnerability of populations living in the Sonoran Desert Border Region.

The 1964 Border Industrialization Program (BIP) initiated the first period of rapid and uneven growth across the border, greatly affecting Ambos Nogales, but also influencing the Agua Prieta/Douglas and Naco/Naco sister cities. BIP established the first incentives for United States industries to take advantage of cheaper labor markets and unequal foreign currency exchange values by moving operations across the border. The result was the explosive development of the *maquila* industry (*maquilar* in Spanish means "to assemble"), in which products are assembled in Mexico and shipped back to meet demand in the United States and abroad. By the 1980s, Nogales, Sonora ranked as Mexico's sixth largest maquila center (Kopinak 1996), was the largest sister city complex along the Arizona–Sonora border, and represented more than 70% of the total Sonoran maquila industry and workforce (INEGI 2005).

Labor demands in the maquila industry initiated a contemporary migration stream that targeted the border region. This demand underpinned the urbanization and heightened the vulnerabilities that characterize the region today (Figure 14.2). Population growth during the BIP years increased dramatically on the Sonora side

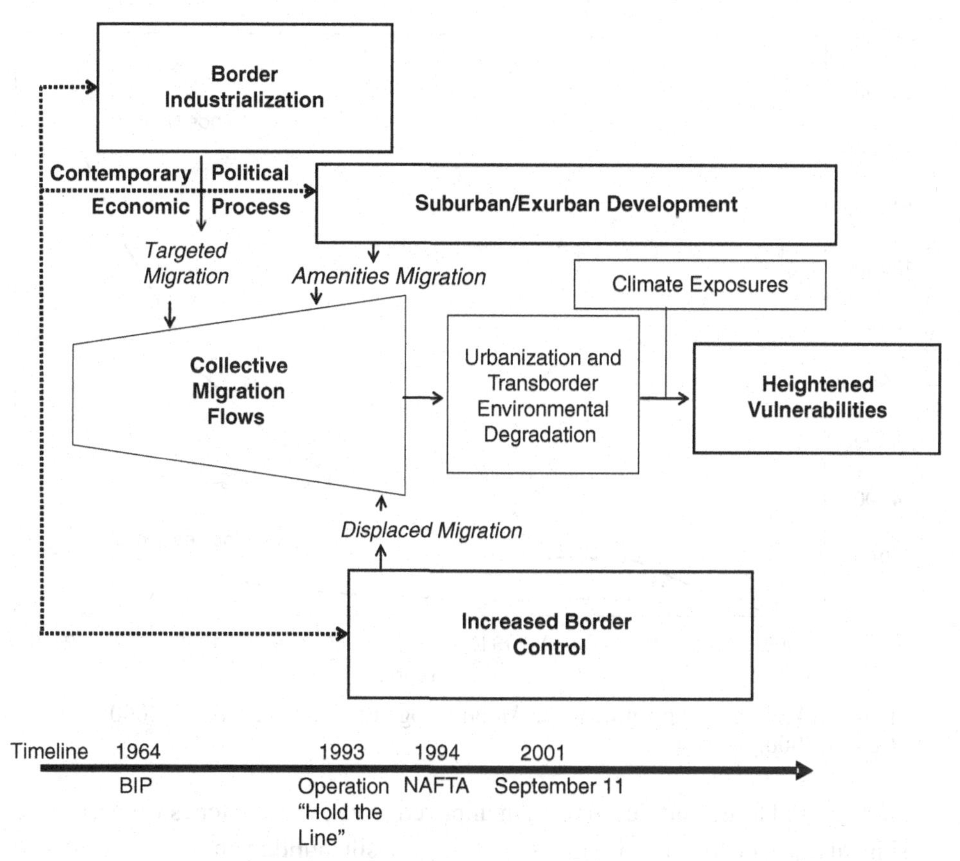

Figure 14.2. Economic development, border control, and vulnerability in the Sonoran Desert Border Region.

of the border in comparison to the Arizona side, especially in the center of maquila activity, Ambos Nogales (Figure 14.3). The urbanized area nearly quadrupled from 1973 to 1985, and doubled again in the next 15 years to 2000 (Figure 14.4). The rapidity of urban growth led to a shortage of adequate housing and the establishment of informal neighborhoods (*colonias*), where migrants coming into the city would invade barren areas and build precarious houses of wood, cardboard, and scrap metal (Ingram *et al.* 1995). Almost all urban amenities and infrastructure did not keep pace with the growth, leading to critical environmental health issues. Water problems were most pressing as access to potable water was limited, sewer lines were often non-existent, and solid waste was disposed of improperly. Migrants built houses on unstable hillsides or within *arroyos* (washes) and therefore were vulnerable to floods. Urban growth stripped vegetation from the hilly landscape, exacerbating soil erosion and – combined with unpaved roads – contributing to dust pollution levels that aggravated respiratory conditions in many border residents.

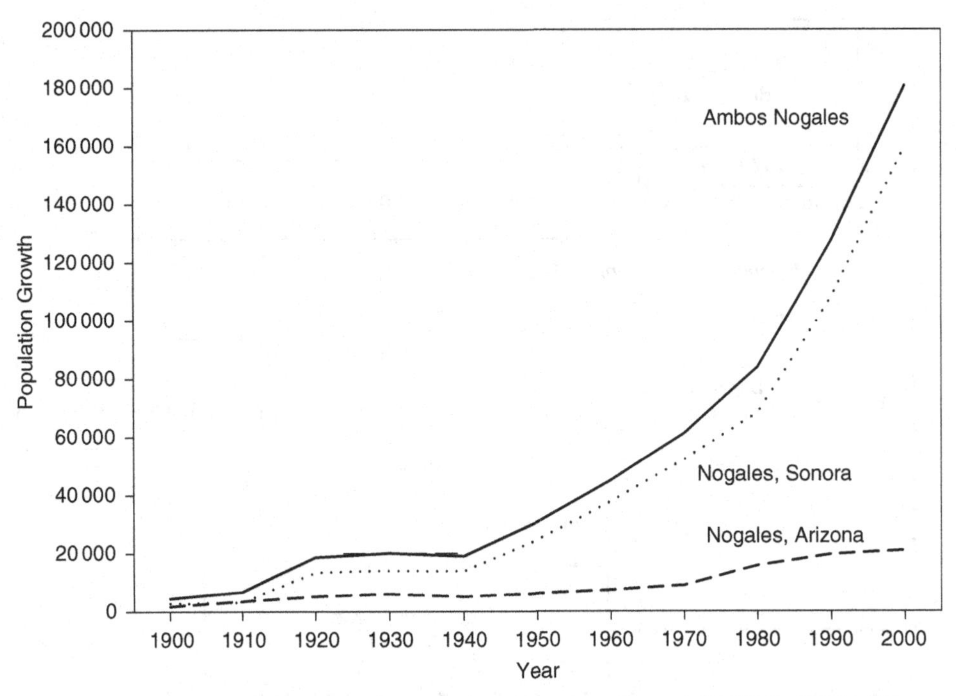

Figure 14.3. Population growth in Ambos Nogales. (Source: INEGI, 2000; US Census 2000.)

Rainfall caused large puddles to form in unpaved roads, while ditches were a source of skin disease in lower extremities and caused silt buildup in sewage systems. Despite these perils, migrants continued to relocate to Nogales, Sonora, believing that circumstances and opportunities there were better than in their places of origin.

Maquilas established in Agua Prieta and Naco provided substantial employment opportunities for residents of both communities. These towns did not attract the overall number of migrants that Nogales, Sonora did and grew at a slower pace in this first period of border industrialization, with 1.57% annual growth in Naco in the period 1950–90, and 1.29% annual growth in Agua Prieta in 1980–90 (INEGI 1990; L. Huntoon and C. Sorrensen, unpublished data). Population growth in Agua Prieta still outpaced its northern counterpart, Douglas, Arizona, as did land-use change to urban cover (Table 14.1; Figure 14.4). These two smaller Mexican border towns also suffered from outdated and limited water supply, sewage, and solid waste systems. They had minimal prospects for generating revenues to support maintenance and upgrades, to implement contingency plans for confronting adverse climatic conditions, or to develop long-range plans to manage resources sustainably. In Naco, only an estimated 65% of the population was connected to sewage lines, while over 400 families were not connected to the city's water supply system (BECC 1996a). The use of septic tanks and latrines instead of sewer connections

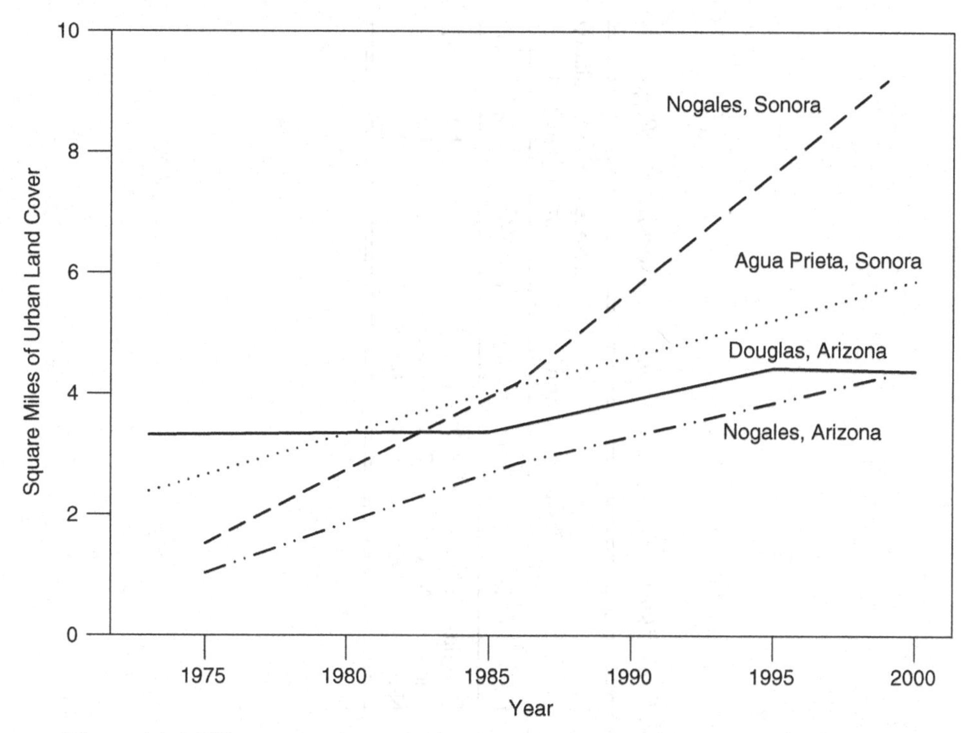

Figure 14.4. Urban cover change in the Sonoran Desert Border Region, 1973–2000.

polluted local soils and aquifers. The municipality of Agua Prieta kept pace with water supply needs, but a lack of urban infrastructure still existed, with solid waste management the most pressing concern.

The signing of the North American Free Trade Agreement (NAFTA) in 1994 introduced a second major surge of trade liberalization between Mexico and the United States, expanding the territory available to maquila development far beyond the border region. While maquila production along the border continued in the 1990s, it dropped to an average of 40% of total Sonoran production (Figure 14.5) while the interior Sonoran capital, Hermosillo, acquired significant maquila production (INEGI 2005). This shift in maquila focus away from Mexico's northern frontier remains an economic concern and source of vulnerability for border towns in the Sonoran Desert Border Region. Maquilas have closed in Nogales, Sonora and in Agua Prieta, some relocating to the interior and others to nations where labor costs are even lower, such as China. Along with the loss of maquilas, employment opportunities decreased. For the smaller two communities of Naco and Agua Prieta, the prospects of competition for maquila development from the interior of the state put them in an even more vulnerable position. They have fewer resources to attract additional maquilas to their area in comparison to the larger Nogales

Table 14.1 *Average annual population growth rates from 1980 to 1990, 1990 to 2000, in the Sonora Desert Border Region*

	Total population			Average annual population growth rate (%)			Total population			Average annual population growth rate (%)	
Mexican cities	1980	1990	2000	1980–90	1990–2000	US cities	1980	1990	2000	1980–90	1990–2000
Agua Prieta	34 380	39 120	61 944	1.3	4.6	Douglas	13 058	12 822	14 312	−0.2	1.1
Naco	4 441	4 645	5 370	0.4	1.5	Naco	N/A	700	833	N/A	1.8
Nogales	68 076	107 936	159 787	4.6	3.9	Nogales	15 683	19 489	20 878	2.2	0.7

Source: US Census Bureau (2000).

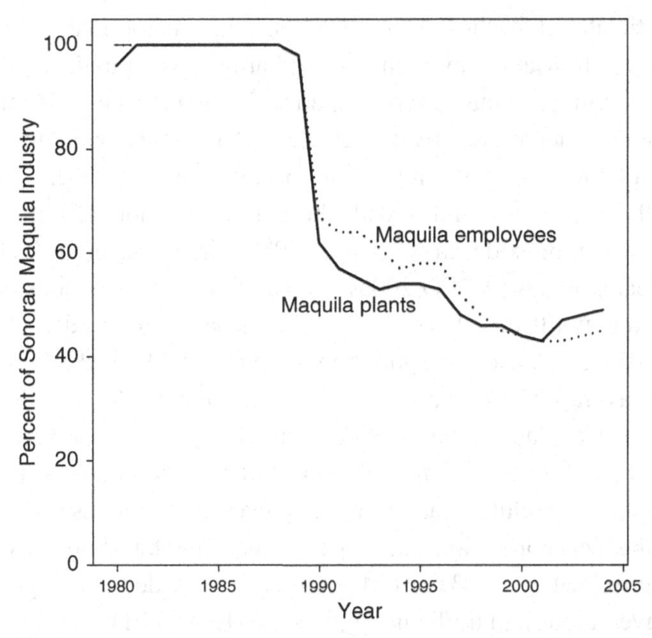

Figure 14.5. Border share of maquila industry, 1985–2004.

industrial complex. At present, private land developers in Agua Prieta are actively trying to attract new maquila development, but thus far, lots remain vacant. Meanwhile migration north to the border continues, contributing to a potentially jobless population along the border and to border issues with the United States.

Other liberalizing measures took place in Mexico during the 1990s to remove protectionist measures and open markets. Privatization features prominently in these measures, including the 1992 modification of Article 27, which allows land managers with collective usufruct rights to public lands (i.e., the legal right to use and enjoy the advantages or profits of another's property) to own and privatize their allotments. On the Mexican side of the Sonoran Desert Border Region, 1 000 000 acres (425 000 ha) are under the *ejido* collective land-tenure regime, a product of the Mexican Revolution and Agrarian Reform of the early 1900s. These ejido lands are now effectively becoming privatized. Private rights to ejido land include rights to all water resources on that land, an important factor in this arid region. In all likelihood, much of this land will continue under agricultural or ranching use. Nevertheless, ejido land that borders urban areas is increasingly susceptible to development as border cities not only expand physically, but also search for more ways to meet increasing demand by increasing their water supply.

Despite the economic uncertainties of maquila operation in a period of heightened trade liberalization, the NAFTA agreement furthered other growth opportunities and environmental challenges. The Deconcini Port of Entry at Nogales became the official port for the CANAMEX corridor, a major transport project aimed at linking Canada to Mexico by highways and at increasing trade, economic development, and growth throughout North America. For the citizens of Ambos Nogales, this designation meant more private vehicle and truck transport traffic flow and increased economic operations related to port-of-entry activities. It also meant increased traffic congestion and – with the high proportion of unpaved streets – dangerous carbon monoxide emissions and PM_{10} levels in both Ambos Nogales and Agua Prieta/Douglas (ADEQ 2005).[1] In addition, during periods of no rainfall, windy conditions uplift dust from the Mexican side of the border, mingle it with vehicular emissions, and carry air pollution north into the United States. In Nogales, Sonora, which averages 71% of all truck traffic crossing the Arizona–Sonora border, PM_{10} ratings exceed the official standards up to 26 times per year, creating an estimated 9800 short tons (8900 metric tonnes) of pollution per year (BECC 2004). In Agua Prieta the air pollution ratings are less severe, but still pose significant health problems for the local population and are part of the larger transboundary air pollution challenge: approximately 81.4% of PM_{10} emissions recorded in Douglas may be the result of unpaved roads and traffic in Agua Prieta (EPA 2001).

The increasing militarization of the United States–Mexico border as a result of heightened national security in the 1990s and the aftermath of 9/11 has had significant impact on the Sonoran Desert Border Region. Specific federal border initiatives have lead to a tightening of the California–Baja California border at San Diego/Tijuana, and the Texas–Chihuahua border at El Paso–Juárez. Consequently, migrants are being pushed further inland from the more temperate climates of the California and Texas coastal/riverine borders to the Sonoran and Chihuahuan deserts, where summer surface temperatures frequently reach over 104 °F (40 °C). This displaced migration stream adds to the collective migration flows that continue to descend upon urban border areas (Figure 14.2). The tripling of apprehensions in the Tucson Border Patrol Sector (which substantially overlaps the study area) from 7.6% of total United States–Mexico border apprehensions in 1993 to almost 37.5% in 2000 (Table 14.2) illustrates this changing flow. The Douglas Station alone

Table 14.2 *Apprehensions by Southwest Border Patrol Sector in 1993 and 2000*

Sector	1993	Share (%)	2000	Share (%)	Average annual apprehensions growth 1993–2000 (%)
Tucson Sector	92 639	7.64	616 346	37.5	27.1
Total US Mexico Border	1 212 886		1 643 679		

Source: GAO (2001).

Table 14.3 *Onboard Patrol Agents 1993–2000*

Sector	1993	1994	1995	1996	1997	1998	1999	2000
Tucson Sector	281	276	400	695	868	1010	1325	1513
Total US Mexico Border	3389	3670	4337	5281	6261	7292	7645	8475
Tucson, AZ percent of total	8	8	9	13	14	14	17	18

Source: GAO (2001).

experienced an unprecedented eightfold increase in apprehensions from 1993 to 2000, with 150 000 more apprehensions than any other station in 2000 (GAO 2001). The number of Border Patrol employees in the sector also increased dramatically, from 281 to 1513 in the same period (Table 14.3).

The privatization efforts in Mexico combined with displaced migration has had a significant impact on the region's second largest urban complex, Agua Prieta/ Douglas. Much of the land bordering the current city limits of Agua Prieta is under ejido land management. After witnessing the land invasions and chaotic growth of Nogales, Sonora in the 1970s and 1980s, ejido leaders took proactive measures towards privatization of urban ejido land by planning urban expansion and development in an organized grid-like fashion. Although growth in Agua Prieta/ Douglas during the 1990s outpaced their counterparts in Ambos Nogales, water infrastructure development in Agua Prieta lagged by only a few years (in contrast to a ten-year lag in Nogales, Sonora). Slightly over 10% of Agua Prieta's population over age 5 arrived in the municipality after 1995, which – considering the overall size of the urban area – signifies a significant migrant population (Table 14.4). While ejido leadership is attempting to attract additional maquila development from American entrepreneurs, it is difficult to determine how much of the population growth is the result of migration by those pursuing maquila jobs. It is likely that displaced migration is also contributing to population increase.

In Arizona, border towns remained small during the border industrialization period, growing minimally in contrast to their southern neighbors. Instead of strong industrial sectors, the economies relied largely on commercial and retail sectors dominated by Sonoran demand and on activities related to border flows, such as warehousing, transport, distribution, and border control.

Despite the relatively stagnant character of Arizona's border towns, exurban and suburban development is burgeoning in other areas of Santa Cruz and Cochise counties, compounding both local and transboundary resource-management issues. A significant impetus for this development comes from the popularity of southern Arizona as a choice for retiring Americans. This amenities-based migration flow (Figure 14.2) brings people to the area seeking a dry climate and recreation

Table 14.4 *Migration to Sonora border municipalities 1985–2000*

Municipality	Total population age 5 or older in 1990	Population age 5 or older that resided in Mexican state other than Sonora in 1985	Population age 5 or older that resided in a different country in 1985	Percent of total population age 5 or older that migrated to the municipality between 1985 and 1990	Total population age 5 or older in 2000	Population age 5 or older that resided in Mexican state other than Sonora in 1995	Population age 5 or older that resided in a different country in 1995	Percent of total population age 5 or older that migrated to the municipality between 1995 and 2000
Agua Prieta	33 512	2 285	158	7.3	53 408	4 885	657	10.4
Naco	3 986	242	41	7.1	4 606	131	65	4.3
Nogales	92 262	10 481	491	11.9	137 480	13 199	1 100	10.4

Source: INEGI 1990, 2000.

opportunities, many of whom want to maintain lawn landscapes and golf courses, which both contribute to increased water consumption rates. The towns of Sierra Vista and Rio Rico have experienced extensive growth, while in unincorporated areas exurban development pressures have outcompeted the struggling ranching industry for private land. The amount of public-owned land in both counties is over 60%, limiting exurban and suburban development to floodplains where most individual and corporate property exists. By 2000, 41.1% of Sierra Vista's population over age 5 resided outside the state five years earlier, whereas in the unincorporated area around the city, the figure was 21.6% (US Census 2000). In Rio Rico, which the Census divides into four unincorporated areas, in 2000, 20.5% of residents in the northeast Census Designated Place resided outside of the state five years earlier. At the county level, Santa Cruz County has experienced over 6% annual growth since 2000, while Cochise County, which is significantly larger in size, also has shown substantial annual growth of more than 5% since 2000.

"Wildcat" subdivisions, in which land-owners split and sell lots without paying for urban amenities, plague much of the exurban growth in southern Arizona. These land-owners avoid many of the responsibilities that developers incur, such as ensuring proper access to properties, attending to land development codes and resource regulations, and providing paved roads, curbs, sewer lines, and drainage systems. Inhabitants of wildcat subdivisions usually rely on individual wells for water and septic tanks for sewer treatment. Because these residents are not part of any formal water management systems, they do not figure in contingency plans that water managers must maintain or in long-term water planning. Outside of Sierra Vista in unincorporated Cochise County, an estimated 3000 unregulated wells are operating, many along the floodplain of the San Pedro River. During summer 2005, the river ran dry for the first time in 70 years. Individual wells are not the sole reason for the river's dry period, but the wildcat subdivisions house a population that utilizes water without regulation and that is unprepared to face hydroclimatic events affecting the resource.

Hydroclimatic challenges in the Sonoran Desert Border Region

The political, economic, and development circumstances described in the previous section sets the stage for understanding the impacts of hydroclimatic events in the Sonoran Desert Border Region and the region's capacity to plan for and respond to future hydroclimatic change. Residents of the Sonoran Desert Border Region frequently experience significant hydroclimatic events. Socioeconomic conditions, which have brought about uneven growth in the border region and varying degrees of vulnerability among the populations, exacerbate these events. The Sonora side of the border feels the impacts of hydroclimatic events more severely because the population is larger and poorer, municipal resources are more limited, and infrastructure lags population needs.

Nevertheless, there are hydroclimatic conditions under which Arizona residents also find themselves in dire circumstances. Despite representing a smaller, richer population, these residents are also vulnerable to hydroclimatic variation and change.

Prolonged drought, adaptations, and water resources in Ambos Nogales

Drought is one of the most common hydroclimatic events experienced in the Sonoran Desert Border Region, with critical implications for the water resources upon which communities rely (see Figure 14.6 for droughts as depicted by the

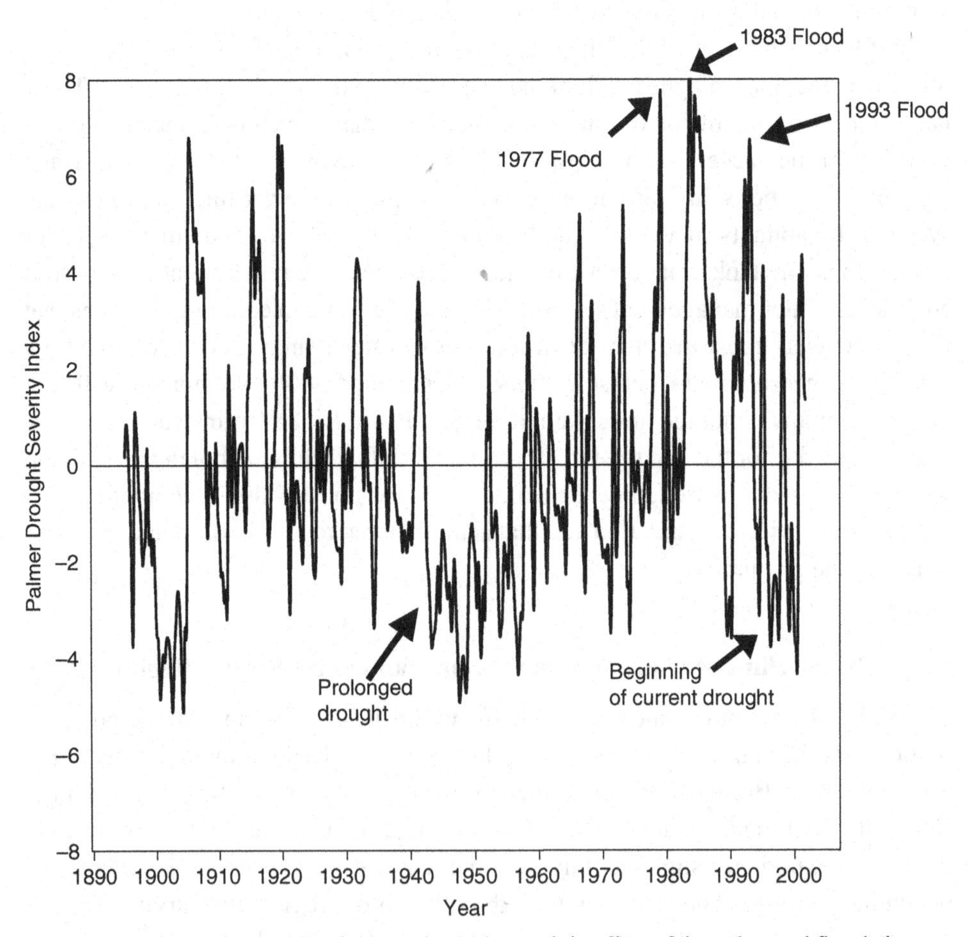

Figure 14.6. Palmer Drought Severity Index and timeline of droughts and floods in southeastern Arizona (Sorrensen 2005).

Palmer Drought Severity Index). The 1950s marked the nadir of an extended drought that dramatically affected ranching and changed the nature of agriculture in the region. Since industrialization began in the 1960s, border cities have – to varying degrees – been able to secure water during dry periods, but they are not in a position to relax strict control of the resource.

Alluvial and low-yielding bedrock aquifers intersperse the region and provide most groundwater resources. Recharge of these aquifers relies heavily on rainfall and, thus, drought affects water supplies. The aquifers on the Arizona side are less susceptible to surface conditions than those on the Sonora side, and wells are less prone to running dry during peak pumping periods. Nogales, Arizona has been more successful than Nogales, Sonora at meeting the water demands of its changing population through adaptation (Figure 14.7). The city built its first pumping facility in 1914, which serviced the population until the extended drought of the 1950s. In response to that drought, however, the city over-pumped the aquifers near the river's infiltration galleries, causing the groundwater table to drop significantly, recharge rates to decline, and the system to fail in meeting demand (Logan 2002). To ensure such shortfalls do not occur again, the city put additional well fields into operation, constructed a dam on Sonoita Creek, and built the Patagonia Lake reservoir.

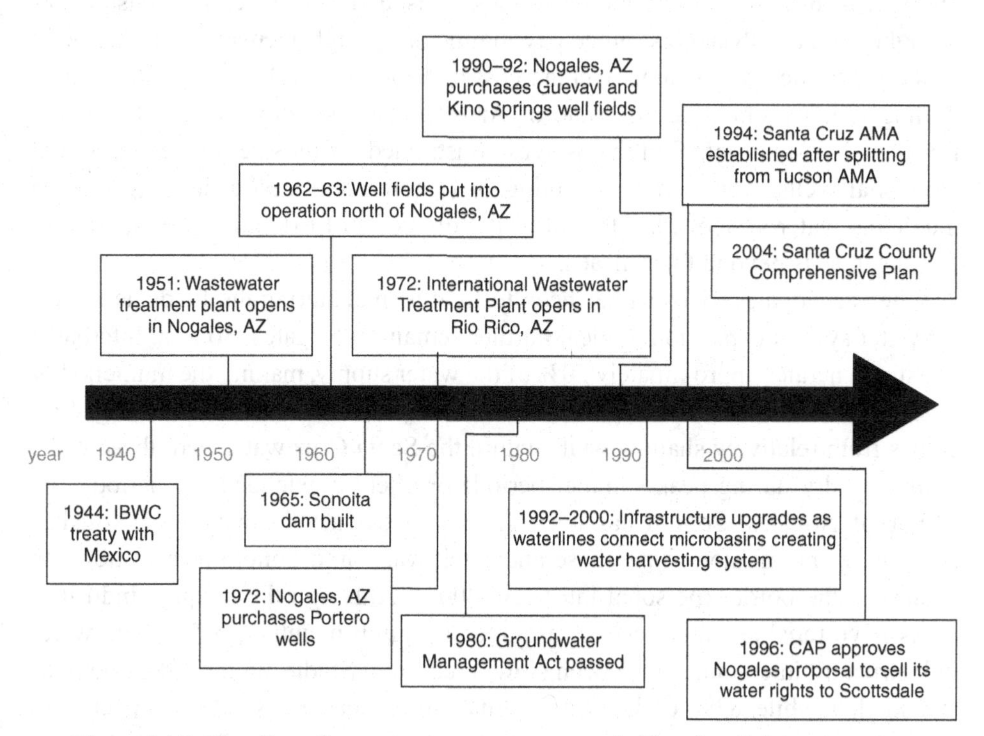

Figure 14.7. Timeline of water-system management in Nogales, Arizona.

The most recent set of adaptations to maintain Nogales' water resources and buffer drought impacts occurred within the context of the Central Arizona Project (CAP), which delivers water to Arizona from the Colorado River. Because the Santa Cruz River is a tributary of the Colorado River, Nogales obtained rights to CAP allocations. However, the canals that distribute CAP water terminate on the southern outskirts of Tucson, approximately 50 miles (80 km) north of Nogales. Delivery of CAP water to Nogales, which is higher in elevation than Tucson, would require an expensive pumping system. Instead of funding such a project, Nogales sold its CAP water rights to Scottsdale, a suburb of Phoenix, and used the resources to search for and purchase water sources closer to the city. Consequently, Nogales established its third and fourth well fields at Guevavi Ranch and Kino Springs in the early 1990s. Now the city has the capacity to stagger its pumping to prevent overextraction and allow for recharge, further buffering itself from drought.

While the Nogales has adapted by fortifying its water supply, the situation for many citizens residing in the unincorporated areas of Santa Cruz County is more precarious. During summer months, when water levels drop in exurban and low-density subdivisions, the energy costs of pumping water increase and make water more expensive for these residents. Personal wells are vulnerable to running dry during peak drought periods and few options exist to access water inexpensively as droughts persist. Residents either pay signficiantly higher energy bills, truck in water, or dig new wells at an average cost of $6000 per well. Even if the cost of drilling is not prohibitive, the state of Arizona has restrictions on well spacing. Residents on small lots find themselves constrained by lot sizes too small to add additional wells. Many of those living in remote areas desire hook-ups to the municipal water systems, but the infrastructure costs of extending the pipeline to remote areas is beyond their means.

Whereas Nogales, Arizona has been fortuitous in acquiring the funds necessary for water system expansion, critical hurdles remain in Nogales, Sonora. Interbasin transfer generates approximately 60% of the water supply, making the municipality vulnerable to management regimes outside its jurisdiction. The remaining 40% comes from relatively shallow wells within the Santa Cruz watershed; these wells easily run dry during peak summer periods or after an extended dry period. The mayor of Nogales, Sonora has been known to contact the mayor of Nogales, Arizona for permission to purchase and truck water into Sonora from water taps located at the border (personal interview 2003). Although the maquila industrial parks have reliable water access, resources to build, maintain, or upgrade water infrastructure in colonias have been slow in coming. Studies in the 1990s demonstrated that while 83% of households may have been physically connected to the water system in Nogales, Sonora, they did not necessarily receive water

(Ingram *et al.* 1995). Even today, when tap water does run, it often does so for limited periods, forcing households to stock up during those times, which magnifies the inconsistency of water flow. In marginalized colonias, the percentage of households with running water has been found to be as low as 30.8% (Sadalla *et al.* 1999).

Access to water is an urgent priority, so the municipality of Nogales, Sonora remains focused on expanding the water supply system and pays less attention to preparing for droughts. Since the mid-1990s, the municipality has been coordinating with the Comisión Nacional del Agua (CNA; the National Commission for Water) to implement upgrades and extensions of the water supply system and to insure 24-hour access (BECC 1996b). Known as the Acuaférico Project, the intent is to repair an estimated 7000 leaks in water pipes, replace old asbestos pipes, and increase distribution of water to all city residents by tapping into the Los Alisos basin (which is not connected to the Santa Cruz basin). It also aims to establish a meter and billing system to recoup initial costs. The project has evolved over the last decade to secure binational authorization and funding from agencies set up during the NAFTA period to support environmental infrastructure in the border region. There is great concern, however, over how most residents of Nogales, Sonora, who are working minimum-wage jobs, will be able to pay for this water. Paralleling this concern is the demand that the maquila industries subsidize the cost of water because of the stress they put on the system. In the interim, water supply is expanding at a sluggish pace, and without attention to long-term management or drought planning.

The international agreement on wastewater treatment made by the International Boundary Water Commission (IBWC) in 1943 compromises the recharge capacity of aquifers serving Nogales, Sonora. The IBWC is a binational institution mandated by the United States and Mexico to address and negotiate all transboundary water issues. As a result of IBWC negotiations, the main treatment facility for Ambos Nogales is located north of the border, where it returns treated effluent downstream in Arizona. Thus, Arizona aquifers and communities benefit from this recharge while Sonora loses precious water for its aquifers. At present, Nogales, Sonora contributes an estimated two-thirds of the total effluent to the facility and legally has the right to reclaim it, but the costs of pumping effluent upstream to recharge areas in Sonora are prohibitive (IBWC 1967; Sprouse 2005). To the dismay of Santa Cruz County residents who have come to rely on that effluent, discussions are under way to build a separate wastewater treatment plant in Sonora to retain effluent. Such an effort would support aquifer sustainability and buffer drought in Sonora.

Another plan to combine wastewater management with energy development would further compromise the amount of effluent returned to Sonora for aquifer recharge, and thereby increase vulnerability to drought. A consortium of Santa Cruz

County citizens proposes the construction of a combined cycle gas turbine (CCGT) plant on the Arizona side, siphoning some of Mexico's wastewater effluent as a cooling mechanism for plant operations (Maestros Group 2004). Although the plant would supply electricity to citizens on both sides of the border, it would also insure that at least 3 million gallons per day of Mexico's effluent would remain in Arizona. For Sonorans, the trade off is clear – the plant would supply needed power, but would not provide a viable recharge mechanism for Sonoran aquifers.

Urbanization and desert floods

Despite the semi-arid environment, the Sonoran Desert Border Region experiences extreme rainfall events and subsequent floods. The most significant regional floods occurred in 1977, 1983, and 1993, each leaving significant damage and a Federal Emergency Management Agency (FEMA) disaster zone declaration. (See Figure 14.6 for a timeline showing the relationship between these floods and the Palmer Drought Severity Index.) All three floods damaged bridges and electrical lines, cost millions of dollars in both urban and farm damage, and, in 1983, caused ten fatalities on the Arizona side of the border. The 1977 and 1983 floods were classified as 100-year floods.

These occasional major floods bring national attention to the region, but seasonal floods that occur with the summer monsoon and winter storm systems have greater impact on everyday life in the three biggest urban areas. Urbanization intensifies the impacts of these smaller floods by removing vegetation from the landscape, increasing the amount of impermeable surfaces, channeling flow, and raising the toxicity of runoff. For Ambos Nogales, urban growth has been particularly challenging for flood mitigation. The Nogales Wash, a tributary of the Santa Cruz River, passes through the very center of each city as it flows north from Sonora into Arizona. The Wash captures runoff from the steep hillsides surrounding the sister cities and then flows through underground channels as it passes the border. Flash floods during the summer monsoon season routinely kill people attempting to cross the border illegally through the underground portion of the wash. Above ground, the floodwaters detain traffic on the main streets, delay border crossings, and interrupt economic activity in the central business areas.

Most of the channelized portion of the Nogales Wash runs directly under the Deconcini Port of Entry, which remains vulnerable to a major flood. Some estimates put channel capacity at 33% below levels needed to handle severe flows from Nogales, Sonora during extreme precipitation events. Damage to the channel could compromise the tunnel's structural integrity and induce a collapse of ground cover, including the Port of Entry, above it. Officials in Santa Cruz County claim that severe flood related damage is highly likely in the future (interview 2004) but fortification of the port remains problematic. Temporary closure would seriously impede transborder economic activity and affect not only local border communities, but also larger regional transactions crossing through the Port of Entry. The Nogales Wash and the border control complex are under the jurisdiction of the federal government, thus leaving local governments on both sides of the border unable to remedy the problem.

In Nogales, Arizona, floods regularly threaten three residential areas – Pete Kitchen, Monte Carlo, and Chula Vista – situated within the confluence of two or more ephemeral washes. Sections of the washes run directly through these neighborhoods, traversing roads and passing within 6 m of many homes. The neighborhoods have only one or two access roads, which historically are the first areas to wash out, isolating the residents from emergency responders. Residents exacerbate the situation by constructing cement block walls into flood easements to divert water from their properties, thus reducing the channel available to carry floodwaters. To illustrate the gravity of the problem, heavy rains in October 2000 inundated homes with 3–6 feet (1–2 m) of water.

The flood problem is even more chronic and severe on the Sonora side of the border, where colonias perch on steep hillsides. Many residents' homes are built of flimsy materials that cannot withstand even minor hillside instabilities and inundations. Residents do what they can to protect themselves, constructing mounds to raise their houses, earthen walls around their properties, and tire walls to prevent erosion. The colonias El Represo, Colosio, and Ferrocarril, as well as houses and businesses along the major thoroughfare, Avenida Instituto Tecnológico, experience the worst problems, with regular floods during the monsoon season that leave behind sediment, long-standing large puddles, and debris in the streets. Other colonias, such as Buenos Aires and Lomas de Nogales I and II, plus houses near the major street Cinco de Febrero, also frequently experience damage to homes and streets. Emergency management officials rank the flood problems here as the worst of three major problems facing the city.

Chronic floods compound water pollution issues in Ambos Nogales because many households in Nogales, Sonora do not connect to the sewage system. Throughout the 1990s, 69% of households in some colonias lacked adequate sewage facilities (Sadalla *et al.* 1999). During heavy rains, waste from these households flows into the Nogales Wash, along with trash, debris, and sediment. Fecal coliform counts from human waste have been measured at levels that exceed both the Mexican Criterios Ecológicos de Calidad de Agua (Quality Criteria for Drinking Water Sources) and the Arizona Aquifer Water Quality Standards. Other contaminants have been recorded in the Wash on both sides of the border, including nitrates, arsenic, and tetrachloroethylene, a suspected carcinogen (IBWC 2001). When floodwaters recede, polluted sediments deposited on streets and in houses and businesses remain to cause respiratory problems.

Floods aggravate wastewater pollution in Ambos Nogales, much of which is without a stormwater drainage system. Residents often open manhole covers to the wastewater drainage pipes to siphon rising floodwaters at street level and reduce the threat to their homes and businesses. The leaky sewage system also allows considerable stormwater incursion into the sewage pipes, which severely stresses Ambos Nogales' main treatment plant. Rainfall during the summer monsoon season can increase influent to 30 million gallons per day, 12.8 million gallons per day over treatment facility operating capacity (personal interview 2004). Winter rains have put daily influent at 20 million gallons per day over a three-month period, again forcing the plant to operate over capacity. When influent exceeds the capacity rating, treatment is compromised, and inadequately treated effluent is deposited in the Santa Cruz River as it heads towards Rio Rico. In addition, the high sediment load of wastewater influent during heavy rains damages equipment and increases maintenance hours. The persistent floods leave Nogales, Arizona drenched in sewage with smells reeking from manholes and even toilets (Vanderpool 2006).

Floods create similar transboundary issues in the Naco/Naco and Agua Prieta/Douglas areas. Border sanitation problems in Naco/Naco, particularly during high-rainfall periods, have occurred since the late 1980s. A faulty sewage pumping station located near the border in Naco, Sonora means that raw sewage frequently overflows and runs through both cities. In addition, Naco, Sonora utilizes stabilization ponds to oxidize the town's wastewater naturally in a closed system that contains no receiving stream for effluent. Unfortunately, the amount of incoming sewage often exceeds the outgoing water (through evaporation, infiltration, and agricultural use), thereby causing overflows of insufficiently treated water. The fugitive water flows through a drainage basin that not only traverses the international boundary, but also crosses a well field used by the town of Bisbee, Arizona for its municipal water supply. Excessive rainfall worsens this overflow, sending even more insufficiently treated wastewater northward.

The situation in the Agua Prieta/Douglas region is no less critical. There, in contrast to the other urban border complexes, the waters of the Agua Prieta wash drain south from Douglas, Arizona into Sonora. The wastewater system of Douglas is comprised largely of vitrified clay pipes originally laid in 1906. Over 30% of the system is deficient because of leaks and overloaded lines. In many places, sewer flows are less than 1.3 miles per hour (0.6 m s^{-1}), rendering the system prone to blockages (BECC 2001). When storm waters increase, additional sediment enters sewer lines, pipes clog, and sewage spews into surface waterways and flows south into Agua Prieta, Sonora. The reliance on individual wells and substandard septic systems in two residential areas of Douglas – Sunnyside and Farview – causes further problems. Repeated septic failures cause sewage to flow through residential yards, threaten local groundwater, and pose health risks to colonia residents. Lot sizes in Sunnyside and Farview are relatively small, making it hard to find space to replace failing septic systems or drill additional wells when water sources get contaminated. The extensive depth to reliable groundwater sources presents additional challenges for these residents.

In downstream Agua Prieta, inadequate solid waste management compounds the city's flood issues. Growth in the 1990s resulted in the need for collection and disposal of 33 tons (30 metric tonnes) of domestic solid waste and 12 tons (11 metric tonnes) of supposedly non-hazardous, maquila-generated solid waste (BECC 1996c). City revenues and waste management capacity has not kept pace with growth so that the estimated 4 tons (4 metric tonnes) of solid waste deposited illegally across the city eventually mingles with storm waters during heavy rains. In addition, the Agua Prieta River is less than 100 yards (90 m) away from the city's main legal open-air landfill. Present city resources are insufficient to enforce environmental controls, technical or sanitary requirements, and basic administrative operations at the landfill. Leachates produced during decomposition of the waste seep into the groundwater, which is the sole source of the city's water. Moreover, during rainy periods, solid waste at the landfill washes into the river where it mingles with runoff coming south from Douglas.

Binational influence on adaptation

In response to rising concerns about pollution in the larger border region, NAFTA carried with it a Side Agreement on the Environment which established linked binational institutions to respond to environmental infrastructure needs: the Border Environmental Cooperation Commission (BECC) and the North American Development Bank (NADBank). The purpose of BECC is to approve environmental infrastructure projects aimed at offsetting pollution and resource demands driven by NAFTA economic policies, while NADBank functions as a financial institution to seek external funding for approved projects, lend its own money to specific projects, and manage funding for all BECC-approved projects. Together, the BECC–NADBank arrangement provides a mechanism for local border communities to prioritize and tackle environmental issues. Unfortunately, there is no mention in BECC–NADBank proposal guidelines for certain types of hydroclimatic hazard mitigation projects, such as drought planning and flood mitigation, even though NAFTA–driven industrialization and urbanization have intensified the impacts of these hazards. Despite this shortcoming, the BECC–NADBank framework provides the best option to communities in the Sonoran Desert Border Region for facing hydroclimatic challenges and creating viable adaptive strategies.

Both larger urban areas and smaller communities in the Sonoran Desert Border Region are actively engaged in the BECC–NADBank process. Presently, there are five BECC–certified projects in the Ambos Nogales area, three in the Agua Prieta/ Douglas area, and two in Naco, Cochise County that lay within the NAFTA border zone have submitted a handful of other projects. All of these projects reflect priorities established by local governments and specifically seek remediation needed to overcome environmental problems that resulted from rapid industrialization and urban growth. To this extent, they represent attempts to mitigate the vulnerabilities that are present. Top priorities are projects to enhance water supplies and upgrade wastewater and sold waste treatment, while newer projects address air pollution issues through street paving. As mentioned earlier in this chapter, these problems are extensive, often require high tech solutions, and would be too costly to fund at the municipal level.

Despite a historic climate record that demonstrates decade-long droughts and a recent disaster record of floods that come every monsoon season and often with winter rains, none of the present BECC–NADBank projects in the Sonoran Desert Border Region focuses on drought planning or flood mitigation. There are several practical reasons for this lack of support. First, building water supply and waste management infrastructure is indeed a high priority in these communities because they are still trying to catch up with uncontrolled development in the border zone. Drought planning is at least in part based on assuring water supply systems and therefore overlaps with proposals already set forth. In addition, drought planning is also perhaps more procedural – as opposed to infrastructural – in character, as it often concerns implementing rationing, conservation measures, or interbasin transfer agreements that do not necessarily require the construction of large physical systems, which tends to be the focus of BECC–NADBank projects. Turning to floods, borrowed funds spent on flood mitigation do not directly create new revenue sources and could be perceived as a riskier loan arrangement by the community borrowing money or NADBank. In contrast, funds spent on water supply or wastewater treatment projects generate public revenues through user fees, thus making loan repayment less risky. Many flood management projects also protect a relatively small, discrete portion of the population and therefore are lower priority within municipal governments. In sum, these and other reasons work against funding drought planning and flood mitigation. BECC–NADBank projects give border communities access to critically needed resources for addressing serious environmental and resource problems, but any mitigation these projects provide against hydroclimatic impacts is coincidental. Moreover, these projects remedy existing problems and do not consider planning for future hydroclimatic variability and change.

Although these projects clearly do not go far enough, the United States Treasury Department and Mexico's Finance Minister are considering the closure of NADBank (La Reforma 2006; Taj 2006). Criticisms include accusations of the bank's bureaucratic inefficiencies and high administrative costs, its inability to provide low-interest loans, and the idea that commercial banks would do a better job (La Reforma 2006). While influential border politicians have successfully countered intimations of closure, the volatility of the situation illustrates the vulnerability of border communities. They are not self-reliant, but depend on the very binational mechanisms that caused their resource management and environmental challenges in the first place. Municipalities on the Mexican side of the border particularly rely on BECC–NADBank to ameliorate NAFTA–fueled resource stress and environmental degradation. The centralized nature of Mexican governance means that these municipalities legally have no access to the municipal bond measures that their United States counterparts could initiate to finance environmental infrastructure. The removal of the BECC–NADBank mechanism will heighten the vulnerability of border populations.

The region is dependent on BECC–NADBank support, but the overall success of – and even the financial logic behind – these projects is in question. Although projects are in process in all three transborder communities, the layers of authority involved in decision-making over transboundary resource management and the lack of consensus among governing bodies have slowed project implementation (Sprouse 2005). In addition, the nearly $200 million in NADBank loans approved for border environmental infrastructure projects in the Sonoran Desert Border Region raises questions about increasing indebtedness in the area, especially on the Sonora side of the border.

Conclusions

Vulnerability to hydroclimatic variation and change in the Sonoran Desert Border Region is a function of the political–economic and development structures that frame border economy, control, and environmental degradation. Vulnerability is also a function of existing climate exposures and the physical demands of migration flows. Because of rapid industrial development and population growth, border communities face tremendous challenges in managing their water resources, waste, and air pollution at all times; these challenges are worse during floods and droughts. Projects to ease many of the chronic environmental problems are under way, but the combination of the institutional constraints and the criticality of the circumstances dominate priority setting, and bureaucratic barriers slow the pace of project implementation. Although the binational presence of the BECC–NADBank is encouraging, it is not pushing border communities to undertake integrated long-term planning to prepare for hydroclimatic change. Important adaptation is occurring, but whether that adaptation will be sufficient or will be sufficiently focused on the prospects of increased climate variability and climate change remains to be seen.

Notes

1. The notation PM_{10} indicates atmospheric particulate matter – fine particles of solid or liquid suspended in a gas – of 10 micrometers or less in diameter. Primary human sources of fine particles are the burning of fuels in motor vehicles and power plants and the blowing of dust from lands denuded of vegetation. Heart and lung disease and other health hazards are associated with increased levels of fine particles in the air.

References

Adams, D., and A. Comrie, 1997. The North American monsoon. *Bulletin of the American Meteorological Society* **78**(10): 2197–2213.

Arizona Department of Environmental Quality (ADEQ), 2005. *Plan of Action for Improving Air Quality in Ambos Nogales*, Developed by the Border 2012 Ambos Nogales Air Quality Task Force and the Border Liaison Mechanism Economic and Social Development Subgroup, adopted by ADEQ and the State of Sonora's Secretariat for Urban Infrastructure and Ecology. Phoenix, AZ: ADEQ. Accessed at www.azdeq.gov.

Border Environment Cooperation Commission (BECC), 1996a. *Water Supply and Wastewater Collection and Treatment System Project for the City of Naco, Sonora*. Ciudad Juárez, Mexico: BECC.

Border Environment Cooperation Commission (BECC), 1996b. *Water Supply and Distribution Project, Nogales, Sonora*. Ciudad Juárez, Mexico: BECC.

Border Environment Cooperation Commission (BECC), 1996c. *Comprehensive Municipal Solid Waste Collection and Final Disposal Project for Agua Prieta, Sonora*. Ciudad Juárez, Mexico: BECC.

Border Environment Cooperation Commission (BECC), 2001. *Water and Wastewater System Improvements in Douglas, Arizona*. Ciudad Juárez, Mexico: BECC.

Border Environment Cooperation Commission (BECC), 2004. *Air Quality Improvement for Nogales, Sonora*. Ciudad Juárez, Mexico: BECC.

Brown, D. B., and A. C. Comrie, 2004. A winter precipitation 'dipole' in the western United States associated with multidecadal ENSO variability. *Geophysical Research Letters* **31**: doi:10.1029/2005GL022911.

General Accounting Office (GAO), 2001. *INS Southwest Strategy Resources and Impact Issues Remain after Seven Years*, GAO 01–842: Report for Congressional Committee. Washington, D.C.: US Government Printing Office.

Ingram, H., M. Laney, and D. Gillilan, 1995. *Divided Waters: Bridging the U. S.– Mexico Border*. Tucson, AZ: University of Arizona Press.

Instituto Nacional de Estadística Geografía e Informática (INEGI), 1990. *Censo General de Poblacion y Vivienda*. Accessed at www.inegi.org.mex/.

Instituto Nacional de Estadística Geografía y Informática (INEGI), 2000. *Censo General de Población y Vivienda 2000*, Municipal data. Accessed at www.inegi.org.mex/.

Instituto Nacional de Estadística Geografía y Informática (INEGI), 2005. *Industria Maquiladora de Exportación*. Aguascalientes, Mexico: INEGI.

International Boundary and Water Commission (IBWC), 1967. *Minute 227: Enlargement of the International Facilities for the Treatment of Nogales, Arizona and Nogales, Sonora Sewage*. El Paso, TX: International Boundary and Water Commission.

International Boundary and Water Commission (IBWC), 2001. *Binational Nogales Wash United States/Mexico Groundwater Monitoring Program, Final Report*. El Paso, TX: International Boundary and Water Commission.

Kopinak, K., 1996. *Desert Capitalism: Maquiladoras in North America's Western Industrial Corridor*. Tucson, AZ: University of Arizona Press.

Maestros Group, 2004. *Ambos Nogales Generating Station*. Accessed at www.maestrosgroup.com/.

Sadalla, E., T. Swanson, and J. Velasco, 1999. *Residential Behavior and Environmental Hazards in Arizona–Sonora Colonias*. San Diego, CA: Southwest Center for Environmental Research and Policy. Accessed at www.scerp.org/projects/sadalla98.pdf.

Sheppard, P. R., A. C. Comrie, G. D. Packin, K. Angersbach, and M. K. Hughes, 2002. The climate of the US Southwest. *Climate Research* **21**: 219–238.

Sorrensen, C., 2005. Adapting to drought and floods in the semi arid landscape of Santa Cruz County, Arizona. *The Geographical Bulletin* **47**: 101–118.

Sprouse, T., 2005. *Water Issues on the Arizona–Mexico Border: the Santa Cruz, San Pedro, and Colorado Rivers*. Tucson, AZ: Water Resource Center.

Taj, M., 2006. Possible shutdown of NADBank worries some U. S. lawmakers. *Tucson Citizen*, March 16.

United States Census Bureau, 2000. *The US Census 2000*. Accessed at http://factfinder.census.gov.

United States Environmental Protection Agency (EPA), 1998. *Climate Change and Arizona*, EPA 236-F-98–007c. Washington, D.C.: Climate and Policy Assessment Division, USEPA.

United States Environmental Protection Agency (EPA), 2001. *Douglas, Arizona Wastewater Collection and Potable Water Distribution Improvement Project: Environmental Assessment*. San Francisco, CA: USEPA.

Vanderpool, T., 2006. NADBank blues. *Tucson Weekly*, April 13.

Watson, R., M. Zinyowera, and R. Moss, 1998. *The Regional Impacts of Climate Change: An Assessment of Vulnerability, 1997, A Special Report of the IPCC Working Group II*. Cambridge: Cambridge University Press.

Part VI

15

Lessons learned from the HERO project

BRENT YARNAL, JOHN HARRINGTON, JR., ANDREW COMRIE, COLIN POLSKY, AND OLA AHLQVIST

The HERO vision revisited

This book started with the premise that to develop sustainable communities on a sustainable planet, an infrastructure should exist that enables scientists to monitor local human–environment interactions, to share and compare data, analyses, and ideas with scientists at other locales, and to participate with colleagues and stakeholders in a global network dedicated to community-level sustainability.

The book recounted the Human–Environment Regional Observatory (HERO) project's attempt to take first steps in developing such an infrastructure and the concepts and research behind that infrastructure. As such, the project did not produce – and never intended to produce – definitive research results about, for example, vulnerability or the causes and consequences of land-use and land-cover change. Consequently, this book has concentrated on conceptualizing the elements needed to make human–environment infrastructure work, and on exploring those elements by proof-of-concept testing.

This chapter summarizes HERO's efforts (and therefore the book) by revisiting a set of questions posed in Chapter 1. The most important part of the chapter is the discussion of lessons learned during the HERO team's attempts to answer those questions. The chapter concludes by trying to support the project's (and book's) claim that there is a need for HEROs.

Answers to and lessons learned from HERO's guiding questions

Chapter 1 reported two fundamental questions that were central to the HERO effort. One overarching question guided the research and addressed infrastructure development via three less-encompassing questions (Table 15.1). The second basic question aimed to prove that it is possible to achieve the HERO vision discussed above by trying to answer one important, complex human–environment research

Sustainable Communities on a Sustainable Planet: The Human–Environment Regional Observatory Project, eds. Brent Yarnal, Colin Polsky, and James O'Brien. Published by Cambridge University Press.

Table 15.1 *HERO's guiding questions and proof-of-concept question*

Guiding (infrastructure development) questions
- How do we understand and monitor local human–environment interactions across space and time?
- How do we collaborate across space and time?
- How do we build networks of human–environment collaborators?
- How do we benefit from collaboration across academic generations?

Proof-of-concept (human–environment) question
- How does land-use/land-cover change influence vulnerability to hydroclimatic variation and change?

question. This section addresses those questions. After addressing the overarching infrastructure question and then discussing each of the three more focused guidance questions, we present lessons learned in trying to follow that guidance. Similarly, after talking about the components of the proof-of-concept question, we cover lessons learned in pursuing that question.

How do we understand and monitor local human–environment interactions across space and time?

The goal of the HERO project was to explore the possibility of developing the infrastructure needed to build a network of sites devoted to understanding and monitoring human–environment interactions across space and time. Infrastructure was taken to mean both the basic framework of the network and the human, computational, and other resources required to make such a network function successfully. HERO used three strategies to achieve this goal: it tried to develop research protocols for studying and monitoring human–environment interactions at individual sites and for comparing and generalizing across sites; it built a computer- and Internet-based "collaboratory" to help investigators collaborate by sharing data, analyses, and ideas from remote locations; and it tested the protocols and collaboratory in a prototype network by investigating vulnerability to hydroclimatic variation and change induced by land use.

Consequently, the HERO research design had two components. The first component was the Web-based networking environment, which not only developed ways to handle the diverse data generated by human–environment research, but also created a collaboratory to enable researchers to interact scientifically and socially while working at their local sites. Chapters 2–4 discussed the theoretical basis of the networking environment and the tools that comprised the collaboratory, as well as examples of the application of geographic information science to human–environment studies. The second component of the research design sought to prove that it was possible to build a network of local human–environment observatories (i.e., HEROs) connected by the collaboratory and common data standards and protocols. The network consisted of the biophysically and socioeconomically diverse Central Massachusetts HERO, Central Pennsylvania HERO, High Plains–Ogallala HERO, and Sonoran Desert Border Region HERO, and these four HEROs focused on answering a single question ("How does land-use change influence vulnerability to droughts and floods?") to test the infrastructure.

To answer this question, the HEROs assessed floods and droughts in the context of land-use/land-cover change and vulnerability – giving the most attention to vulnerability. Researchers used the collaboratory to evolve a framework protocol that defined and characterized vulnerability. Investigators at each HERO placed vulnerability into the local physical and human contexts through archival work, fieldwork, and consultation with local experts and stakeholders. Subsequent analysis sought commonalities and differences across the four study sites.

Chapters 5–7 covered conceptual and methodological questions surrounding the concept of vulnerability, environmental history, and land-use/land-cover change, respectively, and thereby laid the foundation for the proof-of-concept work to follow. Chapters 8–10 applied the frameworks presented in Chapters 5–7 by executing cross-site studies of vulnerability associated with hydroclimatic variation and land-use/land-cover change. Chapters 11–14 presented detailed insights into each of the four HEROs, thus demonstrating that networks of HEROs must simultaneously generalize human–environment interactions across sites while maintaining the unique human–environment characteristics of each place.

How do we collaborate across space and time?

HERO sought to build the infrastructure needed to analyze, assess, monitor, and compare local human–environment interactions across space and time. To know how to build this infrastructure, we needed to answer a two-part question. First, in a network of HEROs, how do we collaborate across space? Second, at any given HERO, how do we collaborate across time?

Considerable time in the early years of the HERO project was devoted to figuring out how to collaborate with our colleagues at the other HEROs. Collaboration is difficult when working with colleagues at one's home institution. Already busy schedules and interminable, often escalating demands on those schedules make it hard for any individual to carve out the time needed to do research. Trying to match limited open times in multiple schedules to find the time to work together compounds this problem. Collaboration is even more difficult when working with colleagues across institutions: trying to match the schedules of collaborators who face the same challenges, but in other time zones and at institutions with different timetables and calendars makes simultaneous collaboration substantially more complex. Moreover, not having the motivation provided by daily face-to-face interaction makes it easier to be drawn into other tasks and not to collaborate with colleagues hundreds or thousands of miles away.

Another challenge facing HERO concerned collaboration over time. For a single HERO site to collect human–environment interactions data at one place over decades would involve multiple generations of researchers with multiple technologies and multiple research paradigms. Comparing these data to detect trends over time would be impossible unless systems were instituted that could archive both quantitative and qualitative data equally well. Moreover, these systems would need to be adaptable so that they could keep up with – and even anticipate – technological changes so that data not only would not be lost, but also would be available in flexible formats to anybody with a desire to use them. Such systems could not be unique to any one location, but would need to work the same everywhere.

For the HERO project, the solution to these dilemmas was to build the collaboratory (Chapter 3) and develop representational approaches for the archival of both qualitative and quantitative knowledge (Chapter 4). The collaboratory aimed at providing investigators with a one-stop Website accessible to any networked computer. At that Website were tools that enabled them to work together synchronously or asynchronously, at the same places or at different places. The videoconferencing tool allowed the investigators to work in real time, simulating face-to-face interaction and sharing text, graphics, and video clips via the Internet. The e-Delphi tool helped the researchers explore questions and ideas anonymously and asynchronously, perhaps coming to consensus. Codex was the most sophisticated tool of all, giving researchers a receptacle for quantitative and qualitative data, an apparatus to manipulate and analyze those data in a mixed-methods format, and an underlying structure intended to provide the flexibility needed to evolve as technologies and science change. The development of support for representation and analysis recognized human–environment research as more than just data archival and retrieval, but a nexus of tangible and intangible aspects (Chapter 2). The intangible aspects of research, such as the hypotheses, concepts, methods, and procedures was of particular interest to us. A particular focus was to find ways to model knowledge about both the conceptual understanding of human–environment interaction and the process of decision-making. To that end a general representation for geographic categories was developed to support formal analysis of hard to define concepts such as vulnerability. Additionally, a GeoAgent-based Knowledge System (GeoAgentKS) was built to help understand the dynamic interactions among humans and their environment.

Lessons learned: collaborating across space and time

Collaboration in the HERO project came to focus on the medium – the collaboratory. By the end of the project, we used videoconferencing routinely, e-Delphi when needed, and Codex experimentally because of its later inception. Complementary to these tools, face-to-face collaboration was essential to the success of the project.

Our vision of videoconferencing – that is, an inexpensive, Web-based system that does not rely on pricey specialized equipment and studios – was ahead of its time and consequently filled with technical challenges. During the first three years of the project, tool development involved adopting scattered, immature hardware and software and using it on relatively underpowered personal computers. This approach consumed inordinate resources, resulted in too many failed meetings, and produced massive frustrations with the tool. During the last years of the project, we dropped the "video" portion of videoconferencing and used a combination of telephone conference calling and screen-sharing. This approach proved to be nearly as good as live video for our purposes: it appears that if all participants are looking at the same image on their personal computers and hearing each other speak to that image, it is similar to them all being in the same room and working with an image projected on a screen at the front of the room.[1] Point-to-point videoconferencing hardware and software are common features built into today's personal computers, which have the power to make these features work well, so this technology has caught up to HERO's original vision. Advanced multi-nodal capabilities are also readily available at modest cost.

While videoconferences and desktop sharing are useful for shorter, instantaneous communication when all participants are available, there is also a need for collaboration that requires more time, that can transcend different schedules or time zones, or when conflicting views may hamper progress. HERO's e-Delphi tool was developed for such asynchronous discussions – those in which participants contribute at different times throughout the course of a day, week, or longer. The Delphi format also has some added advantages of allowing participants time to reflect and providing anonymity to help level the discussion and make the final product representative of the group instead of a few opinionated or dominant members. One drawback of the Delphi approach is that it requires a moderator to move the process through its various stages. Orchestrating an e-Delphi exercise required a modest commitment by all participants, but necessitated a more significant obligation by the moderator. To maintain anonymity, the tool also calls for relatively strict control by the service provider (i.e., HERO). Thus, although HERO wanted all its tools to be open for public use, all users of e-Delphi had to work through a HERO project member to conduct an exercise. The tool can consequently address collaboration needs where complex problems and lack of easy defined agreements call for more planning, a longer time frame, and some control over the discussion tool. This naturally caused e-Delphi to be used on an occasional basis, rather than the everyday basis characteristic of videoconferencing.

Codex is a powerful, multifaceted Web-based tool aimed at helping researchers to store, generate, and share data, information, and knowledge. At its simplest, Codex is a place to store raw quantitative and qualitative data in an easy-to-enter and easy-to-retrieve, flexible format designed to stand changes in software and hardware over time. For all its seeming advantages, however, the tool was little used by HERO researchers. Perhaps the most important reason was that the tool resulted from observing the needs of the HERO collaborators; hence, the concept emerged midway through the project and the tool matured during HERO's later stages. By that time, researchers had learned ways to use (or work around) existing technologies and were unwilling to invest time into a new tool.

The general representation for geographic categories was never integrated with Codex but its development demonstrated several examples related to vulnerability and land-use/land-cover change. Some aspects of the hard to define vulnerability concept could be tackled by using fuzzy and rough set theories by which vague and inexact knowledge can be made explicit and formally represented. This also enabled detailed descriptions of the different classifications used in data on land use and land cover allowing for cross-site comparison or across-time change assessment without the need for standardized data. The GeoAgent-based Knowledge System made it possible to generate models of decision-making processes from documents and interviews with experts. The models integrate data, human knowledge, and knowledge-driven actions to simulate possible scenarios of environmental changes. This system can be used to make direct suggestions in various types of emergency/disaster situations or for long-term environmental management.

The collaboratory taught us some valuable lessons. First, the technological immaturity of videoconferencing, and the work-in-progress status of other collaboration tools deterred HERO investigators from using those tools. In other words, dealing with the technology derailed collaboration and slowed progress on the human–environment research. But once the investigators came to grips with the collaboratory and its weaknesses and strengths, the human–environment component of HERO took off.

It is important not to blame the collaboratory development team for the limited progress made by the human–environment team in the first half of the project. The collaboratory specialists worked diligently on the tools and tried hard to engage the human–environment researchers, while the human–environment team was just being human: in the absence of "bulletproof" collaboratory tools, they hungered for interaction with researchers at other HERO sites. Given limited time and money, however, it was difficult to have many face-to-face meetings.

Thus, another valuable lesson is that collaboration means contact, ideally in person. Humans are social beings and require facial and voice cues to achieve successful collaboration.

How do we build networks of human–environment collaborators?

Providing tools does not ensure that a collaborative network will be successful. At least six elements are essential to building a successful collaborative network: good architecture, strong leadership, effective management, flexible participants, establishing and building a shared knowledge and social base, and sufficient time.

When setting up a collaborative network, a key consideration is its architecture, which influences the functioning and performance of the system. If the architecture is a chain (i.e., work flow is linear, passing through each link in the chain), then the chain is only as strong or quick as the weakest or slowest link. If the architecture is a net (i.e., work flow is multidirectional, passing through multiple nodes), although a missing, weak, or slow node will create a hole in the network, it is possible to work around that spot. Thus, it is important to avoid chains and build nets.

Strong leadership is critical to the success of the collaborative network. A successful leader has a clear vision of the network's goals and how it can achieve them. An effective leader campaigns for funds to initiate and sustain the network. A good leader persistently connects members, reminds them of the network's mission, keeps them on task, and helps them see the value resulting from the daily grind. An able leader contacts other networks and lets them know about the network he/she represents, thereby creating opportunities for the people in his/her network to work with – and learn from – others.

Effective management is imperative to the success of the collaborative network. Good architecture and strong leadership cannot offset the ravages of poor management. Tasks need to be set, workloads need to be apportioned equitably, deadlines need to be set, funds need to be disbursed, and so on. If such management decisions are made well and in a timely way, the network has a good chance of succeeding; if these decisions are poor and late, it will be difficult for the network to function effectively and for collaboration to flower.

Scientists working in collaborative networks need to be flexible. Researchers need not only to share their knowledge, but also to learn from their collaborators. They need to invest time in getting to know their colleagues socially as well as scientifically. They must be willing to speak in plain language and to leave the jargon of their specialties at the door. Collaboration means that scientists need to stick with the project beyond the parts that utilize their expertise and provide direct benefit to them.

Even if the network has excellent architecture, strong leadership, effective management, and flexible participants, it cannot work unless it establishes and builds a shared knowledge and social base. Although all collaborators and research sites want to contribute to the enterprise, investigators in a multidisciplinary setting come with different disciplinary knowledge and language, intellectual traditions, and social customs. Not only do they need to develop a shared scientific understanding and language, but also they need to build personal relationships, a social network, and trust.

Collaborative networks take time to mature. Although experienced researchers clearly understand their science, they need time to understand their role in a network – that is, how their expertise contributes to the overall pool of knowledge represented by the various collaborators. As noted above, the investigators need a period to figure out how to relate to the other researchers not only scientifically, but also socially. If there are several disciplines or research traditions involved, researchers must invest time to understand these ideas and perspectives. If there is specialized technology, sufficient time is required for members to learn these tools. If students are involved, they need time to learn the subject, the technologies, and how to do research. In sum, any collaborative project requires time for all the members of the project to come to a dynamic equilibrium and for the mature project to emerge. The bigger the project and the more people involved, the more time that is needed.

In conclusion, collaborative networks are evolving social constructs that involve an architecture plus leaders, managers, and team members. The architecture is important, but networks are only as good as the people in the networks. Each network is a unique combination of people that with time reaches maturity and equilibrium; when personnel change, another unique entity emerges that may function better or worse than the preceding combination of people.

Lessons learned: building networks of human–environment collaborators

As noted above, our experiences with the HERO project enable us to view several elements as essential to building successful collaborative networks, including good architecture, strong leadership, effective management, flexible participants, and sufficient time. HERO taught us some noteworthy lessons.

The knowledge that nets of researchers are less prone to failure than chains has important implications for funding agencies and collaborative networks. Lower funding levels mean fewer personnel can participate and necessarily result in collaborative chains. Higher funding levels result in nets, which appreciably increase the chances of success. Moreover, higher levels of funding support result in more time invested by the team. Thus, insufficient funding of collaborative research projects and networks suggests a much lower likelihood of success, so it might be better for funding agencies to fund fewer collaborative enterprises at higher levels than more such entities at lower levels. For large projects and networks, this lesson indicates that inadequate funding might doom the enterprise from the beginning, so it might be wiser to seek additional support (including leveraged funds) or to join an existing network, rather than to accept the proffered funds without supplemental support.

For these more expensive collaborative research projects and networks, leadership and management are critical: without strong leadership and effective management, even the best-funded projects and networks will founder. Incipient collaborative enterprises should start by identifying a leader with energy, vision, and social skills; experience at leading a large collaborative effort is a definite advantage. Funding agencies and grant review panels need to pay special attention to the choice of leader, refusing to support even the best-conceptualized projects if they have weak leadership. Ongoing projects and networks must also be careful when selecting replacement leaders. The same ideas apply to effective management: new projects and networks need concrete management plans; funding agencies and grant review panels must demand robust management plans; and ongoing projects and networks need to assess their management constantly, improving it as they go along.

When putting together collaborative research teams, either for large projects or networks, organizers should search for participants who are flexible and socially adept, who are interested in sharing and learning, who can talk in plain language, who are productive and who follow through with their commitments in a timely manner, and who are generous with their time, ideas, and work. In other words, collaborative teams need team players that are willing to roll up their sleeves and get the job done. Experience counts, but a great danger comes with recruiting senior colleagues who are overextended and cannot engage in the collaboration.

Finally, limited time is the enemy of collaborative research. Effective projects and networks need three years or more to spin up and reach equilibrium, especially in such complex topics as human–environment interactions. Most products that come from this period are likely to be (1) individual, non-collaborative efforts, (2) continuations of collaborations that started before the new collaborative work, or (3) technical, methodological work. The new collaboration can be expected to produce new published research by the fourth or fifth year. The problem with this scenario is that funding agencies expect and need to see products after the first year of work, which is reasonable for single-investigator, reductionist research. Moreover, a five-year research grant is a long grant in most cases, yet large collaborative enterprises require that much time to enter their productive years. Consequently, there is a conflict between the needs and expectations of the funding agencies. Researchers who become involved in five-year collaborative projects or networks such as HERO face a dilemma: they either work for many years after the project with no funding to generate publications or they continue to the next project and do not publish results from the big collaborative effort. The former solution is unsatisfactory to the researchers who need to support their efforts financially and the latter is unsatisfactory both to the funding agency (which needs products to justify funding) and to the researchers (who need publications for career advancement). A longer view is needed for large collaborative research enterprises distributed over space.

The HERO project needed one year to spin up the network (e.g., find graduate students, fill key postdoctoral and staff positions, institute work routines, and tackle fundamental research issues) and a second year to establish the social dynamics of the team. One way that the project accelerated the process of building the team was to hold all-hands meetings at the various research sites, which was helpful not only socially, but also in learning about the local human–environment relationships confronting the others. By the third year, however, the HERO collaborators only had three years left in which to tackle the research as a fully functioning team. Again, the lesson here is that collaborative research projects need more time. It is imperative that attention be paid to developing ways for interdisciplinary collaborative research projects – especially projects with limited time horizons – to speed up the spin-up and team-building phases inherent to all such projects.

How do we benefit from collaboration across academic generations?

The HERO project included a full range of academic generations in its research. Over 100 faculty members, postdoctoral researchers, graduate students, undergraduate students, and others contributed significantly to HERO during the five years of the project. Various combinations of funds from HERO and related projects supported these individuals. In some cases, faculty and students contributed their time without financial support.

One of the notable features of HERO was the success of its faculty. During the course of the grant, of the original 18 senior investigators, one became a department chair, four became school, institute, or center directors, and four became deans or higher university administrators. These numbers do not include the several individuals who were already in such positions when the project started or those individuals who rose to such positions after the close of the project. Other important achievements of the HERO faculty included promotions from Associate Professor to Professor (two) and tenure track Assistant Professor to tenured Associate Professor (two). One senior investigator retired.

With an eye to a future with permanent human–environment observatories, HERO cultivated a cadre of postdoctoral, graduate, and undergraduate students specializing in human–environment interactions. From the beginning of the project, an important aspect of the project was postdoctoral and graduate training. Of the eight postdoctoral research associates who worked on the project, one is now a tenured Associate Professor, five are tenure track Assistant Professors or Lecturers, one is a Visiting Assistant Professor, and one is owner of an environmental consulting company. Of the 14 doctoral students who contributed to the project, nine achieved their Ph.D. degrees (with more on the way) and four are now tenure track Assistant Professors. Of the 11 Masters students who took part in HERO, at least three have gone on to doctoral degrees and two of those three have completed the higher degree.

Another crucial focus for HERO, which developed after the project started, was research participation by undergraduates. In Year 1, HERO applied for and received a National Science Foundation Research Experiences for Undergraduates (REU) Supplement. This supplement supported the activities of five women undergraduates – two at Penn State and one each at Clark, Kansas State, and Arizona. Four were Honors students and the fifth was a McNair Scholar. In addition to the HERO REU Supplement, the Central Massachusetts HERO established its own undergraduate Fellowship Program. In its first year of that program, six undergraduate students (one of whom was supported by the HERO REU Supplement) were chosen from a pool of highly qualified applicants after a campus-wide recruitment campaign. A core effort of the Central Massachusetts HERO, this program continues annually.

Spurred by the success of the HERO REU Supplement and Central Massachusetts HERO efforts, the project developed the HERO REU Site (Yarnal and Neff 2007). HERO devoted a considerable portion of its effort to prepare for and run this site. In the second through fifth years of the project, 12 to 16 students – three or four each from universities and colleges in Massachusetts, Pennsylvania, Kansas, and Arizona – attended a two-week short course at which they received training in the theory and method of human–environment interactions. After that, the students returned to their respective HEROs for six weeks to apply the human–environment training, thereby testing the HERO methods and protocols and suggesting improvements to them. The three- or four-person student teams not only collaborated with their mentors and colleagues at the four HEROs, but also interacted with the other REU students through the collaboratory.

HERO even engaged high school students in the research by participating in Penn State's Summer Experience in Earth and Mineral Sciences (SEEMS) program. SEEMS is available to high school students participating in the Upward Bound Math and Science (UBMS) Program, a federally funded program designed to support and motivate students from disadvantaged backgrounds. SEEMS provides academic enrichment throughout the year as well as an intensive six-week summer program. HERO contributed to SEEMS by providing hands-on research experiences for a few of these students, who worked directly with senior faculty, graduate students, and undergraduate students.

Because collaborative research is a social activity, members of the HERO project benefited from the rich interactions of old and young investigators. One important idea was to have multiple levels of research mentoring so that every academic level mentored somebody at a lower level. Hence, faculty mentored all levels, postdoctoral scholars mentored graduate and undergraduate students, doctoral students mentored Masters and undergraduate students, and so forth. Everybody benefited from "teaching research," thereby turning research into an interactive social process instead of a cold, analytical task.

Another important idea was that if a smart undergraduate student could understand and implement a HERO protocol, collecting and analyzing field data with few problems, then it was an effective protocol. By working directly with the younger students, HERO investigators could see the weaknesses in the protocols and ways to improve them.

Lessons learned: benefiting from collaboration across academic generations

The HERO project learned many lessons through its multi-generational collaboration, with most of them being positive. Ironically, perhaps the most negative lesson came from the great success of the faculty, with a surprisingly high percentage moving to important administrative posts. This success speaks highly of the senior faculty's quality, but also suggests that maintaining the attention and contribution of these busy people was difficult. Several of HERO's key senior personnel formally dropped out early in the project and others maintained their presence in name only. Still others tried to continue their research, but by project's end contributed little. On the positive side, this shuffling of personnel created opportunities to bring more junior faculty, postdoctoral scholars, and graduate students into the project, making HERO less "top heavy" and giving younger scientists greater responsibility earlier in their career. The lesson to be learned from this experience is that although the intellectual capital provided by senior personnel is invaluable during conceptualization and start-up of a big, long-duration collaborative project, mid-level and junior personnel are more significant in the long run. Thus, when putting together such a project, although it is important to recruit well-established senior personnel, it is even more crucial to select the best mid-level and junior personnel available (with the appropriate skills needed for collaboration[2]), who will end up doing the lion's share of the work. It is also critical that the management plan and budget has sufficient flexibility to swap out senior personnel for mid-level and junior personnel or to use senior personnel salary to hire more junior investigators.

One of the most surprising, yet positive lessons learned by HERO concerned the importance of undergraduate research to the collaborative research enterprise. The HERO REU program was conceived and added after the start of the project, but drove much of the human–environment research in the later years of the project. There were many reasons. First, planning, administering, and running the REU program provided the HERO researchers with a concrete annual cycle, definite targets, and firm deadlines, thereby keeping the researchers working as a well-coordinated, highly communicative team at all times. Second, teaching novices about human–environment interactions made the investigators think more deeply about the nature of research (what it is and how to teach it), interdisciplinarity (why it is important and how to facilitate it), collaboration (how it works and how to promote it), and quantitative versus qualitative research (when to use one over the other and how the two approaches complement one another in a mixed-methods design). Third, the students tested and provided feedback on the collaboratory tools and protocols, forcing the developers to design them not for people with doctorates, but for educated non-specialists. Finally, because all of the HERO team members above the undergraduate level were educators or aspiring educators, they put the same care and pride into their work with the REU students as they put into their classroom teaching. The lesson we learned from the REU experience was that in a complex multi-investigator project such as HERO, it is vital to provide all project members with a focal point and to make sure that focus has a regular cycle with hard targets, drives the research intellectually and practically, and is something that team members care about deeply. For HERO, an undergraduate research program was ideal, but for other projects the focal point might be annual retreats with formal presentations to distinguished guests, annual internal competitions with meaningful prizes, or annual side events at national meetings.

Perhaps the most important lesson learned from the HERO experience was the simple acknowledgement that collaborative research is an inherently social activity, and that collaborating across the academic generations creates a feeling of family. The more that the investigators engaged in mentoring and collaborating on HERO topics with somebody younger or older than themselves, the deeper their involvement with the HERO project became.

Answers to and lessons learned from HERO's proof-of-concept question

How does land-use/land-cover change influence vulnerability to hydroclimatic variation and change?

The HERO project tried to demonstrate that its vision for local human–environment observatories was viable by addressing a complex human–environment proof-of-concept question. Tackling this question demanded that the participants in the project used data standards specifically designed for human–environment research needs, developed and implemented human–environment research protocols, worked as a team at their local HERO, and collaborated with the other HEROs via the collaboratory in a prototype human–environment network. The HERO investigators addressed the question in four places with dramatically different biophysical, socioeconomic, and historical contexts, thereby testing to see if the protocols fulfilled their intended purpose – to ensure that human–environment research, assessment, and monitoring was comparable across time and space, therefore making integration and synthesis possible.

The research question, "How does land-use/land-cover change influence vulnerability to hydroclimatic variation and change?" has three components – hydroclimatic variation and change, land-use/land-cover change, and vulnerability – that require investigation individually and integration with the other two components. The next section of this chapter discusses some of the specific challenges involved in addressing hydroclimatic variation and change and land-use/land-cover change. The vulnerability analysis is covered in most of the remainder of this section.

As noted in Chapters 5 and 8, it is possible to assess the complex phenomenon of vulnerability in myriad ways. Moreover, although dozens of treatises have conceptualized vulnerability and its assessment, few papers have actually tried to develop vulnerability assessment methodologies for application in the field. Thus, the HERO investigators had to spend considerable effort building practicable protocols that could be used at the four HEROs. Upping the ante, HERO tried to develop methods that could be applied by undergraduate students with little to no previous exposure to human–environment theory and research. One tool developed by the HERO team to help both novices and experts understand vulnerability and quickly identify the data they need to collect was the Vulnerability Scoping Diagram (VSD) presented in Chapter 5. This diagram focuses the user on (a) the three dimensions of vulnerability – exposure, sensitivity, and adaptive capacity – and how the dimensions interact when exposed to a specific hazard in a particular place, and (b) measurements associated with the various components of the dimensions, that might be helpful for monitoring vulnerability across space and time.

The HERO investigators also conceptualized two approaches to assessing vulnerability – Rapid Vulnerability Assessment and Grounded Vulnerability Assessment – which run on short and long timescales, respectively. The investigators applied the Rapid Vulnerability Assessment methodology presented in Chapter 8 at all four sites. Chapter 9 reported the results of that work, with careful looks at exposure, sensitivity, and adaptive capacity across the four HEROs. Grounded Vulnerability Assessment is a much longer process that fell outside the time available to the HERO team and consequently was never tested.

The HERO investigators also wanted to develop a way to validate the results of a vulnerability assessment; they consequently developed the idea of Vulnerability Assessment Evaluation and presented that method in Chapter 8. This approach is not an objective model validation in the classic sense, but it does provide a means to conduct a post-hoc evaluation of the research design, motives, methods, and assumptions. The results from the Vulnerability Assessment Evaluation were presented in Chapter 10.

To understand how and why vulnerability operates the way it does in the four HEROs, and how land-use/land-cover change influences vulnerability to floods and droughts, it was necessary to place the vulnerability assessments in the larger human–environment context. Consequently, an important element of the human–environment research of HERO was a careful appraisal of that context. Chapter 6 developed historical portraits of landscape change that help explain today's landscape, and Chapter 7 characterized today's landscape by comparing aerial photo and satellite imagery. Sections of Chapter 9 presented encapsulated views of contemporary physical and human landscapes, using a Rapid Vulnerability Assessment lens, and Chapters 11–14 explored each of the four HEROs in greater depth, using research conducted without the "rapid" mandate, focusing primarily on hydroclimatic variations and changes and their interactions with people.

Taken together, Chapters 5–14 of this book therefore provide a window on the HERO project's attempt to answer a complex human–environment question and, by doing so, to demonstrate that it is possible to build a viable network of human-environment regional observatories using infrastructure components described in Chapters 2–4. It is important to emphasize that although the research addressed hydroclimatic variation and change, land-use/land-cover change, and vulnerability, those topics were not the heart of the research. Instead, the research focus was on infrastructure development – building and using a suite of collaboration tools, a set of data standards and protocols, and a network of local human–environment researchers. The research question was simply a way to develop that infrastructure in a real-world instance. Nonetheless, we believe that our land-use/land-cover change and vulnerability results are comparable to the findings of other projects in nature and quality. For instance, the Vulnerability Scoping Diagram provides a method to couple natural and human systems across scales from local to continental.

Lessons learned: the influence of land-use/land-cover change on vulnerability to hydroclimatic variation and change

There were many snags associated with the proof-of-concept question and developing the human–environment infrastructure needed to address the question. One problem was the complexity of the proof-of-concept question. By itself, vulnerability is one of the most complex of all human–environment problems. Yet, compounding vulnerability with land-use/land-cover change, which is also a complex human–environment phenomenon, and with hydroclimatic variation and change, which is a multifaceted physical phenomenon with major human implications, produced an extremely complicated problem that severely tested the HERO research team. As noted above, however, it was not necessary for the investigators to "answer" the question definitively; it was only important that they develop and test the infrastructure. Nevertheless, it would have been more satisfying to investigators, sponsors, and the human–environment community if HERO could have presented much improved understanding of some element of human–environment research by project's end. It also might have made it more likely that the project would end with a finished human–environment product (see below).

The HERO investigators needed to reconstruct the historical hydroclimatic variation and change for their individual study sites. Assembling and analyzing the data for this analysis was a mixed exercise. At one extreme, using the National Climatic Data Center climatic division data and applying the Standardized Precipitation Index to these data was straightforward and quick. At the other extreme, collecting data on community water systems (CWSs), local water tables, and reservoir levels, as well as putting together time series for floods and droughts, was much more challenging and a much slower process – especially for the Sonoran Desert Border Region HERO, which constantly had to negotiate two bureaucratic systems and incompatible data formats.[3] A surprising aspect of this research was the struggle in obtaining comparable local hydroclimatic data across the other HEROs. For example, CWS data were freely available on the Web until the September 11, 2001 terrorist attacks; after that, these same data were only obtainable by formally invoking the US Freedom of Information Act. Even after filing the appropriate paperwork, state-level discretion affected data acquisition: two states produced the data quickly at no charge; one state dragged its feet, but shared the data months later at no charge; and one state charged $600 for the same data, delivered it months late, and left out key data fields that it had agreed to provide.

Creating comparable land-use/land-cover change data and maps was challenging because of the different data available and approaches used at the various HEROs. The Central Massachusetts HERO used visual photo interpretation on 1971 and 1999 aerial photographs to produce its data and maps. The Central Pennsylvania HERO produced its data and maps by employing a Landsat Multi-Spectral Scanner image from 1972, a Landsat Thematic Mapper image from 1993, and a Landsat Enhanced Thematic Mapper Plus image from 2000, with manual interpretation and extensive ground-truthing. The High Plains–Ogallala HERO also used Landsat Thematic Mapper images from 1985 and 2001, but applied supervised automated classification to produce its data and maps. Finally, the Sonoran Desert Border Region HERO used Landsat Thematic Mapper images from 1985 and 1999, but used manual classification of the satellite images with further aerial photographic interpretation to improve their classification accuracy. Once the data and maps were assembled, however, the analysis of change was easily compared because one of the HERO investigators had already developed a methodology for measuring and comparing land-use/land-cover change (Pontius *et al.* 2004), which was readily turned into a protocol. We learned that – although the area to be assessed varied and, as a result, the methods selected for studying land-use/land-cover change varied – what mattered most was that all sites were able to assess land-use/land-cover change locally and to produce locally relevant answers that were generalizable to other places. Invoking methods demonstrated in Chapter 4 could possibly provide even more detail and nuance to the cross-site analysis. Unfortunately this methodology to compare heterogeneous datasets was not available in time for the cross-site analysis, but it was recently demonstrated in a land-use/land-cover change analysis of the National Land Cover Database from 1992 and 2001 (Ahlqvist 2008).

As noted above, vulnerability is an extremely complex and contested idea. The concept and the way to approach it evolved substantially over the five years of HERO. The investigators spent considerable time during the first year of the project determining which definition of vulnerability they would employ and developing a protocol for quantitative assessment of vulnerability. The second year saw much effort on applying that quantitative vulnerability assessment tool to the HEROs, but the results were unsatisfactory when compared across HEROs because the tool did not perform well in those study areas with large sizes and low population densities.[4] Consequently, as the project neared its halfway point, there was a concerted effort to retool the approach to vulnerability. The upside to this change in course was a rich, complex approach to a rich, complex concept; the downside was that time had been spent that did not directly contribute to achieving our goals (of course, the experimental nature of research means that sometimes a path turns out to be a dead end). Moreover, although the intricacy of the new approach to vulnerability was more suited to the topic and better tested the project, it required more interaction, effort, and time. The result was that the research went beyond the five-year project period; beyond those five years, key investigators were committed to other projects, while others moved to different positions in new locations. Keeping their attention and finishing the research became increasingly difficult. The important lesson here is that by choosing an extremely complex topic and changing approaches in mid-project, the research could not be completed in the study period (and some aspects of the research could not be completed whatsoever). It is clear, however, that this mid-stream change was not a fatal flaw and that the contributions of the project to the conceptualization and practice of vulnerability assessment far outweighed the negative implications of this decision.

The need for HEROs

At its inception in the late 1990s, the motivation of the HERO project came from the realization that many local research sites and networks devoted to studying human–environment interactions across the United States were under way but were not coordinating with each other. The National Science Foundation (NSF) established a set of human dimensions of global environmental change research centers with some explicitly focusing on local and regional processes. The Department of Energy (DOE) funded the regional National Institute for Global Environmental Change (NIGEC) centers. The National Oceanic and Atmospheric Administration (NOAA) started its Regional Integrated Sciences and Applications (RISA) centers program. The National Aeronautical and Space Administration (NASA) created the Regional Earth Science Applications Centers (RESACs). The National Assessment of Climate Change, an inter-agency project coordinated by the US Global Change Research Program (USGCRP), had nearly 20 regional assessments under way and contemplated building approximately eight permanent regional centers. Finally, the NSF-supported Long-Term Ecological Research (LTER) network added a human–environment component to some of its existing sites and added two urban sites in Baltimore and Phoenix to look at the integration of ecology and human systems. The concern of what were to become the HERO investigators was that these efforts would produce unique, non-comparable data, would not work together to standardize the data, and would not emphasize integration and synthesis across scales, thereby making it difficult for science to develop generalizations about local- and regional-level environmental change and sustainability.

Despite general acceptance of climate change and other global environmental changes, the local and regional agenda lost ground in the United States during the ensuing decade. NSF funding of the human dimensions of global environmental change research centers ran out. DOE changed NIGEC to the National Institute for Climatic Change Research (NICCR), reconfigured the network, and turned more strongly to ecosystems functioning rather than human–environment interactions. The Bush Administration put the National Assessment of Climate Change under siege, unjustly discrediting and burying it (Mooney 2007), while the Administration reconfigured USGCRP into the US Climate Change Science Program and abandoned all support for regional assessments of climate change. On a brighter note, NOAA doubled the number of RISAs from four to eight and expanded support to existing centers. In addition, LTER scientists continued exploring ways to expand the study of human–environment interactions to the network, although NSF has not been able to add significant funding to support such activities.

What did not change over this decade was the need for human–environment protocols and data standards, the need for human–environment researchers to use the growing power of computers and the Internet to share data and ideas, and the need for these small networks to band together into a larger network of researchers dedicated to monitoring and understanding human–environment interactions and promoting local and, ultimately, planetary sustainability. From their synthesis, Liu *et al.* (2007) also suggest that coupling human and natural systems is a complex, but necessary endeavor that will generate new questions and help identify new patterns and processes. In other words, during the period from the inception of this research project to today, the need for HEROs did not change and, in our opinion, became more apparent and greater.

The future of the HERO vision

The HERO team is not the only group of scientists to have recognized the value of developing collaborating research sites to address human–environment interaction. For instance, the LTER network has recently produced two important volumes engaging this topic. In the first, Magnuson *et al.* (2006) produced a long-term synthesis of the North Temperate Lakes LTER of Wisconsin. Although the human–environment interactions were not the original motivation for the LTER, they became a significant augmentation of the original research site and a focus of much current research. In the second, Redman and Foster (2008) cover six case studies from the LTER system, including ones from New England, the Appalachian Mountains, Michigan, Kansas, Colorado, and Arizona. They focused on how human activities influence agrarian landscapes and how these influences vary over time and across biogeographic regions (with four of their environs being the same as the HERO sites and studied in a manner somewhat similar to Chapter 6 in this book).

Especially important is the workshop report produced by Vajjhala *et al.* (2007) in which the authors sought to improve social science integration into various NSF Environmental Observatories, including the LTERs. They realized that social sciences are central to resolving many of the major science questions addressed by the observatories and that coordinated research efforts across the social and biophysical sciences are necessary for solving these human–environment questions. Thus, they proposed:

for NSF and the broad community of social and natural scientists to:

(1) Initiate a demonstration or test-bed project for integrated observation
(2) Develop a cross-observatory advisory committee to guide integration efforts
(3) Coordinate with the Social, Behavioral, and Economic (SBE) Sciences Directorate at NSF to align incentives for social scientists to participate in observatory planning and agenda setting

(4) Engage NSF and other funding agencies to design programs and funding vehicles to sustain collaborative observatory research in the long term
(5) Establish a center, modeled on the National Center for Ecological Analysis and Synthesis (NCEAS), to encourage and strengthen observatory collaborations.

(Vajjhala et al. *2007, 2)*

One of the clearest calls from Vajjhala *et al.* was for infrastructure and data standards. Such efforts (see, for instance, US Long-Term Ecological Research Network 2007) increasingly indicate that science recognizes the need for networks of HEROs to integrate across the biophysical and social sciences at one site, facilitate knowledge transfer across sites, and synthesize that knowledge into understanding of the Earth's human–environment condition. HERO is one of the first projects to attempt a tight integration of human–environment researchers with a technical infrastructure.

So, with this growing recognition and consensus, where does the HERO vision suggest the human–environment research community should go from here? First, the human–environment community should build networks of HEROs. Although these sites could be freestanding local human–environment observatories, it would make good practical sense for them to be integral parts of already standing sites (e.g., LTERs) or networks now under construction (e.g., the Water and Environmental Research Systems – WATERS – Network). It would be highly desirable for all human–environment observatories, whether freestanding or associated with other efforts, to form a super-network of HEROs, thereby giving more impetus to developing and adhering to data standards and protocols, and expanding the number and geographical range of HEROs. This super-network should include nodes in the developed world and developing world and in diverse biophysical and socioeconomic regions.

Second, the HERO vision suggests that these HEROs should embark on continuous building and refining of infrastructure and testing of that infrastructure. The collaboratory built for the HERO project demonstrated that technologies change so rapidly that constant collaboratory development would be necessary to facilitate the storage, retrieval, and analysis of data, enable remote interactions, and keep the network alive, vibrant, and evolving. But more importantly, we have recognized that data standards and integration alone are only a partial solution. Researchers need tools that help them understand the local context of any particular dataset, as well as the motivations, methods, and workflows that created it and shaped the science that produced the data.

Third, interdisciplinary collaborative networks are constructs built on social capital – i.e., connections within social networks built on goodwill, social intercourse, and trust. Building social capital, and therefore putting in place the mortar that holds together the bricks of the physical network, takes time.

Finally, the HERO vision suggests that HEROs should increasingly involve stakeholders in the work – not only as subjects of studies, but as true collaborators in the research enterprise. As human–environment research has matured, stakeholder interaction has become an integral element of the agenda. The inclusion of stakeholders is essential because the HERO vision ultimately seeks the sustainability of their communities on their planet.

Notes

1. One reason for the effectiveness of this approach might have been the familiarity that the researchers had gained with one another through years of virtual and real contact.
2. Junior members of the team can learn collaborative skills through observation, e.g., by studying how the leaders demonstrate teamwork by keeping their egos in check during the establishment of the effort.
3. It was clear that going into the research, procuring data from Sonora, Mexico would be challenging. In fact, a major reason for choosing the Sonoran Desert Border Region HERO was the cross-border challenge, which we thought would be an interesting test of data standards and protocol development. It proved to be both interesting and challenging.
4. The results produced by using this vulnerability assessment protocol can be excellent, depending on the local context. For example, Wu *et al.* (2002) and Kleinosky *et al.* (2007) successfully applied the protocol to coastal counties in the eastern United States. Others continue to use and refine it (e.g., Frazier *et al.* 2008).

References

Ahlqvist, O., 2008. Extending post-classification change detection using semantic similarity metrics to overcome class heterogeneity: a study of 1992 and 2001 US National Land Cover Database changes. *Remote Sensing of Environment* **112**(3): 1226–1241.

Frazier, T., N. Wood, and B. Yarnal, 2008. Current and future vulnerability of Sarasota County, Florida to hurricane storm surge and sea level rise. *Solutions to Coastal Disasters Conference Proceedings,* April 2008. Reston, VA: American Society of Civil Engineers.

Kleinosky, L., B. Yarnal, and A. Fisher, 2007. Vulnerability of Hampton Roads, Virginia to storm-surge flooding and sea-level rise. *Natural Hazards* **40**: 43–70.

Liu, J., T. Dietz, S. R. Carpenter, M. Alberti, C. Folke, E. Moran, A. N. Pell, P. Deadman, T. Kratz, J. Lubchenco, E. Ostrom, Z. Ouyang, W. Provencher, C. L. Redman, S. H. Schneider, and W. W. Taylor, 2007. Complexity of coupled human and natural systems. *Science* **317**: 1513–1516.

Magnuson, J. J., T. K. Kratz, and B. J. Benson (eds.), 2006. *Long-Term Dynamics of Lakes in the Landscape: Long-Term Ecological Research on North Temperate Lakes.* Oxford: Oxford University Press.

Mooney, C., 2007. An inconvenient assessment. *Bulletin of the Atomic Scientists* **63**: 40–47.

Pontius Jr., R. G., E. Shusas, and M. McEachern, 2004. Detecting important categorical land changes while accounting for persistence. *Agriculture, Ecosystems and Environment* **101**: 251–268.

Redman, C., and D. R. Foster, 2008. *Agrarian Landscapes in Transition: Comparisons of Long-Term Ecological and Cultural Change.* Oxford: Oxford University Press.

United States Long-Term Ecological Research Network, 2007. *Integrative Science for Society and the Environment: A Plan for Science, Education, and Cyberinfrastructure in the US Long-Term Ecological Research Network.* Accessed at cwt33.ecology.uga.edu/ecology/web_learning/LTER_Integrated_Research_Plan.pdf.

Vajjhala, S., A. Krupnick, E. McCormick, M. Grove, P. McDowell, C. Redman, L. Shabman, and M. Small, 2007. *Rising to the Challenge: Integrating Social Science into NSF Environmental Observatories.* Washington, D. C.: Resources for the Future. Accessed at www.rff.org/Events/Pages/RFF-NSFWorkshop.aspx.

Wu, S.-Y., B. Yarnal, and A. Fisher, 2002. Vulnerability of coastal communities to sea-level rise: a case study of Cape May County, New Jersey. *Climate Research* **22**: 255–270.

Yarnal, B., and R. Neff, 2007. Teaching global change in local places: the HERO Research Experiences for Undergraduates program. *Journal of Geography in Higher Education*, **31**: 413–426.

Index

Printed in the United States
by Baker & Taylor Publisher Services